THE ENERGY SAVER'S HANDBOOK

FOR TOWN AND CITY PEOPLE

About the Author

The Massachusetts Audubon Society is one of the oldest and largest environmental organizations in the United States. Since its founding in 1896 as an organization to protect endangered birds, the society has expanded its programs to include a wide variety of environmental concerns including the energy transition, toxic substances, water resources, coastal zone protection, conservation of ecological resources and environmental education.

The Massachusetts Audubon Society is the only state environmental organization in the United States to employ a full-time research staff of professional scientists. The Scientific Staff conducts and stimulates original research on important environmental problems, provides technical information and advice on environmental issues and identifies future environmental concerns. Acid rain, pesticides, endangered and threatened species and indoor air pollution are current research topics, in addition to energy issues.

The society became involved with energy issues in the early 1970s, when it became evident that few factors would have as great an impact upon future environmental quality as the production, distribution and consumption of energy resources. In response to the need for environmentally conscious energy planning, MAS embarked on a program to promote an energy transition based on the development of renewable energy sources and the efficient use of remaining fossil fuel resources. Demonstration and research projects have included the weatherization of society buildings and the installation of active and passive solar systems. In its educational programs, the society has compiled and distributed technical information on energy conservation and renewable energy technologies through workshops, seminars and publications.

THE ENERGY SAVER'S HANDBOOK

FOR TOWN AND CITY PEOPLE

by the Scientific Staff of the Massachusetts Audubon Society

Ronnie D. Lipschutz

Deborah L. Bleviss

William R. Alschuler

David W. Conover

Illustrations by David W. Conover

Rodale Press, Emmaus, Pa.

Printed in the United States of America on recycled paper containing a high percentage of de-inked fiber.

Book design by Jerry O'Brien
Book layout by Darlene Schneck

Library of Congress Cataloging in Publication Data
Main entry under title:
The Energy saver's handbook.
     Includes index.
       1. Energy conservation—Handbooks, manuals, etc.
I. Lipschutz, Ronnie D. II. Massachusetts Audubon Society.
TJ163.3.E548   1981   333.79'16   81-19244
ISBN 0-87857-373-9 hardcover   AACR2
ISBN 0-87857-374-7 paperback

2 4 6 8 10 9 7 5 3   hardcover
2 4 6 8 10 9 7 5 3   paperback

Table of Contents

List of Figures

List of Tables

Acknowledgments

Numerous people deserve credit for their part in seeing *The Energy Saver's Handbook* through to its completion. Among them are Dick Henry, Bill Newbury, Bob Saltonstall and Tom Urquhart who, with one of the authors, first conceived of what eventually became known as the Riley Project. Ian Nisbet, Jim Aldrich and John Fitch, all of whom directed the Scientific Staff of the Massachusetts Audubon Society over the duration of the project, are the recipients of our heartfelt thanks for their oversight and advice.

The many individuals who reviewed drafts of various parts of this book and offered technical assistance and advice must also be thanked. Among them are Pentti Aalto, Tim Johnson, Bob Timmerman, Eric Terini, Don Jackson, Henry Lee, Arnold Wallenstein, David Johnson-Wint, Barbara Brandt, the Urban Solar Energy Association, Ambrose Spencer, Jack Gleason, Richard Zeidman, Paul Lorris, Hal Mahon, Henry Joseph, Ann Brian Murphy, P. Nicholas Elton, Connie Springer, Phil Simmons of the Massachusetts Audubon Society and the many people in the Riley Project Advisory Group.

Credit must also be given to those who provided help in the revision of this book for publication. Among them are Carol Stoner of Rodale Press, Tom Wilson of Star Route Studios and Rick Diamond, Dave Krinkel, Mark Modera, Bruce Dickinson, Dave Grimsrud and others in the Energy Performance of Buildings Group at Lawrence Berkeley Laboratory.

People deserving of special thanks are Sue Blake, who administered the project, Charlotte Bensdorp-Wilson, who stormed into the Scientific Staff building, woke everyone up, made sure that things got done and nursed an earlier version of the book from manuscript to print and Marsha Rockefeller, who labored over various drafts leading to publication. The authors are also grateful for the help of Betty Taylor and Lee Malloy, who did much of the typing of the various drafts and manuscripts of the book, and Joan Wittig and Adrian Gold, who also helped with the preparation of the manuscript.

Finally, our greatest appreciation goes to the Mabel Louise Riley Charitable Trust and its trustees, who thought this project worthy of funding and made *The Energy Saver's Handbook* a reality.

Introduction

Many Americans think the use of energy to heat homes, drive cars and run machinery is one of their inalienable rights. In the past ten years, however, the price of this "right" has risen dramatically, and the reliability of energy supplies has become questionable. The availability of oil supplies is now subject to political events in foreign countries, and natural gas reserves are beginning to run out. As new, more costly electrical generating plants begin to operate, the price of electricity will rise astronomically.

In many parts of the United States, the effects of the rising cost of energy are most directly felt in home energy use. In the northern states, residential energy use is as much as 30 percent of the total regional energy consumption as opposed to 20 percent on a national basis. In these areas, a much larger percentage of residential energy use is consumed by home heating. In New England, for example, 60 percent of a home's energy consumption is for space heating.

The only ways to cope with the energy crisis are to conserve the fuel that is already being used and to substitute renewable energy sources for traditional ones. Energy conservation is the best short-term solution to the crisis. It involves using fuels more efficiently and using each fuel for the work to which it is best suited. The longer-term solution, the replacement of traditional fuels with renewable ones, holds much promise because the supply of renewable fuels (solar, wind, biomass) cannot be depleted.

The Energy Saver's Handbook addresses both energy conservation and the use of renewable fuels in urban homes. Urban residences have a high rate of occupancy by lower-income families who can ill afford to pay for the escalating cost of home heating and whose homes are very much in need of energy conservation improvements. Homes in cities have unique features and problems that are not characteristic of suburban buildings, which are the focus of most texts on energy conservation and renewable energy systems. Solar energy systems have been traditionally thought of as unsuitable and too costly for urban housing. Yet, it is in the city that renewable energies can, in fact, deliver important benefits to city people's homes, to their neighborhoods and to their communities.

The Energy Saver's Handbook is divided into eight chapters:

Chapter 1: "Making an Energy Survey of a Building" describes how heat is lost from a building and how this loss can be detected;

Chapter 2: "The Building Envelope" describes the process of reducing heat loss through the exterior of a building with caulking, weather stripping and insulation;

Chapter 3: "Heating and Cooling Systems" describes ways to increase the efficiency of conventional heating and cooling systems;

Chapter 4: "Managing Energy in the Home" describes how home energy use can be reduced with more efficient lighting and appliances, with better load management and with changes in life-style and energy-use habits;

Chapter 5: "Renewable Energy Systems" discusses solar space and domestic water heating and natural cooling systems, photovoltaic systems and wind power;

Chapter 6: "Large-Scale Energy-Conserving Systems" describes community-scale heating systems such as cogeneration, district heating and seasonal energy storage;

Chapter 7: "Institutional Issues: Lifting the Barriers to Energy Conservation" addresses the questions of solar and wind access rights, legal restrictions on renewable energy systems, tenant-landlord relations and energy pricing and rate structures;

Chapter 8: "Financing Energy Conservation and Renewable Energy
   Systems" addresses the problem of how individuals and groups can
   get financial assistance for making energy improvements to a building.

   These chapters speak directly to the occupants of the buildings
concerned and to those who are doing structural renovations on an
urban residence and wish to make some energy improvements to
it as well.

   As you read these chapters, keep in mind the following points:

   *A building should be viewed as a total system when making energy
conservation improvements.* Waste in energy use can be reduced both
by improving the exterior of a building and by improving its heating or
cooling system.

   Energy conservation improvements should be cost-effective. These
improvements may be expensive, but they will immediately reduce your
oil, gas or electric bills, and the cost of these improvements will be
recovered within a few months to a few years by the accrued value of
the energy you save.

   *Decisions on energy conservation improvements should be
forward-looking.* As the prices of conventional fuels rise in the coming
years, renewable fuels will become more attractive financially. It is
important to consider this as improvements are made. If decisions are
made wisely so that building improvements are compatible with
renewable energy systems, it will be easier to retrofit a building with
these systems in the future.

   Energy conservation and renewables go hand-in-hand. Many people
would like to add solar systems to their houses and forget about
weatherization, but the fact is that the less energy a home requires for
heating, the greater is the contribution that can be made by renewable
energy. A weatherized home requires a smaller solar system, too,
which means lower installation costs.

   *Renewable systems have relatively high first costs but small
continuing fuel costs.* The relatively high first costs of retrofitting a
renewable energy system to your home may deter you from going
ahead with the project. It's important to compare the costs of such
systems to the long-term operating costs of a conventional heating
system. The fuel for a solar or wind system is, of course, free for the
taking, while the oil, gas or electricity for your present system is
increasing in cost. As the costs of these conventional fuels rise, the
economics of your renewable system will look better and better in terms
of how long it will take for the system to pay for its first cost with the
accumulated savings in conventional fuel.

   Complicated technology does not always give better results. The

**Table 0-1
Degree-Days for the Four Climatic Zones**

Zone	July	Aug.	Sept.	Oct.	Nov.	Dec.
I	0	0	0	48	272	460
II	0	0	43	232	552	752
III	0	10	92	372	728	1,055
IV	26	50	211	511	943	1,324

SOURCE: Degree-days are from J. Leckie et al., *More Other Homes and Garbage* (San Francisco: Sierra Club Books, 1981).
NOTE: Degree-days for the following cities were used to calculate average monthly degree-days for each climatic zone:

Zone I: Los Angeles, Calif.; Waco, Tex.; Macon, Ga.; Alexandria, La.; Vicksburg, Miss.
Zone II: Ashville, N.C.; Roanoke, Va.; Portland, Oreg.; Albuquerque, N. Mex.
Zone III: Boston, Mass.; Akron, Ohio; Hartford, Conn.; Chicago, Ill.
Zone IV: Green Bay, Wis.; Helena, Mont.; Burlington, Vt.; Rochester, Minn.

problem of high first cost is of particular concern to low-income families, but solar systems need not be expensive or complicated to work well. We describe many systems that are fairly simple and can operate as efficiently as more complicated systems, but with lower first costs and with fewer problems.

Your results from implementing some or many of the energy or weatherization improvements described in this book may be different from the results described here. No two families or houses are alike. Indeed, studies show that energy use by similar families in identical houses can vary by as much as a factor of two! If you weatherize your house but become careless in managing your energy use, you cannot expect to realize significant energy savings. If you install a solar water heater and expect it to provide hot water on demand whatever the weather, you will be disappointed.

HOW TO USE THIS BOOK

The calculations for *The Energy Saver's Handbook* were originally prepared for climate conditions that occur in the Greater Boston area. Where possible, text, charts and tables have been revised to include climate data for other cities in the United States. The reader should consult the following map to determine the appropriate climatic zone for his or her city. At the bottom of many of the tables in this book, there

Jan.	Feb.	Mar.	Apr.	May	June	Total Annual
496	380	276	64	4	0	2,000
792	642	572	282	112	21	4,000
1,130	981	847	506	238	42	6,000
1,484	1,244	1,103	662	341	101	8,000

are correction factors for the four climatic zones. These factors should be applied to the numbers in the tables in order to derive the right data for a particular climatic zone. Degree-day distributions for the four zones were determined by averaging together monthly degree-day figures for several cities in each zone. These monthly averages are presented in table 0-1.

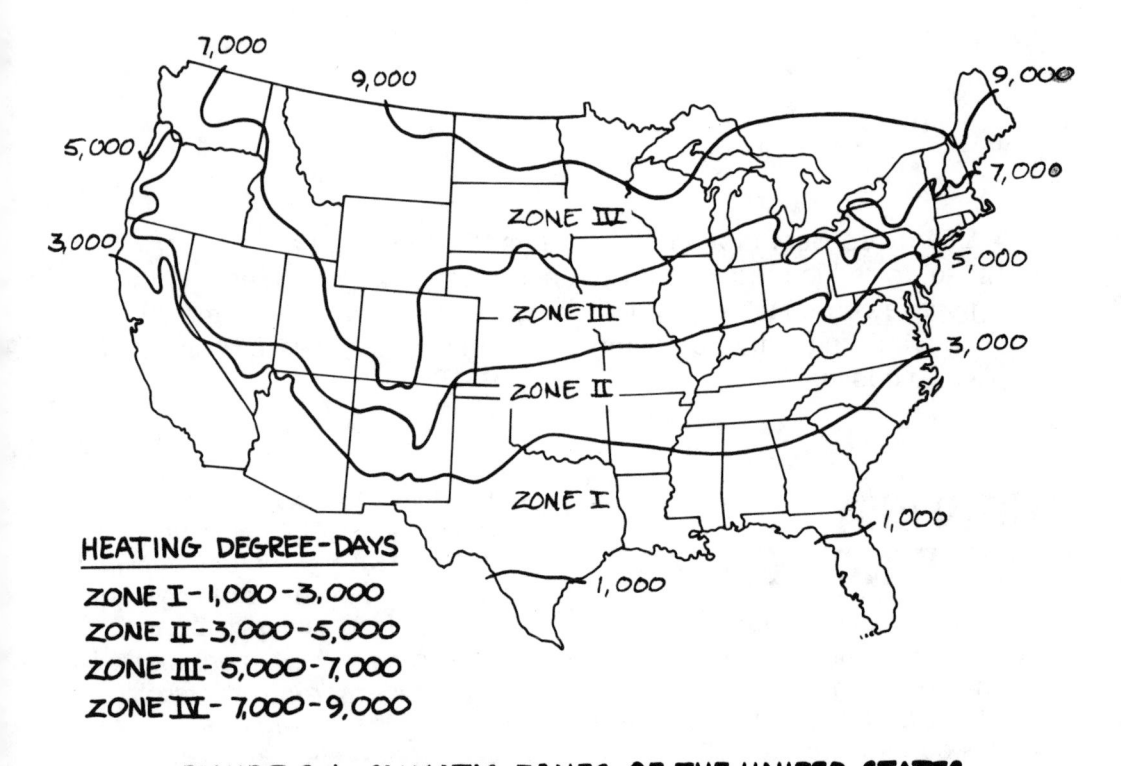

FIGURE 0-1: CLIMATIC ZONES OF THE UNITED STATES

Making an Energy Survey of a Building

Energy conservation improvements to a building have one major purpose: to reduce the amount of energy used while still keeping the living space comfortable. The following discussion applies primarily to heating a building, although some of the principles discussed can also be applied to cooling.

HOW IS A BUILDING HEATED?

In most homes, warmth is provided by burning a fuel (e.g., natural gas, heating oil, wood) or using electric resistance heaters. This heat eventually passes out of the building through its exterior, or *envelope* (the walls, roof and basement floor). To ensure that fuel is not used wastefully, it is important that the heating system operates efficiently and that the building retains heat for as long as possible.

What can be done to ensure that a heating system runs efficiently? To answer this question, it is first necessary to explain how such a system works. Most buildings in the United States use

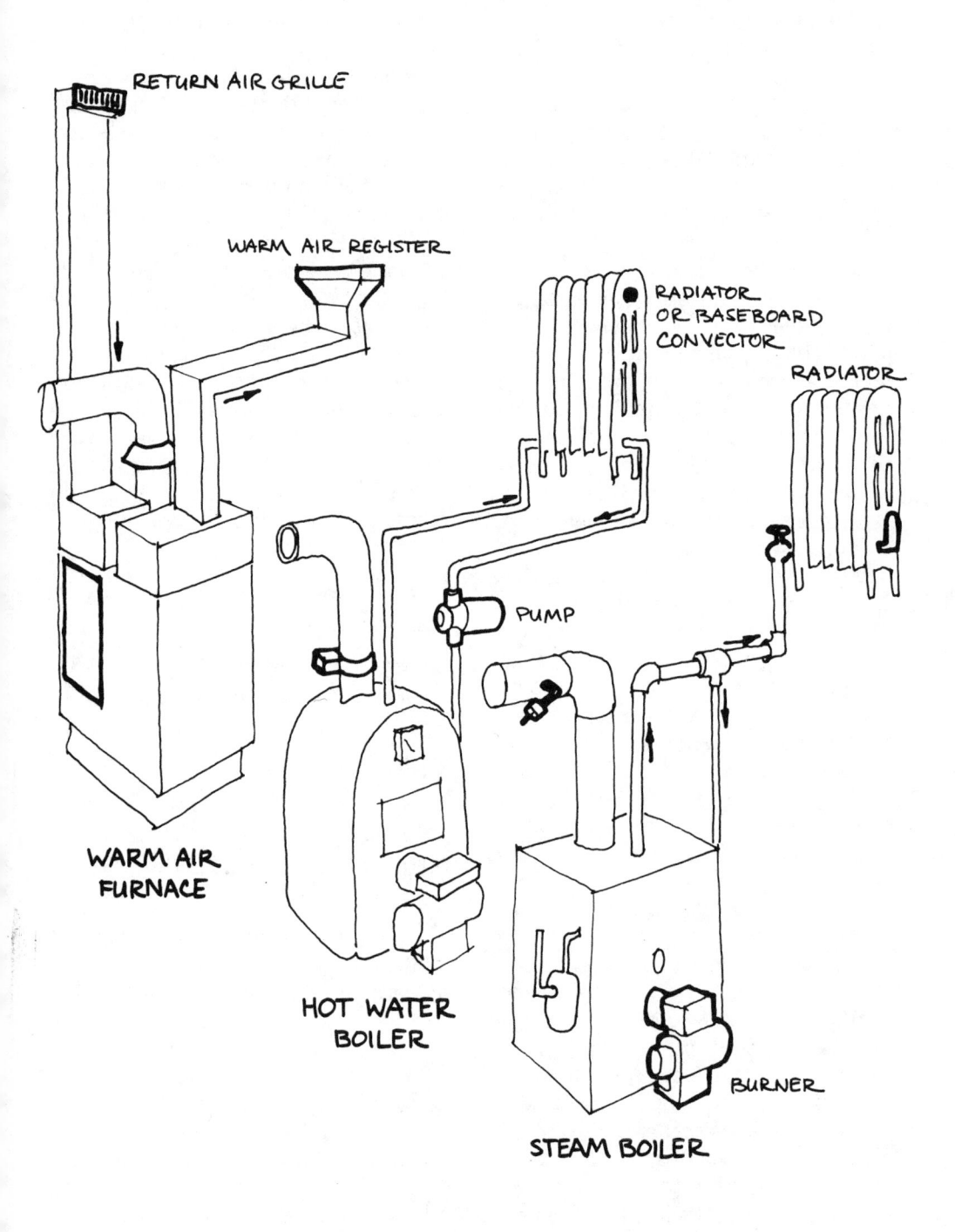

RETURN AIR GRILLE

WARM AIR REGISTER

RADIATOR
OR BASEBOARD
CONVECTOR

RADIATOR

PUMP

WARM AIR
FURNACE

HOT WATER
BOILER

BURNER

STEAM BOILER

FIGURE 1-1: DIFFERENT HEATING SYSTEMS

oil or natural gas as a fuel for heating. When this fuel is burned, it heats either air or water, or it generates steam. These fluids are then moved throughout the building where they transmit their heat to the structure. Heating systems are composed of several parts. Gas or oil is mixed with air in a *burner*. It is then ignited by the burner in the *combustion chamber* of a *furnace* or *boiler*. The fluid contained within the furnace (air) or boiler (water or steam) picks up the heat of combustion, carries it to the building through pipes or ducts and then releases the heat through *baseboard heaters* if the fluid is water, through *radiators* if the fluid is water or steam or through *air grilles* or *registers* if the fluid is air. If the heating system is efficient it will deliver a large amount of heat for every gallon of fuel used. To maintain this efficiency, the system should be tuned regularly, and worn-out parts should be replaced. Some buildings, particularly newer ones, use electric resistance heaters to provide warmth. These heaters are less desirable than oil or gas because electricity is more expensive, and the process of making this energy in large oil, coal or gas generating plants wastes about two-thirds of the original fuel's heat content.

Assuming that the heating system in a building operates efficiently, what can be done to retain that heat for as long as possible? Any building warmer than the outside temperature will gradually lose that warmth to the outdoors. Heat is lost primarily in two ways: by *conduction* and by *infiltration*. A third way of losing heat, *radiation*, is significant only with respect to windows and is usually included in conduction losses. Conduction is the direct transfer of heat through a material. Brick walls, for example, allow for substantial heat

loss by conduction, while wood walls allow for less. Infiltration is the loss of heat through leaks in the building that let warm air out and cold air in. Loose windows and doors and cracks in the foundation are some points that allow heat loss through infiltration.

A building should be *weatherized* or improved so that its heat loss is reduced, but before this can happen, you must estimate how much heat is being lost from the building and from where it is being lost. There are several ways to do this. The measure of a material's resistance to heat loss by conduction is known as the *R-value*. The higher the R-value, the more a material resists heat flow. The following list shows the R-values for various parts of a building before and after it has been weatherized.

Unweatherized Building	R-Value
Single-pane window	0.88
Uninsulated wood frame wall	3.46
Uninsulated attic (ventilated to outdoors)	1.69

Weatherized Building	R-Value
Window with storm window	1.85
Wall with 3 inches blown-in cellulose	13.66
Ceiling with 6 inches fiberglass batt	21.69

The measure of heat loss due to infiltration is the number of cubic feet of warm air escaping from the building per hour, known as the *air change rate*. Another goal of weatherization is to minimize this rate as much as possible. Of course, this amount of airflow cannot be reduced to zero, because every building must be ventilated in order to remove water vapor, pollutants and odors produced in cooking and other domestic activities.

Heat loss from conduction can be

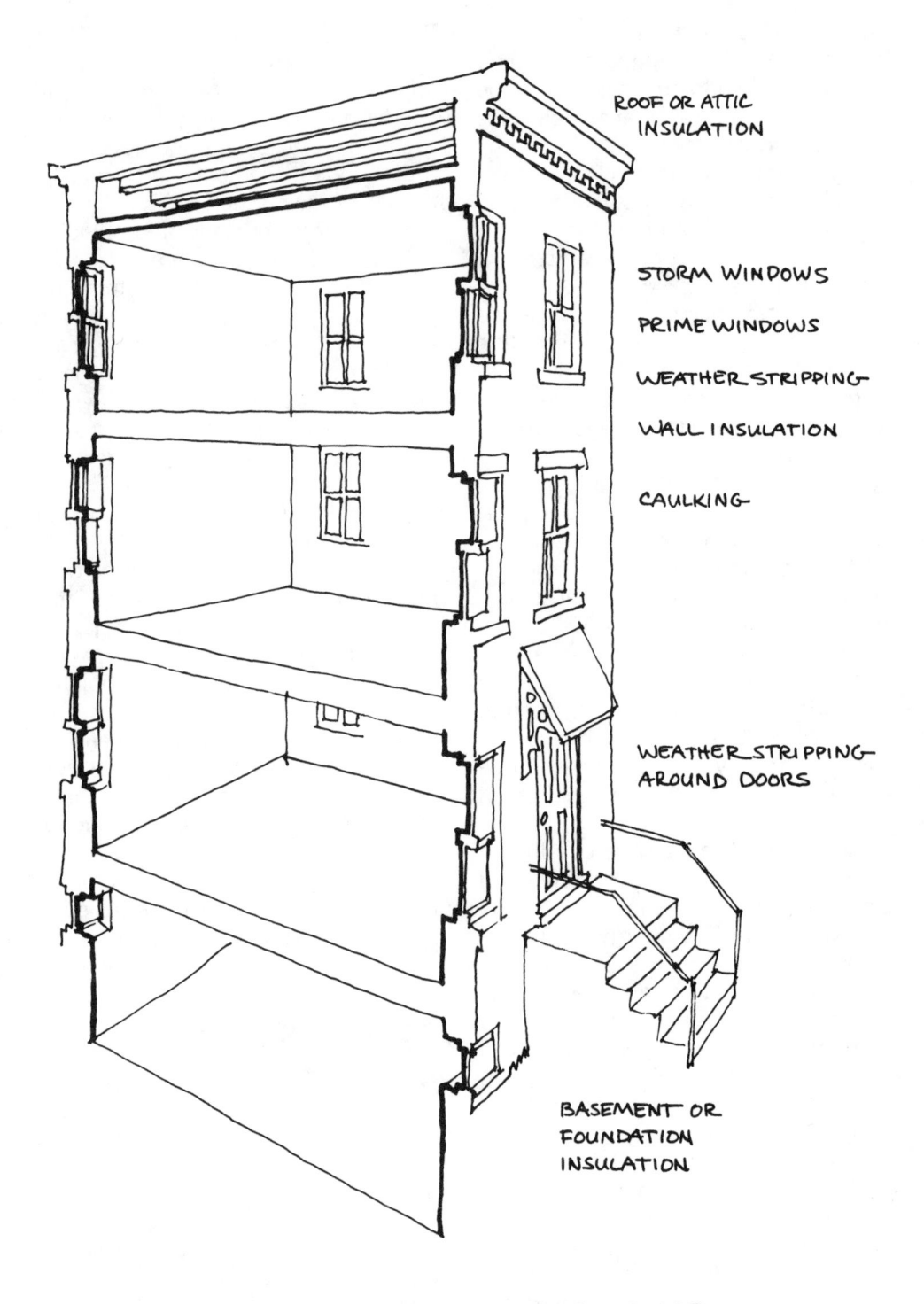

ROOF OR ATTIC
INSULATION

STORM WINDOWS

PRIME WINDOWS

WEATHER STRIPPING

WALL INSULATION

CAULKING

WEATHER STRIPPING
AROUND DOORS

BASEMENT OR
FOUNDATION
INSULATION

FIGURE 1-2: THE BUILDING ENVELOPE

reduced by adding storm windows and storm doors to existing windows and doors, and by adding insulation to the roof, walls and basement. Heat loss from infiltration can be reduced by plugging and caulking cracks, weather-stripping doors and windows and adding storm windows and doors.

How do you decide what step to take first? The cost of replacing a heating system or weatherizing a house can be high, especially for older houses. It is helpful to have a set of priorities to determine what should be done first. There is one basic way of establishing these priorities. Consider how long it will take for the cost of the fuel saved by weatherization to equal the original cost of the improvement. This length of time is called the *simple payback period.* For example, if adding storm windows to a house costs $500, and the result is an annual fuel savings of $100, the payback period will be five years. If you set priorities for heating system alteration or weatherization by looking at payback periods, those options with the fastest payback should have highest priority.

An *energy audit* can be conducted to determine payback periods and potential fuel savings for various improvements. An energy audit is a thorough review of the building's characteristics from an energy per-spective. First, the parts of the building through which heat flows are studied to determine their R-value and the magnitude of air leaks through them. These parts, collectively known as the envelope, include the roof, the outside and foundation walls, windows, doors and the basement ceiling if the basement is unheated or the basement floor and walls if the basement is heated. Second, the heating systems for space and domestic water heating are examined.

Finally, an auditor may inspect the lighting system for possible areas of improvement. From these observations, the auditor determines where the greatest inefficiencies and heat losses are and what the costs and paybacks are for reducing them.

Since there are many different kinds of buildings, made of different materials and with different heating systems, the priorities for heating system replacement and weatherization vary widely. Each building should receive an individual audit to establish these priorities. To illustrate basic problems peculiar to certain building types and to give you an idea of what priorities might be established by an energy auditor, this book presents an energy survey that is applicable to four different housing types representative of urban housing. These are the *row house,* which is similar to a brownstone or townhouse, the *triplex,* which is a three-story frame building with three apartments (one on each floor), the *single-family house* and the 24-unit *multi-unit apartment.* If your building is one of these types, the survey results may provide you with enough information to set your energy priorities. But making a survey of your actual building will always be time well spent in uncovering special energy problems.

THE ENERGY SURVEY

To complete a basic energy survey of your building, start in the basement and work your way up through the structure. Use the Energy Survey form that follows to record your building's energy-use characteristics. If your building resembles one of the four common urban housing types, you can apply the results of your energy survey to the information included in the section "Model Energy Surveys of the Four Housing Types" later in this chapter, which lists the costs of various weatherization improvements, the amount of fuel saved by these improvements and the payback period.

The Basement

The Heating System

What type of heating system do you have? You should know what type of fuel you use. Oil is delivered by your local oil company; gas is piped into your home from the gas company. Locate your boiler or furnace.

Are there any obvious leaks or corrosion spots in the boiler, furnace or connecting piping? If there are, you should contact your oil or gas company, because leaks reduce a heating system's efficiency. Gas leaks are, of course, dangerous and require immediate action. You can smell out a gas leak by its odor. Your local gas utility may have "scratch and sniff" cards that duplicate the smell of natural gas so you can recognize this distinctive odor.

Are the ducts or pipes carrying the heat away from the boiler insulated? Uninsulated ducts or pipes will cause some of the heat they carry to be lost to the basement.

Locate the temperature regulator (called an airstat or aquastat) on your boiler or furnace and read the temperature. Often, this regulator is set higher than is necessary and can be turned down. Water systems can often be turned down to 150°F, and air systems down to 120°F. Steam systems usually cannot be turned down. By turning down this regulator, fuel consumption can be reduced.

Check the on/off time for your furnace or boiler. When you make the survey during the heating season, sit next to your furnace or boiler for an hour and see how many minutes it is on. Then check the outside temperature with a thermometer and consult table 1-1 to determine how long the furnace should have been on depending upon the outside temperature. If the on-time of your burner is much less than the on-time in the chart, your heating system is oversized and needlessly using fuel. You may want to consult your oil company about reducing the nozzle size or your gas company about reducing the firing rate. A constantly operating system during milder weather may also indicate inefficient operation, leakage or poor heat delivery. Fewer firing cycles or continuous burning during the coldest weather usually means that the burner is properly sized.

Locate your hot water heater. Bring your hand close to the heater. If you can feel the warmth, it is probably not adequately insulated or the thermostat is set too high. Lack of sufficient insulation means that heat is lost from your hot water to the basement.

Are the hot water pipes insulated? Why pay for heat that is lost through pipes? If hot water is used throughout the day, the amount of heat lost through uninsulated pipes can be considerable. Even if hot water is used only in the mornings and evenings, insulated pipes will conserve fuel. Insulating pipes is an easy job that most people can do without the aid of a plumber.

Look for the temperature regulator on your heater and read the temperature. If you do not have a dishwasher, this temperature setting should be no more than 110 to 120°F. If you have a dishwasher with a separate heating element, the setting can still be 110 to 120°F. Otherwise, the setting should

Table 1-1
On-Time for Boiler or Furnace

Outdoor Temperature (°F)	Minutes On in an Hour (for an indoor temperature of 70°F)			
	Zone I	Zone II	Zone III	Zone IV
−20	60			
−10	54	. . .	. . .	. . .
0	47	60	. . .	. . .
10	40	51	60	. . .
20	34	43	50	60
30	27	34	40	48
40	20	26	30	36
50	13	17	20	24
60	7	9	10	12

NOTE: Ellipses indicate unlikely temperatures for certain zones.

be 140°F. Turn the power off to an electric water heater before adjusting the thermostats. (There are usually two, one for each heating element.)

The Basement Envelope

Is the basement heated? That is, is it a finished basement that is used as living space? Most basements are warmed to some degree by distribution losses from the heating system. Our calculations presume a relatively cold basement.

If the basement is heated, are the outside walls insulated? To determine this, remove the plate from a light switch or electric outlet on the wall (if the wall is finished), and look inside. Be sure to remove the appropriate fuse or turn off the circuit breaker before doing this. Since the outside wall is usually masonry (concrete, stone or brick), interior walls are usually insulated either with fiberglass or rock wool batts which have the texture of cotton or with rigid foam boards (polyurethane, Styrofoam).

Is the floor insulated? Usually, the basement floor slab is not insulated but the foundation may be. The only way to determine this is to dig down to the foundation from the outside. On all but very new homes, you can assume the foundation is uninsulated. A layer of carpet on the basement floor provides some insulation as well as increased comfort.

Are the basement windows weather-stripped? Weather stripping comes in various forms (see "Caulk and Weather-

Stripping" in chapter 2.) Its purpose is to seal out cold air infiltration.

Do storm windows cover the basement windows? If they do, check how well they fit in their frames. Feel for a draft. Do they allow much infiltration of air? If so, they require some work and possible replacement.

Are cracks caulked? If there is major infiltration through cracks in the foundation or between the foundation and a frame wall, you will be able to feel a draft coming through. If so, caulking will be necessary.

If the basement is not heated, is there insulation in the ceiling? This insulation slows the flow of heat from the house into the relatively cold basement. It is usually suspended from the ceiling in the form of fiberglass batts, either with an R-value of 11 (approximately 3 inches of fiberglass) or 19 (approximately 6 inches). In colder climates, insulation in the basement ceiling, while reducing heat loss, may make the basement so cold that water pipes freeze. You may want to consult a professional contractor before you install such insulation under the first floor.

The Living Space

The Heating System

How well is heat delivered to the building?

If you have a forced air system, are the grilles or registers clean? Dust can reduce the efficiency of the heating system.

Are there any objects blocking grilles or registers? Heat cannot flow through blocked outlets.

If you have a hot water system with radiators, are any objects on top of or *blocking the radiators?* Radiator covers might look nice, but they block convection currents and the distribution of heat.

Are the valves in steam or water radiators working properly? If the valves are not working properly, air may be getting into the pipes and blocking the flow of heat (chapter 3 tells how to correct this). This is often a problem with old heating systems.

Are there any reflectors behind the radiators? These can often aid in the delivery of heat to a room by reflecting heat back that would normally be conducted through the exterior walls, where radiators are usually installed.

If you have hot water baseboard heaters, are there any objects blocking the baseboards? Baseboards are convectors. Don't block the inlet or outlet vents, because heat flow to the living space will be reduced.

Are the fin tubes in the baseboard clean? Dust can reduce the efficiency of these baseboards.

If you have electric resistance heaters, are any objects blocking these heaters? Like hot water baseboards, these are also convectors and should be clean and not blocked.

Is the distribution system in the building well balanced? Check for any hot or cold rooms. If there are any balance problems, consult your oil or gas company or a heating contractor.

Check your thermostat settings. Daytime settings of 65°F and settings of 60°F at night or when the building (or room, if you have individual room thermostats) is unoccupied can save up to 20 percent of your fuel bill compared to a conventional setting of 72°F. If dressed warmly, most infants and pregnant women can tolerate temperatures as low as 62 degrees,

derly persons should consult their
icians before drastically lowering
thermostats.

The Living Space Envelope

Are the outside walls insulated? Use
the procedure described in testing basement walls. On wood frame structures,
insulation may have been blown in
between the exterior and interior walls.
This type of insulation may be cellulose
(which looks like pulverized paper),
fiberglass (which looks like pink or yellow
cotton) or urea-formaldehyde (which
is a hardened foam). In relatively new
buildings, rigid board insulation may
have been applied on the other side of
the wood sheathing (see "Insulation,"
chapter 2). In this case, it is difficult to
detect unless you make a hole through
the wall section.

Are all cracks caulked? Check all
over for drafts. You might try lighting
an incense stick and watching the smoke
to see in which direction it moves. If
the smoke moves into cracks, openings
or joints, air leaks exist there and must
be sealed. Don't use an open flame to
check for drafts. Air may be flowing
into a crack or hole, and the flame can
be sucked into the wall and ignite
flammable materials. Another useful
and safe tool for detecting air leaks is
the smoke stick, a glass tube containing
chemicals that produce a smoky vapor
upon contact with air. These tubes can
be ordered from National Draeger, Inc.,
P.O. Box 120, Pittsburgh, PA 15230.
They are called Air Current Tubes and

cost about $ 18 for ten sticks.

*Are storm windows and doors in
place?* If they are, check how well they
fit in the door or window frames. Do
they allow for much infiltration of air?
If so, they require some work and
possible replacement.

*Are doors and windows weather-
stripped?* If they are, is the weather
stripping still effective? Check for drafts,
using the smoke test.

The Attic

Is the attic insulated? The attic can
be insulated in several ways depending
on whether or not it is used for living
space. Figure 1-3 shows some of the
ways possible. Batts of insulation can
be fitted between the roof rafters or
the ceiling joists or mounted on the
floor of the attic. If the attic is used as a
living space, insulation must also be
added to the outside of the short, vertical
walls known as *kneewalls*. Insulation in
the form of loose cellulose or fiberglass
may be blown into the attic, particularly
if it is not easily accessible for installing
batts. In new homes, insulation in the
form of rigid boards may be attached
to the roof.

*If the attic is used as a living space,
are all cracks caulked?* Check for drafts,
particularly around flue pipes.

*If the attic is a living space, are
storm windows and doors in place?*
Check that they fit securely in their
frames and are not allowing unwanted
infiltration.

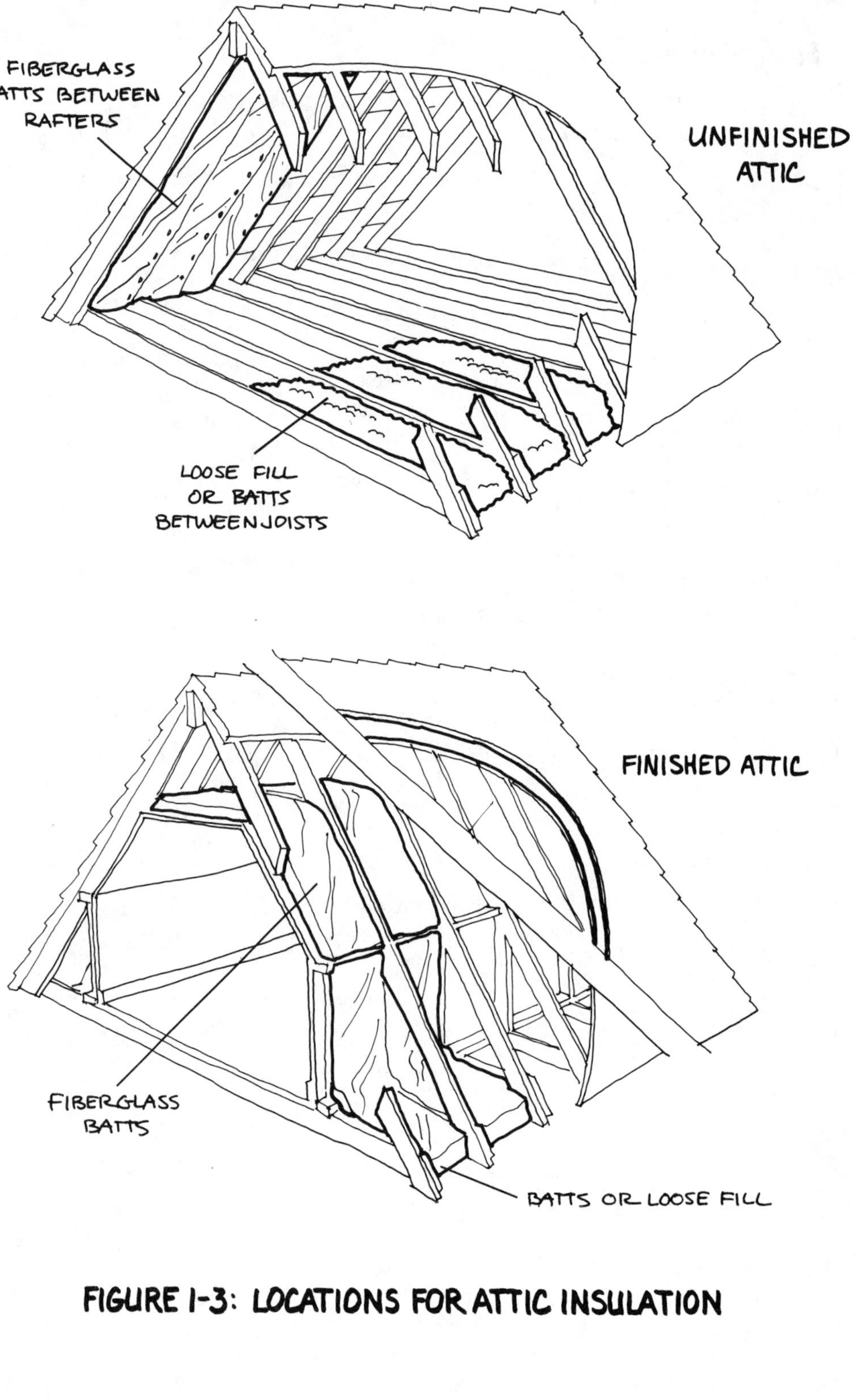

FIBERGLASS BATTS BETWEEN RAFTERS

UNFINISHED ATTIC

LOOSE FILL OR BATTS BETWEEN JOISTS

FINISHED ATTIC

FIBERGLASS BATTS

BATTS OR LOOSE FILL

FIGURE 1-3: LOCATIONS FOR ATTIC INSULATION

Energy Survey

Record the results of the energy survey of your house here.

BASEMENT

- **Heating System**
  1. Fuel
     oil   gas
  2. Distribution
     steam   water
          air
  3. Leaks or corrosion?
     yes   no
  4. Insulated pipes or ducts?
     yes   no
  5. Temperature setting on boiler or furnace:
     ___ °F
  6. Furnace or boiler on-time:
     ___ minutes at an outside temperature of ___ °F

- **Hot Water Heating System**
  1. Heater warm to touch?
     yes   no
  2. Additional insulation added to exterior of heater?
     yes   no

3. Insulated pipes?
   yes   no
4. Temperature setting on heater:
   ___ °F

- **Envelope, If Heated**
  1. Insulated walls?
     yes   no

     If yes, how much?
     _____

     What type?
     _____

     R-value—if you know it:
     _____

  2. Insulated floor or foundation?
     yes   no
  3. Weather-stripped windows?
     yes   no

     If yes, quality?
     good   fair
          poor
  4. Storm windows?
     yes   no

If yes, quality?
   good   fair
        poor
5. Caulked cracks?
   yes   no

- **Envelope, If Unheated**
  1. Insulated ceiling?
     yes   no

     If yes, how much?
     _____

     What type?
     _____

     R-value—if you know it:
     _____

LIVING SPACE

- **Heating System**

  AIR
  1. Clean grilles or registers?
     yes   no
  2. Registers open or closed?
     _____

     If closed, where?
     _____

3. Blocked grilles or registers?
   yes   no

RADIATORS

1. Blocked radiators?
   yes   no

2. Valves operational?
   yes   no

3. Reflective backers?
   yes   no

BASEBOARD HEATERS

1. Blocked baseboards?
   yes   no

2. Clean baseboards?
   yes   no

3. Crushed fin tubes?
   yes   no

ELECTRIC RESISTANCE HEATERS

1. Blocked heaters?
   yes   no

2. Clean heaters?
   yes   no

GENERAL

1. Cold or hot rooms?
   yes   no

   If yes, which ones?

   _____

   _____

2. Thermostat setting:
   ___°F

3. Set back at night?
   yes   no

   If yes, to what?
   ___°F

• Envelope

1. Insulated wall?
   yes   no

   If yes, insulated with:
   batts
   rigid board
   blown-in cellulose
   blown-in fiberglass
   urea-formaldehyde
      foam

   R-value—if you
      know it:

   _____

2. Caulked cracks?
   yes   no

3. Weather-stripped windows and doors?
   yes   no

   If yes, quality?
   good   fair
      poor

4. Storm windows?
   yes   no

If yes, quality?
   good   fair
      poor

5. Storm doors?
   yes   no

   If yes, quality?
   good   fair
      poor

ATTIC

1. Insulated attic?
   yes   no

   If yes, insulated with:
   batts in roof rafters
   batts in ceiling joists
   batts on floor
   rigid board
   blown-in cellulose
   blown-in fiberglass

   R-value—if you
      know it:

   _____

2. If living space, are cracks caulked?
   yes   no

3. If living space, are storm windows in place?
   yes   no

   If yes, quality?
   good   fair
      poor

RESULTS OF THE ENERGY SURVEY

Some Interacting Effects of Energy Conservation

As a general rule of thumb, energy conservation measures installed in one part of a building do not affect the rate of heat loss in another part. For example, attic insulation does not affect heat loss through walls, windows, basements or doors.

There are, however, instances in which such improvements do interact. The presence or absence of weather stripping does affect heat loss through storm doors or windows. The purpose of a storm door or window is to create a layer of "dead" air between the storm and prime door or window that will serve as insulation to impede the conduction of heat from the interior to the exterior. Double-glazed windows serve a similar function. If there is no weather stripping, cold air can enter around the window frame and negate much of the energy savings created by the double glazing.

Windows and doors with storms and weather stripping will save more fuel than those with only one or the other. But the fuel saved by adding both improvements is less than the sum of the fuel saved from storm windows or weather stripping alone. That is, the effects of multiple improvements to one part of a building are not always additive. For example, adding only storm windows may save 100 gallons of oil, while adding only weather stripping may save 50 gallons. Adding both would save more than 100 gallons but less than 150. If you already have storm doors and windows, the additional fuel saved by weather stripping will be less

than it would have been if there were no storms.

A similar point can be made with regard to the interaction between weatherization and heating system improvements. Both weatherizing a building and improving a heating system will save more fuel than making either improvement alone, but the fuel saved will be less than the sum of the savings from weatherization alone and heating system improvements alone. This happens because each time an improvement is made, the heating load of the building is reduced and, consequently, the percentage of fuel saved by a subsequent improvement is applied to a smaller total heating load.

Let's take an example. Suppose you want to make two improvements—a change to the heating system and the addition of storm windows—to a building that consumes 2,000 gallons of heating oil per year. The first improvement by itself is expected to reduce energy consumption by 15 percent, or 300 gallons, per year. The second improvement by itself is expected to save 500 gallons per year. But if both improvements are made together, the combined savings will not total 800 gallons per year. The storm windows will save the same 500 gallons, but the heating system improvement will save only 15 percent of the *reduced* energy consumption, or 225 gallons (15 percent of 1,500 gallons). Hence, the total savings for the two improvements will actually be 725 gallons of oil per year.

This example points out an important distinction in the way energy savings are expressed and calculated in this book. If a particular improvement is described as saving a certain number of gallons or therms or kilowatt-hours of energy, this quantity will remain constant as long as additional improvements are not made to the same building component. On the other hand, if energy savings are expressed as a percentage of total consumption, the absolute value of the savings will decrease (and the payback period for the improvement will increase) as total energy use is reduced through improvements to the building envelope.

Priorities for Weatherization Improvements

Improvements to the Envelope

The single most significant weatherization step that can be taken in any building (whether insulated or partially insulated) is reduction of infiltration by weather-stripping, caulking and the general plugging of major air leaks. Weather stripping is plastic, rubber or metal strips placed over or in the movable parts of windows and doors. Caulk usually comes in a tube that is used with a caulking gun to seal air leaks around window and door frames, plumbing and electrical penetrations through walls and places where unlike materials such as wood and masonry meet.

Often, in addition to cracks, there are large and obvious holes in the building envelope, such as broken windowpanes and doors, cracks in walls or damperless fireplaces. These types of leaks should always be repaired first. Other unseen leaks may exist, including heat by-passes from the basement to the attic through spaces in walls and around vent pipes and chimneys. These are much more difficult to find and fix, but they may be major sources of heat loss.

After minimizing infiltration, the most cost-effective weatherization step that

can be applied to all four building types is to insulate an uninsulated attic. "Cost-effective" means a conservation improvement with a relatively rapid payback (less than three years). If the attic floor or roof pitches aren't very accessible, the insulation may have to be blown in.

In the single-family house and triplex, insulating the walls, although the most expensive step, saves the most fuel. The walls are insulated by blowing cellulose or fiberglass into the space between the interior lath and exterior walls. If the attic in the single-family house has been converted into a living space, both its walls and ceiling should be insulated.

The next most cost-effective step is installing storm windows and doors. The installation of storm windows, in particular, can cut fuel consumption considerably. Usually, storm windows are more important with regard to comfort than either double-door vestibules or storm doors. Vestibules can often be inexpensively made in older city houses by simply replacing a second door that was once removed.

Additions to Existing Insulation

If a building already has some wall or basement ceiling insulation, it sometimes does not pay to add to it, because the cost of the installation is much greater than the small amount of fuel savings realized. Often the only way wall insulation can be increased is by adding wood studs to the walls and putting in fiberglass batts or rigid board insulation, a very costly process.

It may be financial good sense to add to already existing attic insulation if you live in a cold climate and the existing insulation is less than R-11. This is usually done by blowing cellulose or fiberglass into the attic space, or by adding fiberglass batts to the roof or floor of the attic space. The fiberglass batts should be without a vapor barrier if one is already in place. The type of insulation and quantity to be added is dependent upon existing insulation, how much space is available for additional insulation and the presence or absence of a vapor barrier. These matters are best decided in consultation with an energy professional.

Labor Costs in Energy Conservation

Residential energy conservation is labor-intensive, so labor costs can figure prominently in conservation costs. Many conservation improvements can be done by the homeowner, including weather-stripping, installing storm windows, adding attic insulation, caulking cracks and insulating water heaters, pipes and ducts.

Insulating brick walls on the outside or inside is costly and requires the services of more than an insulation installer, because insulation (rigid foam), must be added and then an exterior or interior wall placed over it. It is often not cost-effective to insulate brick walls on the inside unless you are also planning to replace the entire interior finish wall. It is generally agreed, though, that exterior insulation for masonry walls makes the most sense. Either way, insulation of these walls can reduce fuel consumption considerably and increase comfort. Construction of party walls should also be noted. If they are hollow cinder block, there may be bypass leaks from the basement to attic that should be plugged.

The payback on insulating the basement ceiling tends to be longer than other weatherization steps (six to eight years for a row house and apartment building; five to six years for a triplex or single-family house), so it has a lower priority than other weatherization steps. Adding this insulation in very cold climates (over 6,000 degree-days) may cause an unheated basement to become cold enough for water pipes to freeze. If insulation is to be added to the basement ceiling, a professional energy expert should be consulted to see if this danger exists.

Improvements to Space Heating and Domestic Water Heating Systems

If the heating system is old, it may no longer be delivering heat effectively to the house. Or, it may be oversized for the building, especially after the building is better insulated and sealed. The boiler should be cleaned and tuned regularly. The valves on radiators should be checked to find if they still work. Each floor should be checked to determine if heat is being delivered uniformly throughout the building. Your oil or gas company serviceman should be able to assist you in these tests.

You should also check the temperature regulators (thermostats, airstats and aquastats). Ideally the thermostat should be no higher than 65 to 68°F in the day and 60°F at night. The aquastat on a water boiler often can be set to 150°F or lower and on an air furnace to 120°F. On a domestic water heater, the thermostat can also often be set to 120°F. Insulating the hot pipes or ducts used in heating and hot water systems is cost-effective.

In buildings with stand-alone water heaters, it is extremely cost-effective to wrap them with insulation. If the tank is poorly insulated, as most are, the payback period can be as short as several months. If your building has a "tankless" system, in which water is run through the boiler or furnace, it can be very cost-effective to replace or supplement this system with a stand-alone water heater fueled by gas or solar energy (see Oil- and Gas-Fired Domestic Hot Water Systems in chapter 3, and Solar Domestic Hot Water in chapter 5).

If your heating plant is old or inefficient, replacement may be in order. As a general rule of thumb, any furnace or boiler that is more than about 30 years old and has a tuned combustion efficiency of 65 percent or less is a good candidate for replacement. As mentioned above, if the plant is oversized for the heating load of the building as a result of building weatherization, replacement may also be cost-effective. A number of other cost-effective heating system improvements—for example, clock thermostats, automatic flue dampers, flame retention burners and heat pumps—can be made. These possibilities are discussed in chapter 3, "Heating and Cooling Systems."

MODEL ENERGY SURVEYS OF THE FOUR HOUSING TYPES

These surveys tabulate the costs of various weatherization improvements, the amount of fuel saved by these improvements and the payback period. The numbers are representative of the moderately cold (5,000 to 7,000 degree-days) regions of the United States. They have been calculated on the basis of specific energy costs as shown at the top of each column. To determine energy savings and paybacks for your city, use the appropriate correction factor for your climate zone given at the bottom of each table. See the map in the Introduction (figure 0-1) to determine your zone.

The tables are based upon characteristics of one building in each housing type, but buildings within each housing type vary greatly as to number of windows, area of living space, type of building materials and type of heating system. For example, if you live in a row house, the numbers in the table on row houses may not exactly apply to your building, but they should give you an idea of which improvements have the fastest payback or save the most fuel. If you wish to have specific numbers, consult an energy auditor. Your local gas or electric utility may be offering low-cost energy audits, or there may be companies in your city that do more expensive, but more detailed, audits. If your building uses less fuel than the buildings described here, you cannot expect energy savings for each improvement to be as great. The costs given in the tables are approximate and may be high or low compared to actual costs for your area.

When using the tables on the following pages, remember that you cannot always simply add together the listed savings for individual retrofits, especially when several improvements are made to the same building component. Remember, too, that if you try to figure in heating system improvements (generally expressed in this book as percentage reductions), the energy savings must be calculated on the *reduced* heating load.

Row House

The row house used in this analysis is typical of such houses in many cities. It is made of brick and has a flat roof. It has four floors and a full basement, with one stairway connecting them. Two walls of the house are exposed to the outdoors; the other two are attached to the neighboring houses. The house has three exterior doors, two in the front and one in the back.

The house is serviced by one main heater and one water heater, usually an 80-gallon tank heated by oil, gas or electricity. Its heating system is typically quite old, heating the building with steam or hot water radiators. Both heaters are located in the unheated basement.

Specifications of the Row House

Dimensions: 35 feet by 20 feet

Heated floor area: 2,800 square feet

Levels: Basement (7½ feet high) and four floors (9 feet floor to floor)

Walls: Exterior—12 inches masonry, basement and first floor; 8 inches masonry, upper floors; interior—¾-inch air space, lath and plaster

Roof: Flat; tar and gravel on top of 1-inch wood board; 2-foot space between roof and ceiling

Ceiling/Floor: Wood lath and plaster; ¾-inch subfloor and ¾-inch floor (both wood); 16-inch space between floor and ceiling

Basement: 3 feet above grade; 2½-inch concrete slab floor

Windows: Wood sash, double-hung; five on street level, five on main level, six on each two floors above (all 15 square feet); four in basement (all 4½ square feet)

Doors: 1½-inch wood; three (all 20 square feet)

Stairways: one

Heat-loss rate: 12.4 Btu per square foot per degree-day or 34,720 Btu per degree-day

Table 1-2
Row House Monthly and Annual Heat Load
(millions of Btu)

Zone	July	Aug.	Sept.	Oct.	Nov.	Dec.	Jan.
I	0	0	0	1.67	9.44	15.97	17.22
II	0	0	1.49	8.06	19.16	26.11	27.50
III	0	0	3.19	12.92	25.28	36.63	39.23
IV	0	1.74	7.32	17.74	32.74	45.97	51.52

*Assumes 55% conversion and distribution efficiency.
†Assumes 100% distribution efficiency.

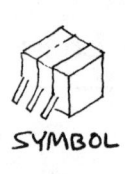

SYMBOL

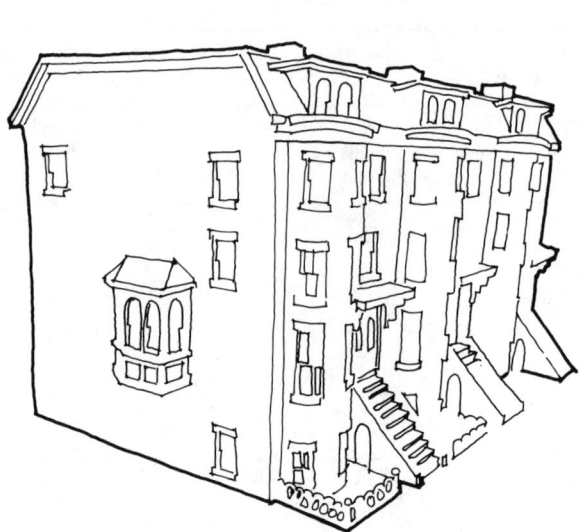

FIGURE I-4: ROW HOUSES

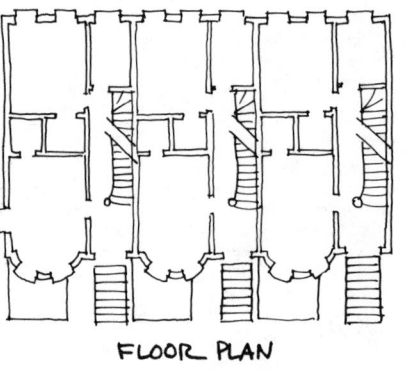

FLOOR PLAN

Feb.	Mar.	Apr.	May	June	Total Annual	Heat load in terms of fuel and electric power		
						Oil* (gal)	Gas* (therm)	Elec.† (1,000 KWH)
13.19	9.58	2.22	0.14	0	69.43	896	1,264	20.34
22.29	19.86	9.79	3.89	0	138.15	1,783	2,515	40.48
34.06	29.41	17.57	8.26	1.46	208.01	2,685	3,786	60.95
43.19	38.30	22.98	11.84	3.51	276.85	3,573	5,039	81.12

Table 1-3
Costs and Benefits for Energy Improvements to a Row House

Improvements made by a contractor	Cost ($)	Heating Oil* (@ $1.25/gal)		Natural Gas* (@ $0.50/therm)		Electricity† (@ $0.07/KWH)	
		Annual Savings (gal)	Payback (yrs)	Annual Savings (therms)	Payback (yrs)	Annual Savings (KWH)	Payback (yrs)
Weather stripping (no existing storms)	560	124	3.6	175	6.4	2,818	2.8
Weather stripping (existing storms)	560	79	5.7	111	10.1	1,795	4.5
Storm doors (no existing weather stripping)	360	51	5.6	72	10.0	1,159	4.4
Storm doors (existing weather stripping)	360	51	5.6	72	10.0	1,159	4.4
Storm windows (no existing weather stripping)	880	444	1.6	626	2.8	10,088	1.2
Storm windows (existing weather stripping)	880	405	1.7	571	3.0	9,202	1.3
Weather stripping & storm doors	920	175	4.2	247	7.5	3,976	3.3
Weather stripping & storm windows	1,440	526	2.2	742	3.9	11,952	1.7
Attic insulation— R-11 fiberglass batts	405	283	1.1	399	2.0	6,430	0.83
Attic insulation— R-19 fiberglass batts	560	317	1.4	447	2.5	7,203	1.1
Attic insulation— R-30 fiberglass batts	720	333	1.7	470	3.0	7,566	1.3
Attic insulation— R-38 fiberglass batts	1,125	342	2.6	482	4.6	7,771	2.0
Attic insulation— R-30 cellulose fill	440	336	1.0	474	1.8	7,634	0.75
Attic—R-19 added to existing R-11 (cellulose)	385	51	6.0	72	10.7	1,159	4.7
Attic—R-11 added to existing R-19 (cellulose)	335	17	15.8	24	28.1	386	12.4
Basement ceiling—R-11 fiberglass batts	440	62	5.7	87	10.1	1,409	4.5
Basement ceiling—R-19 fiberglass batts	600	73	6.6	103	11.7	1,659	5.2
Wall insulation— R-11 fiberglass batts	3,675	362	8.1	510	14.4	8,225	6.4
Basement wall insulation— R-4 rigid board	1,030	257	3.2	362	5.7	5,840	2.5
Hot water pipes— foam wrapped	1.90/ft‡	0.53/ft‡	2.9	0.75/ft‡	5.2	...	...

Table 1-3 — Continued

Improvements made by a contractor	Cost ($)	Heating Oil* (@ $1.25/gal)		Natural Gas* (@ $0.50/therm)		Electricity† (@ $0.07/KWH)	
		Annual Savings (gal)	Payback (yrs)	Annual Savings (therms)	Payback (yrs)	Annual Savings (KWH)	Payback (yrs)
Steam pipes—rigid fiberglass wrap	1.90/ft‡	0.42/ft‡	3.6	0.59/ft‡	6.4	...	...
Ducts—fiberglass duct wrap	1.60/ft²§	0.66/ft²§	1.9	0.93/ft²§	3.4	...	...
Domestic water heater— R-19 insulation (80-gal tank)	38	122	0.25	172	0.42	2,769	0.16
Improvements made by individual							
Weather stripping (no existing storms)	110	124	0.71	175	1.3	2,818	0.58
Weather stripping (existing storms)	110	79	1.1	111	2.0	1,795	0.83
Storm windows (no existing weather stripping)	374	444	0.66	626	1.2	10,088	0.5
Storm windows (existing weather stripping)	374	405	0.75	571	1.3	9,202	0.58
Weather stripping & storm windows	484	526	0.75	742	1.3	11,952	0.58
Attic insulation—R-11 fiberglass batts	121	283	0.33	399	0.58	6,430	0.25
Attic insulation—R-19 fiberglass batts	137	317	0.33	447	0.58	7,203	0.25
Attic—R-11 added to existing R-19 (batts)	121	17	6.4	24	11.4	386	6.5
Attic—R-19 added to existing R-11 (batts)	137	51	1.9	72	3.4	1,159	1.5
Hot water pipes— foam wrapped	0.63/ft‡	0.53/ft‡	0.92	0.75/ft‡	1.6	...	...
Ducts—fiberglass duct wrap	0.23/ft²§	0.66/ft²§	0.25	0.93/ft²§	0.42	...	...
Domestic water heater—R-19 insulation (80-gal tank)	26	122	0.16	172	0.33	2,769	0.25

Correction factors: To find fuel savings, multiply by the appropriate factor for your zone. To find payback, divide by the same factor. Zone I: 0.33; Zone II: 0.67; Zone III: 1.00; Zone IV: 1.33.

* Assumes 55% conversion and distribution efficiency.
†Assumes 100% distribution efficiency.
‡Measure the total length of hot pipes; multiply that number by the cost per foot to find the total cost, and by fuel saved per foot to find total fuel savings.
§Measure the average distance around the cross section of a duct; multiply this number by the length of exposed ducts; multiply that product (total square feet) by the cost per square foot to find the total cost; multiply total square feet the fuel per square foot to total fuel savings.

Triplex

The triplex used in this analysis is typical of such buildings in many cities. It is a shingled or clapboard wood-frame building with a flat roof. It has three floors, each an individual residence, and a basement, with two stairways connecting them. Each residence has two exterior doors, one in the front and one in the back.

This building is serviced by three oil boilers, one for each residence. The heating systems are typically quite old and heat the building with steam or hot water radiators. Domestic hot water is provided by a common 80-gallon tank heated by oil, gas or electricity. All heaters are located in the unheated basement.

Specification of the Triplex

Dimensions: 40 feet by 22 feet

Heated floor area: 2,640 square feet

Levels: Basement (7½ feet high) and three floors (9 feet floor to floor)

Walls: Exterior—½-inch clapboard or shingle, tar paper and ¾-inch wood sheathing; interior— 3½-inch air space, lath and plaster

Roof: Flat; tar and gravel on top of 1-inch wood board; 2-foot space between roof and ceiling

Ceiling/Floor: Wood lath and plaster; ¾-inch subfloor and ¾-inch floor (both wood); 16-inch space between floor and ceiling

Basement: 2½ feet above grade; 2-inch concrete slab floor

Windows: Wood sash, double-hung; 15 on each floor (all 15 square feet); 6 in basement (all 4½ square feet)

Doors: 1½-inch wood; two per floor (all 20 square feet)

Stairways: two

Heat-loss rate: 22.5 Btu per square foot per degree-day or 59,400 Btu per degree-day

Table 1-4
Triplex Monthly and Annual Heat Load
(millions of Btu)

Zone	July	Aug.	Sept.	Oct.	Nov.	Dec.	Jan.
I	0	0	0	2.85	16.16	27.32	29.46
II	0	0	2.55	13.78	32.79	44.67	47.04
III	0	0	5.46	22.10	43.24	62.67	67.12
IV	0	2.97	12.53	30.35	56.01	78.64	88.15

*Assumes 55% conversion and distribution efficiency.
†Assumes 100% distribution efficiency.

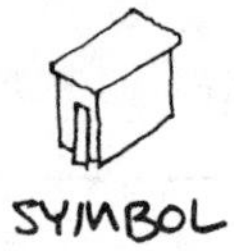

SYMBOL

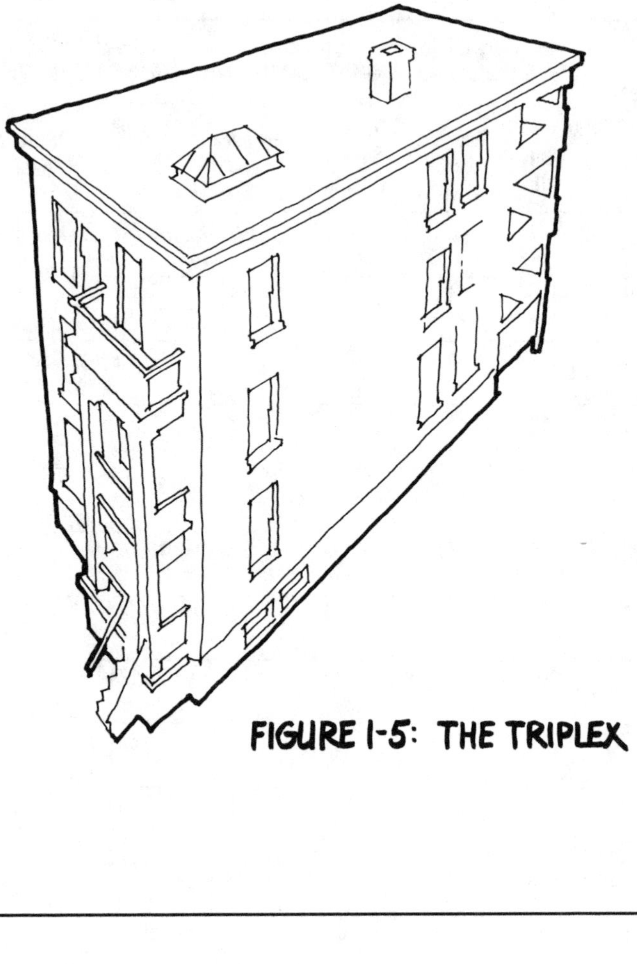

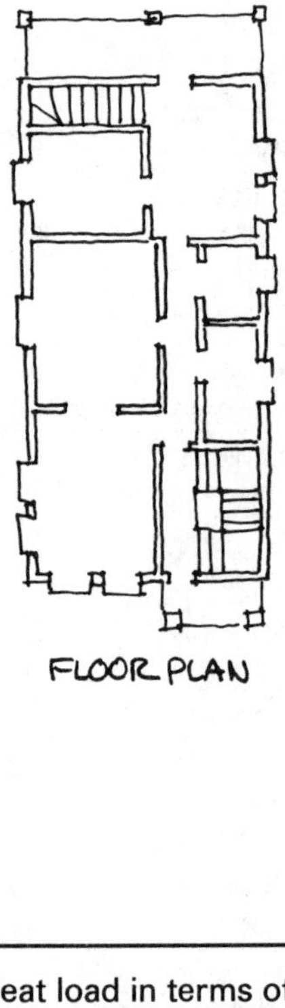

FLOOR PLAN

FIGURE I-5: THE TRIPLEX

Feb.	Mar.	Apr.	May	June	Total Annual	Heat load in terms of fuel and electric power		
						Oil* (gal)	Gas* (therm)	Elec.† (1,000 KWH)
22.57	16.39	3.80	0.24	0	118.79	1,532	2,160	34.80
38.13	33.98	16.75	6.55	0	236.34	3,048	4,297	69.23
58.27	50.31	30.06	14.14	2.49	355.86	4,588	6,470	104.27
73.89	65.52	39.32	20.26	6.00	473.64	6,107	8,612	138.75

Table 1-5
Costs and Benefits for Energy Improvements to a Triplex

Improvements made by a contractor	Cost ($)	Heating Oil* (@ $1.25/gal)		Natural Gas* (@ $0.50/therm)		Electricity† (@ $0.07/KWH)	
		Annual Savings (gal)	Payback (yrs)	Annual Savings (therms)	Payback (yrs)	Annual Savings (KWH)	Payback (yrs)
Weather stripping (no existing storms)	1,140	297	3.1	419	5.5	6,742	2.4
Weather stripping (existing storms)	1,140	195	4.7	275	8.4	4,426	3.7
Storm doors (no existing weather stripping)	720	144	4.0	203	7.1	3,269	3.1
Storm doors (existing weather stripping)	720	130	4.4	183	7.8	2,951	3.4
Storm windows (no existing weather stripping)	1,800	903	1.6	1,273	2.8	20,498	1.2
Storm windows (existing weather stripping)	1,800	815	1.8	1,149	3.2	18,500	1.4
Weather stripping & storm doors	1,860	427	3.5	602	6.2	9,693	2.7
Weather stripping & storm windows	2,940	1,109	2.1	1,564	3.7	25,174	1.6
Attic insulation— R-11 fiberglass batts	505	356	1.1	502	2.0	8,081	0.83
Attic insulation— R-19 fiberglass batts	710	396	1.4	558	2.5	8,989	1.1
Attic insulation— R-30 fiberglass batts	910	421	1.7	594	3.0	9,557	1.3
Attic insulation— R-38 fiberglass batts	1,415	430	2.6	606	4.6	9,761	2.0
Attic insulation— R-30 cellulose fill	555	421	1.0	594	1.8	9,557	0.75
Attic—R-19 added to existing R-11 (cellulose)	485	65	6.0	92	10.7	1,476	4.7
Attic—R-11 added to existing R-19 (cellulose)	425	25	13.6	35	24.2	568	10.6
Basement ceiling—R-11 fiberglass batts	555	105	4.2	148	7.5	2,384	3.3
Basement ceiling—R-19 fiberglass batts	760	119	5.1	168	9.1	2,701	4.0
Wall insulation—R-13 cellulose fill	2,055	791	2.1	1,115	3.7	17,956	1.6
Hot water pipes— foam wrapped	1.90/ft‡	0.53/ft‡	2.9	0.75/ft‡	5.2	...	...

Table 1-5—*Continued*

Improvements made by a contractor	Cost ($)	Heating Oil* (@ $1.25/gal)		Natural Gas* (@ $0.50/therm)		Electricity† (@ $0.07/KWH)	
		Annual Savings (gal)	Payback (yrs)	Annual Savings (therms)	Payback (yrs)	Annual Savings (KWH)	Payback (yrs)
Steam pipes—rigid fiberglass wrap	1.90/ft‡	0.42/ft‡	3.7	0.59/ft‡	6.6	...	...
Ducts—fiberglass duct wrap	1.60/ft²§	0.66/ft²§	1.9	0.93/ft²§	3.4	...	...
Domestic water heater—R-19 insulation (80-gal tank)	38	122	0.25	172	0.42	2,769	0.16
Improvements made by individual							
Weather stripping (no existing storms)	225	297	0.58	419	1.0	6,742	0.45
Weather stripping (existing storms)	225	195	0.92	275	1.6	4,426	0.75
Storm windows (no existing weather stripping)	880	903	0.75	1,273	1.3	20,498	0.58
Storm windows (existing weather stripping)	880	815	0.83	1,149	1.5	18,500	0.66
Weather stripping & storm windows	1,103	1,109	0.79	1,564	1.4	25,174	0.58
Attic insulation—R-11 fiberglass batts	150	356	0.33	502	0.58	8,081	0.25
Attic insulation—R-19 fiberglass batts	170	396	0.33	558	0.58	8,989	0.25
Attic—R-11 added to existing R-19 (batts)	150	25	5.4	35	9.6	568	4.2
Attic—R-19 added to existing R-11 (batts)	170	65	1.8	92	2.3	1,476	1.4
Hot water pipes— foam wrapped	0.63/ft‡	0.53/ft‡	0.92	0.75/ft‡	1.6	...	...
Ducts—fiberglass duct wrap	0.23/ft²§	0.66/ft²§	0.25	0.59/ft²§	0.42	...	...
Domestic water heater—R-19 insulation (80-gal tank)	26	122	0.16	172	0.29	2,769	0.12

Correction factors: To find fuel savings, multiply by the appropriate factor for your zone. To find payback, divide by the same factor. Zone I: 0.33; Zone II: 0.67; Zone III: 1.00; Zone IV: 1.33.

*Assumes 55% conversion and distribution efficiency.

†Assumes 100% distribution efficiency.

‡Measure the total length of hot pipes; multiply that number by the cost per foot to find the total cost, and by fuel saved per foot to find total fuel savings.

§Measure the average distance around the cross section of a duct; multiply this number by the length of exposed ducts; multiply that product (total square feet) by the cost per square foot to find the total cost; multiply total square feet by the fuel saved per square foot to find total fuel savings.

Single-Family House

The single-family house used in this analysis is typical of such houses in many cities. It is a shingled or clapboard-sided wood-framed building with a pitched roof that is 12 feet high at its peak. It has two floors, an attic that may have been converted to living space and a basement, with one stairway connecting them. The building has two exterior doors, one in the front and one in the back, and a doorway leading to the basement.

The house is serviced by an oil boiler for space heating and a 30- to 50-gallon water heater heated by oil, gas or electricity. The heating system is typically quite old and heats the building with steam or hot water radiators. The boiler and water heater are located in the unheated basement.

Specifications of the Single-Family House

Dimensions: 30 feet by 40 feet

Heated floor area: 2,400 square feet

Levels: Basement (7½ feet high), two floors (9 feet floor to floor) and attic (12 feet high at peak)

Walls: Exterior—½-inch clapboard or shingle, tar paper and ¾-inch wood sheathing; interior—3½-inch air space, lath and plaster

Roof: Pitched, wood shingles mounted on paper and wood sheathing; 7½-inch space between roof and attic ceiling (lath and plaster)

Ceiling/Floor: Wood lath and plaster; ¾-inch subfloor and ¾-inch floor (both wood); 16-inch space between floor and ceiling

Basement: 2½ feet above grade; 2-inch concrete slab floor

Windows: Wood sash, double-hung; 15 on each floor (all 15 square feet); 6 in attic—2 dormer windows (all 12 square feet); 8 in basement (all 4 square feet)

Doors: 1½-inch wood; two (both 20 square feet); one bulkhead to basement (30 square feet)

Stairways: one

Heat-loss rate: 21.2 Btu per square foot per degree-day or 50,880 Btu per degree-day

Table 1-6
Single-Family House Monthly and Annual Heat Load
(millions of Btu)

Zone	July	Aug.	Sept.	Oct.	Nov.	Dec.	Jan.
I	0	0	0	2.44	13.84	23.40	25.24
II	0	0	2.19	11.80	28.08	38.26	40.30
III	0	0	4.68	18.93	37.04	53.68	57.49
IV	0	2.54	10.74	26.00	47.98	67.36	75.50

*Assumes 55% conversion and distribution efficiency.
†Assumes 100% distribution efficiency.

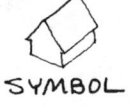

SYMBOL

FIGURE I-6: THE SINGLE-FAMILY HOUSE

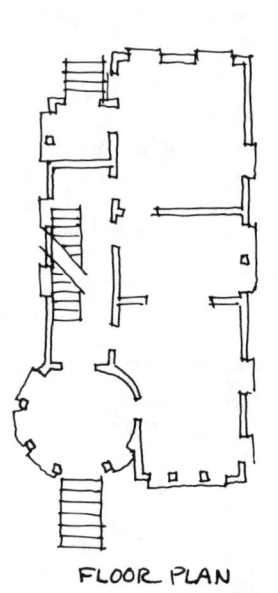

FLOOR PLAN

Feb.	Mar.	Apr.	May	June	Total Annual	Heat load in terms of fuel and electric power		
						Oil* (gal)	Gas* (therm)	Elec.† (1,000 KWH)
19.33	14.04	3.26	0.20	0	101.75	1,312	1,850	29.81
32.66	29.10	14.35	5.70	0	202.44	2,610	3,781	59.32
49.91	43.06	25.74	12.11	2.14	304.78	3,931	5,542	89.30
63.29	56.12	33.68	17.35	5.14	405.70	5,231	7,359	118.87

Table 1-7
Costs and Benefits for Energy Improvements to a Single-Family House

Improvements made by a contractor	Cost ($)	Heating Oil* (@ $1.25/gal)		Natural Gas* (@ $0.50/therm)		Electricity† (@ $0.07/KWH)	
		Annual Savings (gal)	Payback (yrs)	Annual Savings (therms)	Payback (yrs)	Annual Savings (KWH)	Payback (yrs)
Weather stripping (no existing storms)	860	178	3.9	251	6.9	4,041	3.0
Weather stripping (existing storms)	860	93	7.4	131	13.2	2,111	5.8
Storm doors (no existing weather stripping)	240	48	4.0	68	7.1	1,090	3.1
Storm doors (existing weather stripping)	240	34	5.6	48	10.0	772	4.4
Storm windows (no existing weather stripping)	1,140	699	1.6	986	2.8	15,867	1.2
Storm windows (existing weather stripping)	1,140	631	1.8	890	3.2	14,324	1.4
Weather stripping & storm doors	1,100	212	4.2	299	7.5	4,812	3.3
Weather stripping & storm windows	2,300	818	2.2	1,153	3.9	18,569	1.7
Attic insulation—R-11 fiberglass batts	875	489	1.4	689	2.5	11,100	1.1
Attic insulation— R-19 fiberglass batts	1,225	551	1.8	777	3.2	12,508	1.4
Attic insulation— R-30 fiberglass batts	1,460	579	2.0	816	3.6	13,143	1.6
Attic insulation— R-38 fiberglass batts	2,050	593	2.8	836	5.0	13,461	2.2
Attic insulation—R-30 cellulose fill	965	593	1.3	836	2.3	13,461	1.0
Attic—R-19 added to existing R-11 (cellulose)	840	90	7.5	127	13.4	2,043	5.8
Attic—R-11 added to existing R-19 (cellulose)	735	28	21.0	39	37.4	636	16.4
Basement ceiling—R-11 fiberglass batts	760	138	4.4	195	7.8	3,133	3.4
Basement ceiling—R-19 fiberglass batts	1,035	161	5.1	227	9.1	3,655	4.0
Wall insulation—R-13 cellulose fill	1,970	757	2.1	1,067	3.7	17,184	1.6
Hot water pipes— foam wrapped	1.90/ft‡	0.53/ft‡	2.9	0.75/ft‡	5.2	...	...

Table 1-7—*Continued*

Improvements made by a contractor	Cost ($)	Heating Oil* (@ $1.25/gal)		Natural Gas* (@ $0.50/therm)		Electricity† (@ $0.07/KWH)	
		Annual Savings (gal)	Payback (yrs)	Annual Savings (therms)	Payback (yrs)	Annual Savings (KWH)	Payback (yrs)
Steam pipes—rigid fiberglass wrap	1.90/ft‡	0.42/ft‡	3.6	0.59/ft‡	6.4	...	...
Ducts—fiberglass duct wrap	1.60/ft²§	0.66/ft²§	1.9	0.93/ft²§	3.4	...	...
Domestic water heater—R-19 insulation (50-gal tank)	35	80	0.33	113	0.58	1,816	0.25

Improvements made by individual

	Cost ($)	Annual Savings (gal)	Payback (yrs)	Annual Savings (therms)	Payback (yrs)	Annual Savings (KWH)	Payback (yrs)
Weather stripping (no existing storms)	165	178	0.75	251	1.3	4,041	0.58
Weather stripping (existing storms)	165	93	1.4	131	2.5	2,111	1.1
Storm windows (no existing weather stripping)	612	699	0.71	986	1.3	15,867	0.58
Storm windows (existing weather stripping)	612	631	0.75	890	1.3	14,324	0.58
Weather stripping & storm windows	775	818	0.75	1,153	1.3	18,569	0.58
Attic insulation—R-11 fiberglass batts	265	489	0.42	689	0.75	11,100	0.33
Attic insulation—R-19 fiberglass batts	300	551	0.42	777	0.75	12,508	0.33
Attic—R-11 added to existing R-19 (batts)	265	28	8.6	39	15.3	636	6.7
Attic—R-19 added to existing R-11 (batts)	300	90	2.4	127	4.3	2,043	1.9
Hot water pipes— foam wrapped	0.63/ft‡	0.53/ft‡	0.92	0.75/ft‡	1.6	...	...
Ducts—fiberglass duct wrap	0.23/ft²§	0.66/ft²§	0.25	0.93/ft²§	0.42	...	...
Domestic water heater—R-19 insulation (50-gal tank)	23	80	0.25	113	0.42	1,816	0.16

Correction factors: To find fuel savings, multiply by the appropriate factor for your zone. To find payback, divide by the same factor. Zone I: 0.33; Zone II: 0.67; Zone III: 1.00; Zone IV: 1.33.

*Assumes 55% conversion and distribution efficiency.

†Assumes 100% distribution efficiency.

‡Measure the total length of hot pipes; multiply that number by the cost per foot to find the total cost, and by fuel saved per foot to find total fuel savings.

§Measure the average distance around the cross section of a duct; multiply this number by the length of exposed ducts; multiply that product (total square feet) by the cost per square foot to find the total cost; multiply total square feet by the fuel saved per square foot to find total fuel savings.

Multi-Unit Apartment Building

The multi-unit apartment building used in this analysis is typical of such buildings in many cities. It is made of brick and has a flat roof. It has four floors and a basement with two stairways connecting them. Six units are on each floor. The building has two exterior doors, one in the front and one in the back. It may also have elevators or a trash chute.

The building is serviced by one main heater, usually an oil boiler. The hot water needs are provided by a tankless hot water system attached to the boiler (summer-winter hook-up). Heat is supplied to the building by radiators or baseboard heating. The boiler is located in the unheated basement.

Specifications of the Multi-Unit Apartment Building

Dimensions: 50 feet by 90 feet

Heated floor area: 18,000 square feet

Levels: Basement (7½ feet high) and four floors (9 feet floor to floor); six units per floor

Walls: Exterior—12-inch masonry; interior—¾-inch air space, lath and plaster

Roof: Flat; tar and gravel on top of 1-inch wood board; 2-foot space between roof and ceiling

Ceiling/Floor: Wood lath and plaster; ¾-inch subfloor and ¾-inch floor (both wood); 16-inch space between floor and ceiling

Basement: 2½ feet above grade; 2-inch concrete slab floor

Windows: Wood sash, double-hung; 46 per floor (all 13½ square feet); 16 in basement (all 4½ square feet)

Doors: 1½-inch wood with glazing; two (12 square feet wood, 8 square feet glazing); glazing around front door—on top, 3 feet by 1 foot; on sides, 1½ foot by 8 feet

Stairways: two

Heat-loss rate: 9.8 Btu per square foot per degree-day or 176,400 Btu per degree-day

Table 1-8
Multi-Unit Apartment Building Monthly and Annual Heat Load
(millions of Btu)

Zone	July	Aug.	Sept.	Oct.	Nov.	Dec.	Jan.
I	0	0	0	8.47	47.98	81.14	87.49
II	0	0	7.58	40.92	97.37	132.65	139.71
III	0	0	16.23	65.62	128.42	186.10	199.33
IV	0	8.82	37.22	90.14	166.34	233.55	261.78

*Assumes 55% conversion and distribution efficiency.
†Assumes 100% distribution efficiency.

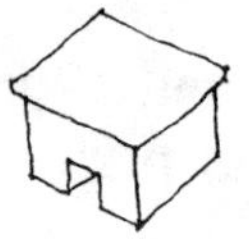

SYMBOL

FIGURE I-7: THE MULTI-UNIT APARTMENT BUILDING

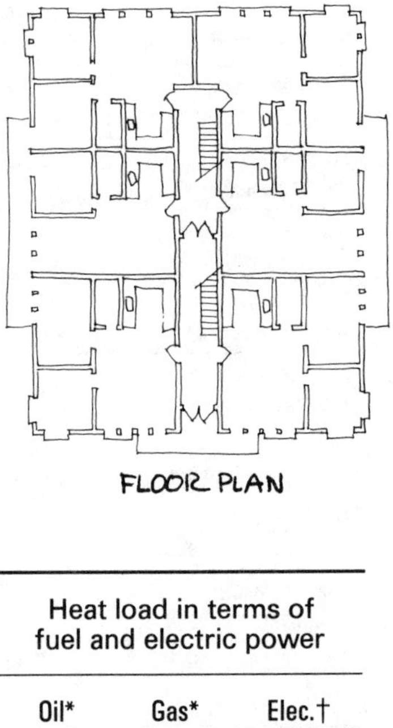

FLOOR PLAN

Feb.	Mar.	Apr.	May	June	Total Annual	Heat load in terms of fuel and electric power		
						Oil* (gal)	Gas* (therm)	Elec.† (1,000 KWH)
67.03	48.69	11.29	0.70	0	352.79	4,549	6,414	103.37
113.25	100.90	49.74	19.76	0	701.88	9,050	12,762	205.65
173.05	149.41	89.26	41.98	7.41	1,056.81	13,625	19,211	309.58
219.44	194.57	116.78	60.15	17.82	1,406.61	18,138	25,574	412.13

Table 1-9
Costs and Benefits for Energy Improvements to a Multi-Unit Apartment Building

Improvements made by a contractor	Cost ($)	Heating Oil* (@ $1.25/gal)		Natural Gas* (@ $0.50/therm)		Electricity† (@ $0.07/KWH)	
		Annual Savings (gal)	Payback (yrs)	Annual Savings (therms)	Payback (yrs)	Annual Savings (KWH)	Payback (yrs)
Weather stripping (no existing storms)	4,265	704	4.8	993	8.5	15,981	3.7
Weather stripping (existing storms)	4,265	370	9.2	522	16.4	8,399	7.2
Storm doors (no existing weather stripping)	240	51	3.8	72	6.8	1,158	3.0
Storm doors (existing weather stripping)	240	45	4.3	63	7.6	1,022	3.4
Storm windows (no existing weather stripping)	7,360	3,329	1.8	4,694	3.2	75,568	1.4
Storm windows (existing weather stripping)	7,360	2,295	2.0	3,236	3.6	52,097	1.6
Weather stripping & storm doors	4,505	749	4.8	1,056	8.5	17,002	3.7
Weather stripping & storm windows	11,625	3,696	2.5	5,211	4.4	83,899	2.0
Attic insulation—R-11 fiberglass batts	2,590	1,823	1.1	2,570	2.0	41,382	0.83
Attic insulation— R-19 fiberglass batts	3,620	2,026	0.92	2,857	1.6	45,990	0.75
Attic insulation— R-30 fiberglass batts	4,660	2,148	1.7	3,029	3.0	48,760	1.3
Attic insulation— R-38 fiberglass batts	7,245	2,196	2.6	3,096	4.6	49,849	2.0
Attic insulation—R-30 cellulose fill	2,845	2,153	1.1	3,036	2.0	48,873	0.83
Attic—R-19 added to existing R-11 (cellulose)	2,485	325	6.1	458	10.9	7,378	4.8
Attic—R-11 added to existing R-19 (cellulose)	2,175	122	14.2	172	25.3	2,769	11.1
Basement ceiling—R-11 fiberglass batts	2,845	345	6.6	486	11.7	7,832	5.1
Basement ceiling—R-19 fiberglass batts	3,880	398	7.8	561	13.9	9,035	6.1
Wall insulation—R-11 fiberglass batts	25,975	2,181	9.5	3,075	16.9	49,509	7.4
Wall insulation—R-4 rigid board	7,369	1,472	4.0	2,076	7.1	33,414	3.1

Table 1-9—*Continued*

Improvements made by a contractor	Cost ($)	Heating Oil* (@ $1.25/gal)		Natural Gas* (@ $0.50/therm)		Electricity† (@ $0.07/KWH)	
		Annual Savings (gal)	Payback (yrs)	Annual Savings (therms)	Payback (yrs)	Annual Savings (KWH)	Payback (yrs)
Hot water pipes— foam wrapped	1.90/ft‡	0.53/ft‡	2.9	0.75/ft‡	5.2	...	...
Steam pipes—rigid fiberglass wrap	1.90/ft‡	0.42/ft‡	3.6	0.59/ft‡	6.4	...	...
Ducts—fiberglass duct wrap	1.60/ft²§	0.66/ft²§	1.9	0.93/ft²§	3.4	...	...

Improvements made by individual

	Cost ($)	Annual Savings (gal)	Payback (yrs)	Annual Savings (therms)	Payback (yrs)	Annual Savings (KWH)	Payback (yrs)
Weather stripping (no existing storms)	770	704	0.87	993	1.6	15,981	0.66
Weather stripping (existing storms)	770	370	1.7	552	3.0	8,399	1.3
Storm windows (no existing weather stripping)	3,128	3,329	0.75	4,694	1.3	75,568	0.58
Storm windows (existing weather stripping)	3,128	2,295	1.1	3,236	2.0	52,097	0.83
Weather stripping & storm windows	3,898	3,696	0.83	5,211	1.5	83,899	0.66
Attic insulation—R-11 fiberglass batts	775	1,823	0.33	2,570	0.58	41,382	0.25
Attic insulation—R-19 fiberglass batts	880	2,026	0.33	2,857	0.58	45,990	0.25
Attic—R-11 added to existing R-19 (batts)	775	122	5.8	172	10.3	2,769	4.5
Attic—R-19 added to existing R-11 (batts)	880	325	1.9	458	3.4	7,378	1.5
Hot water pipes— foam wrapped	0.63/ft‡	0.53/ft‡	0.92	0.75/ft‡	1.6	...	...
Ducts—fiberglass duct wrap	0.23/ft²§	0.66/ft²§	0.25	0.93/ft²§	0.58	...	...

Correction factors: To find fuel savings, multiply by the appropriate factor for your zone. To find payback, divide by the same factor. Zone I: 0.33; Zone II: 0.67; Zone III: 1.00; Zone IV: 1.33.

*Assumes 55% conversion and distribution efficiency.

†Assumes 100% distribution efficiency.

‡Measure the total length of hot pipes; multiply that number by the cost per foot to find the total cost, and by fuel saved per foot to find total fuel savings.

§Measure the average distance around the cross section of a duct; multiply this number by the length of exposed ducts; multiply that product (total square feet) by the cost per square foot to find the total cost; multiply total square feet by the fuel saved per square foot to find total fuel savings.

A Typical Energy Improvement Package

Here are some results, costs and paybacks for what could be a typical package of energy conservation measures applied to the four building types. Such a package would consist of caulking, storm windows and doors, weather stripping, insulation in the attic, walls and basement ceiling and installation of three layers of glass on windows not facing south. South-facing windows would be left double-glazed in order to take advantage of solar gain through them. In most of the climate zones, paybacks for this weatherization package for all four building types using oil, gas or electricity are less than seven years. This is equivalent to a 15 percent annual return on investment (tax-free!).

Table 1-10
Building Heat Loss Reduction

Building Type	Heat Loss Rate Prior to Changes (Btu/ft²/DD)	Heat Loss Rate after Changes (Btu/ft²/DD)	Estimated Cost ($)
Row house	12.4	5.4	5,170
Triplex	22.5	7.6	7,470
Single-family	21.2	6.4	7,310
Apartment	9.8	3.3	27,410

Building Type and Heated Floor Area	Energy Improvements
Row House 2,800 ft²	Caulking, storms Weather stripping Attic: R-30 batts Walls: R-11 rigid board Basement: R-19 batts on ceiling Triple glazing on nonsouth windows
Triplex 2,640 ft²	Caulking, storms Weather stripping Attic: R-30 batts Walls: R-13 cellulose fill Basement: R-19 batts on ceiling Triple glazing on nonsouth windows
Single-Family House 2,400 ft²	Caulking, storms Weather stripping Attic: R-30 batts Walls: R-13 cellulose fill Basement: R-19 batts on ceiling Triple glazing on nonsouth windows
Multi-Unit Apartment Building 18,000 ft²	Caulking, storms Weather stripping Attic: R-30 batts Walls: R-11 rigid board Basement: R-19 batts in ceiling Triple glazing on nonsouth windows

Table 1-11
Simple Payback for the
Total Package*
(in years)

Building Type	Zone	Oil† (@$1.25/ gal)	Gas† (@$0.50/ therm)	Elec.‡ (@ $0.07/ KWH)
Row house	I	8.2	14.5	6.4
	II	4.1	7.2	3.2
	III	2.7	4.8	2.1
	IV	2.0	3.6	1.6
Triplex	I	7.3	12.9	5.7
	II	3.6	6.4	2.8
	III	2.4	4.3	1.9
	IV	1.8	3.2	1.4
Single-family	I	6.4	11.3	5.0
	II	3.2	5.6	2.5
	III	2.1	3.8	1.7
	IV	1.6	2.8	1.2
Apartment	I	7.3	12.9	5.7
	II	3.6	6.4	2.8
	III	2.4	4.3	1.9
	IV	1.8	3.2	1.4

*Divide the total estimated cost of the package by the cost of fuel (in dollars per gallon, therm or kilowatt-hour) in order to find the total amount of fuel saved during the payback period. To find annual fuel savings, divide total fuel savings by the payback period. Remember, fuel savings continue even after the investment has been paid back.

†Assumes 55% conversion and distribution efficiency.

‡Assumes 100% distribution efficiency.

For Further Reference

American Society of Heating, Refrigerating and Air-Conditioning Engineers (ASHRAE). *Cooling and Heating Load Calculation Manual.* New York, 1979.

Carrier Air Conditioning Co. *Carrier System Design Manual.* Syracuse, N.Y., 1972.

Central Mortgage and Housing Corporation of Canada. *The Conservation of Energy in Housing.* Ottawa, Ontario, 1977. Order no. NHA 5149.

Dubin, F. S., and Long, C. G., Jr. *Energy Conservation Standards.* New York: McGraw-Hill Book Co., 1978.

Farallones Institute. *The Integral Urban House.* San Francisco: Sierra Club Books, 1979.

Harrje, D. T.; Dutt, G. S.; and Gadsby, K. J. "Isolating the Building Thermal Envelope." Princeton, N.J.: Center for Energy and Environmental Studies, Princeton University, n.d.

Knight, P. A. *Home Retrofitting for Energy Savings.* New York: McGraw-Hill Book Co., 1981.

Leckie, J. et al. *More Other Homes and Garbage.* San Francisco: Sierra Club Books, 1981.

National Association of Realtors®. *The Realtor's® Guide to Residential Energy Efficiency.* Chicago, Ill., 1980.

Pacific Gas and Electric Co. "Home Use Energy Survey," San Francisco, n.d.

Socolow, R. H., ed. *Saving Energy in the Home.* Cambridge, Mass.: Ballinger Publishing Co., 1978.

U.S. Department of Energy. *Project Retro-Tech Home Weatherization Manual.* Washington, D.C.: U.S. Government Printing Office, 1975.

_____. *Residential Conservation Service Auditor Training Manual.* Washington, D.C.: U.S. Government Printing Office, 1980.

_____. *Saving Money with Energy Conservation: An Energy Audit Workbook for Apartment Buildings.* Washington, D.C.: U.S. Government Printing Office, 1980.

_____.Technical Assistance Program for the Residential Conservation Service Program. *Your Home Energy Audit: The First Step to More Energy Efficient Living.* Washington, D.C.: U.S. Government Printing Office, n.d.

U.S. Department of Housing and Urban Development. *In the Bank . . . Or Up the Chimney?* Washington, D.C.: U.S. Government Printing Office, 1977. (Also published as: *How to Keep Your House Warm in Winter, Cool in Summer.* New York: Cornerstone Library, 1977.)

Wilson, T., ed. *Home Remedies.* Philadelphia: Mid-Atlantic Solar Energy Association, 1981.

Wing, C. *From the Walls In.* Boston: Little, Brown and Company, 1979.

The Building Envelope

The building envelope consists of all the parts of a structure through which heat can pass from the inside to the outside. These include the exterior walls, roof, windows, doors and basement ceiling, if the basement is unheated, or basement walls and floors, if it is heated. Heat moves out of a building either by flowing through the materials making up the envelope (conduction) or by flowing through cracks, joints and seams in the envelope (infiltration). Heat is also radiated directly from surfaces such as windows, but this source of heat loss is generally small in relation to conduction and infiltration losses. Figure 2-1 shows where and how heat flows out of a building and gives the approximate percentages of heat lost through different building components.

CAULK AND WEATHER STRIPPING

One of the most cost-effective conservation improvements that you can make is the plugging of air leaks in the building envelope. Studies have shown that anywhere from 25 to 50 percent of the heat lost from a typical house (even if it is insulated) is the result of infiltration of outside air through the envelope. Caulk and weather stripping are materials made expressly for the purpose of sealing these leaks. Caulk generally comes in a tube as a pliable, sticky, gumlike material that can be worked into cracks and holes when it's forced out of the tube by a caulking gun. Weather stripping is made of metal, plastic, rubber or foam strips that can be placed in window frames and doorjambs. These materials are available in a variety of forms and costs and are of variable durability and effectiveness. Generally, it's best to buy higher-quality caulks and weather stripping in order to get the most durability and the best performance. The first cost will be higher, but with better materials it's a once-and-done job.

Where the Leaks Occur

Air leaks through the building envelope generally occur wherever two different types of material come into contact or where two different building components meet. Examples of the former are points where a brick chimney meets a wood frame wall or where a metal storm window touches the frame of a wood prime (original) window. An example of the latter is the seam between a wood frame wall and a window. Age, shrinkage of materials and foundation settling can cause cracks to open up where none previously existed, and consistent dry heat can cause wood frames to crack and split. Because these cracks can develop on both the exterior and interior sides of a wall, cold air can easily enter a house and warm air can similarly escape.

Other points where infiltration "short circuits" occur are places where walls and foundations meet; corners formed by siding; openings around plumbing and wiring penetrations through walls, floors and ceilings; joints beneath baseboards and finish floors; the point of contact between a porch and an exterior wall; seams between exterior walls and built-in cabinets; seams around fireplaces and mantels; and, of course, the seams around doors and operable windows.

Some points of infiltration are less obvious, or even invisible, but can still be sources of significant heat loss. The hot air rising through a building causes what is called the *stack effect*. Hot air escaping through the roof draws cold air in around the foundation and through cracks in the walls of the lower floors. In some instances, there may be direct passages from the basement to the attic through spaces inside the walls or around chimneys and vent pipes. These can be important sources of heat loss by infiltration. Any reservoir of cold air in the lower part of a house, such as a basement, can act as a feed to the stack effect, and it is important that these leaks be found and eliminated.

How to Find the Leaks

How, then, can you go about locating both the obvious and the invisible air leaks in the building envelope? On a windy day, air leaks in the walls are easy to detect. You can take a piece of tissue paper or light plastic and hold it

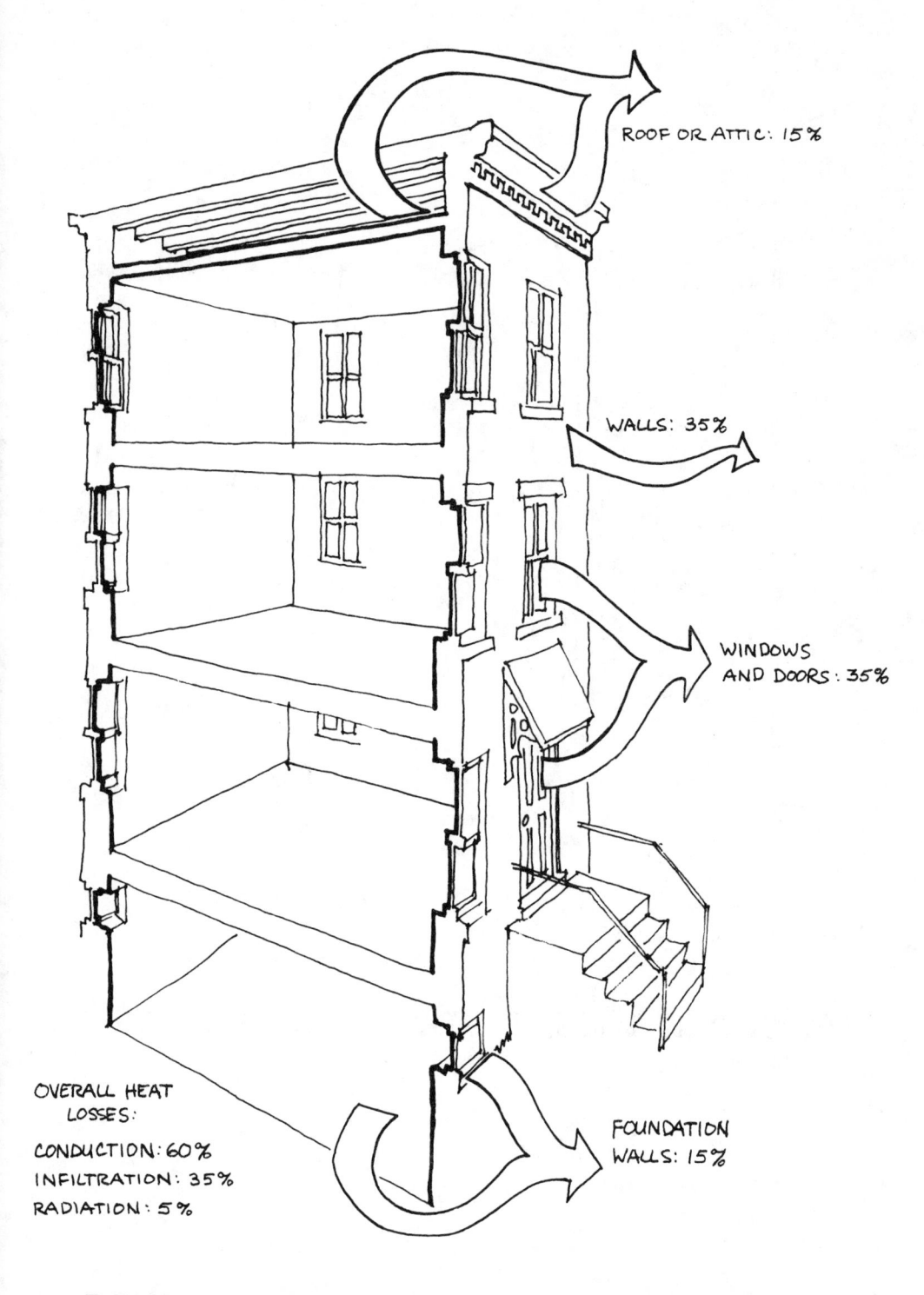

ROOF OR ATTIC: 15%

WALLS: 35%

WINDOWS AND DOORS: 35%

OVERALL HEAT LOSSES:

CONDUCTION: 60%
INFILTRATION: 35%
RADIATION: 5%

FOUNDATION WALLS: 15%

FIGURE 2-1: HEAT FLOW THROUGH THE BUILDING ENVELOPE

FAN IS TEMPORARILY SEALED IN WINDOW FRAME

INCENSE STICK

FIGURE 2-2: FAN PRESSURIZATION FOR FINDING AIR LEAKS

in front of suspected leaks. The piece of paper or plastic will move whenever you find a breeze. This method of leak detection, however, is dependent upon weather conditions and will only enable you to find leaks on the upwind side of the house.

A more dependable approach is to use a source of smoke, such as an incense stick, a punk or a smoke stick. (Do not use an open flame; it can be sucked into the wall and cause a fire.) Depending upon whether the airflow is into or out of the house at a particular location, you will see the smoke sucked into or blown away from leaks. Likely locations for such leaks are around windows and doors; near baseboards, electrical outlets, light fixtures and switches; and around plumbing penetrations, fireplaces, furnace registers and kitchen or bathroom vents. After some practice, you will not even need to use a smoke source; leaks will become self-evident.

A more involved approach to finding air leaks is to "pressurize" the house with a large window fan. The fan is installed in a window, with care taken to seal all openings around the fan. Be sure to install the fan so that it blows air into the house. Close all other windows and exterior doors and cover any bathroom and kitchen vents and furnace registers with plastic, and turn on the fan. The air blowing into the house will cause the air pressure inside the house to be higher than that outside, and air will escape through all the cracks and holes in the building envelope. If your fan is too small to pressurize the entire house, try closing off some rooms. Your smoke source will now reveal the myriad of leaks that occur in any house and will suggest which are the most important to seal first.

Caulking Materials

Caulking materials come in a number of different forms (see table 2-1). They include rope caulk; oil-, latex- and butyl-based compounds; solvent-based acrylic, polysulfide, urethane and silicone sealants; nitrile rubber and polymeric foam sealants. While many caulks are interchangeable, there are usually only two or three types suitable for a specific application.

Caulks vary greatly in cost and durability. For example, oil-based caulk is relatively inexpensive but does not stand up well to external weather conditions. Even when used inside a building, it may dry out and crack after several years. Silicone sealant, on the other hand, is very durable and well-suited for outside use, but it cannot be painted with ordinary paints and is very expensive. Latex- and butyl-based caulks are intermediate in cost and tend to last from 5 to 10 years, depending upon where they are used.

Rope caulk is very inexpensive and convenient to use, particularly as a temporary seal around window joints, but will usually last only one or two years and is difficult to reuse. Foam sealant is very convenient to use, especially for large openings and hard-to-reach places. It is, however, quite expensive. Advertisements commonly claim that one small can of foam is equivalent to as many as a dozen tubes of ordinary caulking compound, but one should be skeptical of such claims. Some materials, such as nitrile rubber, are for use only in high moisture areas, and it is a good idea to check any caulk you may buy for recommended uses. Figure 2-3 shows typical locations where caulk is used.

Table 2-1
Types of Caulking Compounds

Type	Relative Cost	Lifetime (yrs)	How It's Used
Rope caulk	Low	1-2	Easy to use; needs dust and oil-free surfaces; use above 40°F
Oil-based caulk	Lowest	3-5	Comes in tubes; easily applied and tooled; minimal cleaning of joints; use above 40°F
Latex-based caulk	Low to moderate	5-10	Comes in tubes; easily applied and tooled; requires well-cleaned joints; use above 40°F
Butyl-based caulk	Low to moderate	5-10	Comes in tubes; easily applied and tooled; clean surfaces with solvents; use above 40°F
Solvent-based acrylic sealants	Moderate	10-20	Heat to 120°F before using; needs ventilation and minimal cleaning of joints; comes in tubes
Polysulfide sealants	Moderate to high	20-30	Comes in tubes; two-part compound can be messy when mixing; can't be applied and won't cure below 40°F
Urethane sealants	High to very high	20-30	Joints must be very clean; use above 32°F and below 70% relative humidity
Silicone sealants	High to very high	30	Comes in tubes; easily applied and tooled; very clean surfaces required; can be applied down to −55°F
Nitrile rubber	Moderate to high	15-20	Primarily for use in areas requiring seal against liquid water
Polymeric foam sealants	High	20-30	Comes in spray can or cartridge; minimal cleaning of joints; contents must be kept above 60°F when using

SOURCE: U.S. Department of Energy. *Residential Conservation Service Auditor Training Manual* (Washington, D.C.: U.S. Government Printing Office, 1980).

Weather Stripping

Weather stripping is also available in a variety of forms and quality grades (see table 2-2). As a general rule of thumb, inexpensive weather-stripping materials, such as foam, felt and rolled vinyl, are also the least durable and may be difficult to install properly. Some of the more expensive materials, while extremely durable, can require professional installation, thereby reducing their

Performance	Good Features	Bad Features
Moderate to high expansion; little shrinkage; becomes brittle at low temperatures	Available in packages suitable for two or six standard-size windows; easy to remove	Must be replaced frequently; windows cannot be opened
Little expansion: low shrinkage; performs from −10 to 180°F, although poorly at lower range	Good for nonmoving interior surfaces and joints; fast cure; paint after two-three days	Poor for outside or moving joints; should not be used for glazing work
Low to moderate expansion; high shrinkage; best on small joints; good from −20 to 150°F	Can be used on damp absorptive surfaces; fast cure; paint shortly after application	Poor adhesion on moving parts
Moderate expansion; high shrinkage; use on larger joints and cracks; good from −25 to 250°F	Adheres well to all building materials, especially masonry to metal; comes in many colors	Good only for joints with moderate movement; paint after one week
Low to moderate expansion; little shrinkage; good from −20 to 150°F	Good on wide and/or moving joints; comes in colors	Strong odor limits use to well-ventilated spaces or outside
High expansion; very low shrinkage; good for large joints; good from −60 to 250°F	Good on wide and/or moving joints; comes in colors	Primer may be required before application
Very high expansion; very low shrinkage; good for large joints; good from −60 to 275°F	Similar to polysulfide sealants; very good adhesion; abrasion resistant	Can adhere so strongly to masonry as to cause crumbling
High expansion; no shrinkage; very versatile as joint filler; good from −90 to 400°F	Good adhesion to metal and glass; fast cure; comes in colors	Poor adhesion to concrete; must be painted with silicone paint
Moderate expansion; high shrinkage; can't be used on moving joints or cracks	Adheres well to metal and masonry in high moisture areas	Poor performance if applied to wide or moving joints
Little expansion; no shrinkage; not for moving joints; good from −90 to 200°F	Good adhesion; use on large cracks and hard-to-reach spots	Use requires adequate ventilation; may crumble if disturbed

cost-effectiveness. Weather stripping tends to be more visible than caulking, so aesthetic considerations may also enter into your choice of material. Table 2-3 describes common infiltration areas and prescribes remedies for sealing up these leaks.

Materials such as foam and felt strips are good for compression joints and places where little or no abrasion occurs (for example, where a door strikes a jamb). Spring metal is very durable, can eliminate rattling windows and is good in areas of high abrasion. Rolled or reinforced vinyl is suitable for joints with moderate abrasion such as the

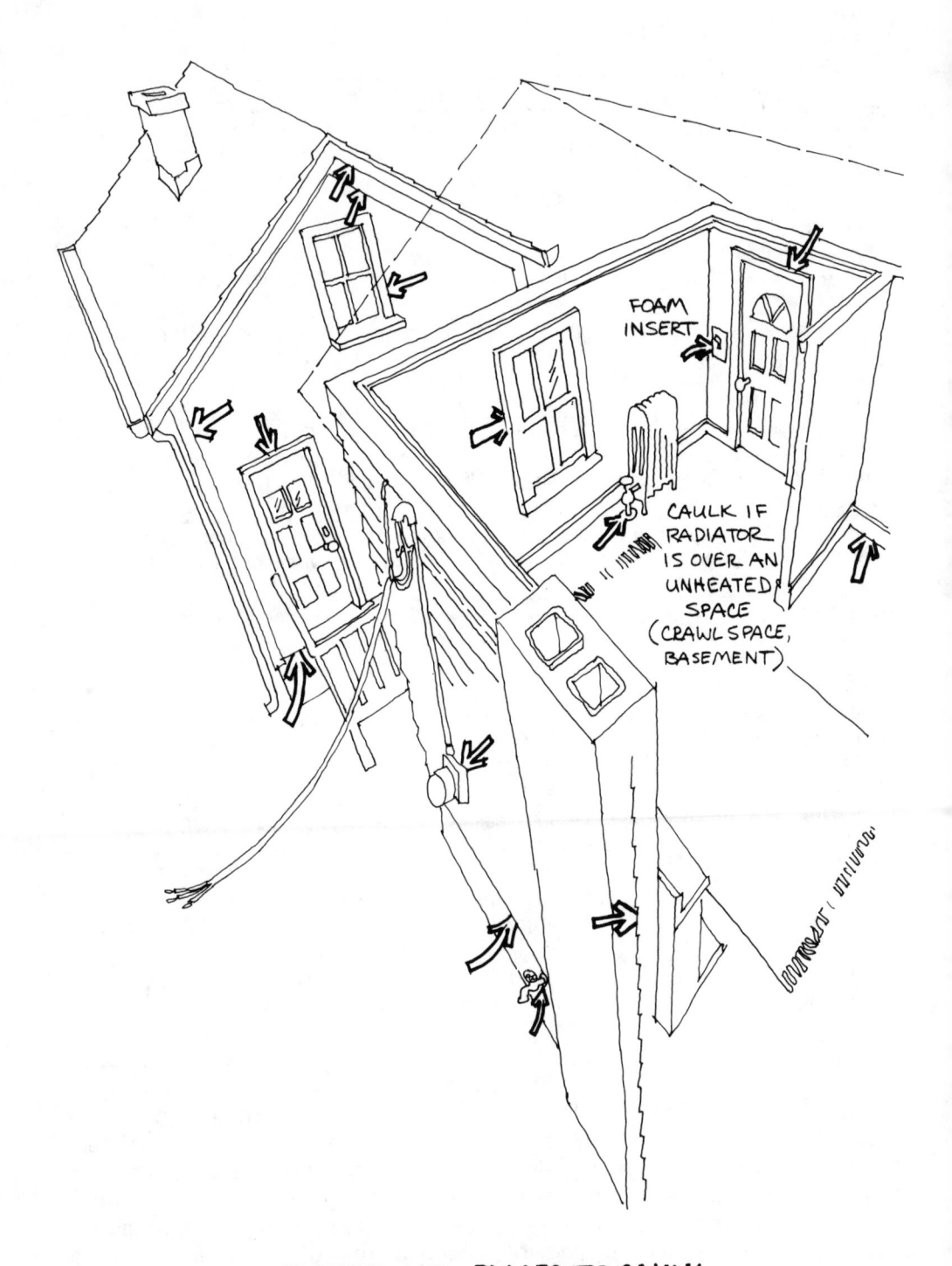

FOAM
INSERT

CAULK IF
RADIATOR
IS OVER AN
UNHEATED
SPACE
(CRAWL SPACE,
BASEMENT)

FIGURE 2-3: PLACES TO CAULK

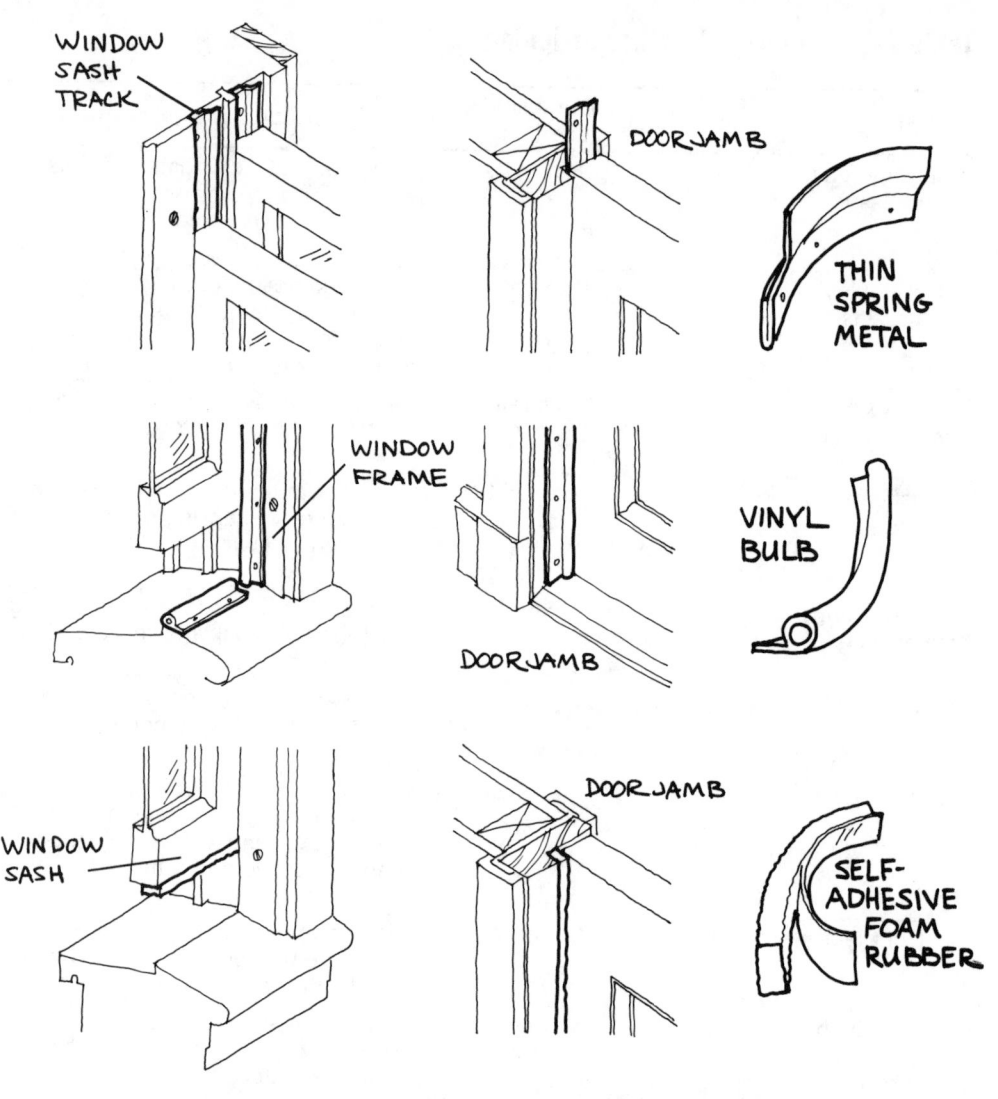

FIGURE 2-4: PLACES TO WEATHER-STRIP

sliding action of a double-hung window. Many types of weather stripping must be nailed or screwed into wood and are highly visible. You may wish to choose those materials that will be less visible or self-adhering. Figure 2-4 shows several types of weather stripping and where they should be installed. Another type of weather stripping material that is very effective is the outlet gasket. These are small pieces of foam rubber, costing a few cents apiece, that fit behind an outlet or switch cover plate. Studies have shown that infiltration through exterior wall outlets and switches can account for a great deal (as much as 5 to 10 percent) of a building's heat loss.

Table 2-2   Types of Weather Stripping

Type	Relative Cost	Durability and Lifetime (yrs)	Description and Performance
Spring metal or plastic	Moderate to high	High; 10-15	Strips of metal or plastic in V shape nailed or bonded to window or door frame; very effective
Rolled vinyl gaskets	Low	Moderate; 2-4	Extruded hollow or filled tubing; fastened to frame so as to provide slight contact pressure; very effective when properly installed
Reinforced vinyl gaskets	Moderate to high	Moderate; 5-7	Similar to above but with flexible metal or plastic reinforcing frame
Plastic foam	Low	Low; 1-2	Strips of open- or closed-cell foam, with adhesive backing or mounted on wood or aluminum
Interlocking metal channels	High	High; 10-20	Two-part strips on frame and moving part that interlock and seal together; very effective
Reinforced felt	Low to moderate	Moderate; 2-4	Felt, wool, cotton, or hair encased in aluminum; used in compression joint
Poly tape	Low to moderate	Moderate; 2-3	Clear, strong plastic tape placed over cracks and joints; seals totally; can be applied and removed seasonally
Duct tape	Moderate	Moderate; 3-5	Strong, nontransparent (silver), adhesive-backed tape covered with plastic or weather seal
Vinyl bulb doorway threshold	High	Moderate; 5-7	Aluminum strip on floor under door, with vinyl bulb in strip that makes seal against door
Door shoe	High	High; 10-15	Aluminum strip with vinyl gasket attached to door bottom; seals by contact over existing threshold
Door sweep	Low	Moderate; 5-7	Vinyl strip backed with aluminum at door bottom; provides light contact with threshold; less effective than above threshold types
Outlet gasket	Low	Moderate; 5-7	Piece of foam with holes for outlet or light switch; installed behind cover plate; very effective

SOURCE: Institute for Local Self-Reliance, *Weatherization Materials Handbook* (Washington, D.C.: Community Services Administration, 1979).

Ease of Installation	Good Features	Bad Features
Requires nails or screws every 10''; must be placed so as to meet moving part but not obstruct	High resistance to abrasion; good for sliding surfaces	More difficult to install; if movement is obstructed, reinstallation may be required
Comes in rolls; can be nailed to frame or may have own adhesive	Can be painted; may be adjusted to seal uneven or warped surfaces	Difficult to install so as to provide equal contact at all points
Comes in rolls; usually has predrilled nail holes in metal frame	Same as above, but more durable	Same as above
Comes in rolls or long strips; can be glued or nailed to window or door frame	Properly installed strip can last long time if only compressed; good for nonuniform gaps	If used on moving surfaces, abrasion greatly shortens lifetime; adhesive often ineffective
Little tolerance in installation; may require assistance of professional	Very durable; can provide good seal if properly installed	May work poorly if improperly installed; expensive
Comes in strips or rolls; nailed to frame or may have own adhesive; used between two surfaces	More durable than other plastic or foam materials	Susceptible to insects, moisture, fungus, abrasion; won't close nonuniform gaps
Use like any tape	Easy to install; useful for temporary repair of cracked windowpanes	May dry out and become ineffective; may cause paint to chip or crack
Use like any tape	Easy to install; very strong adhesion provides good seal	May be difficult to remove without causing damage to paint or wood
Nailed or screwed in place of threshold; door bottom may require beveling to provide good seal	Very effective as new or replacement threshold; replacement bulbs available	Proper installation may require professional
Nailed or screwed in place; door may require planing; contact must not be too strong	Very effective; may come with drip cap that sheds rain; replacement gaskets available	Proper installation may require professional
Requires only that screw holes be made in door in order to install	Inexpensive; easy to install	May not provide good seal if threshold is unevenly worn
Remove cover plate, place gasket over box, reinstall plate	Cheap and effective	Some types may be flammable; be sure to check label; turn off power before installation

Table 2-3
Infiltration Sites and Remedies

Location	Problems	Remedies	Other Problem Locations
Windows	Loose and rattling window sash	Install sash lock, channel guides, spring metal weather strip	Windows in the basement, stairwells, hallways, sunporch, laundry room, attic, front and rear doors, storm windows, even some closets
	Loose windowpanes	Replace window putty and paint	
	Cracked or broken panes	Patch with poly tape or replace	
	Molding-to-wall seam	Caulk	
	Seams between moving and stationary parts	Weather-strip	
	Broken sash ropes	Repair and tape over pulley holes	
	Bent metal frame	Repair or replace	
	Cracked wooden frame	Patch or replace	
	Rotten wood	Replace parts or entire frame	
Doors	Light visible under door or worn or uneven threshold	Install door sweep or new threshold	Basement bulkhead door; kitchen-to-basement door; doors from living space to garage, porch, attic, crawl space; secret doors; plumbing access hatches; storm doors
	Drafty mail slot	Poly tape over slot	
	Drafty, old-style keyhole	Poly tape over keyhole	
	Molding-to-wall joint	Caulk	
	Joint between moving and stationary parts	Weather-strip	
	Cracked door or wood frame	Patch or replace	
	Bent metal frame	Repair or replace	
	Door improperly hung	Rehang or weather-strip	
	Rotten wood	Replace parts or entire frame	
Floors	Furnace registers	Caulk between duct and wood	
	Radiator or baseboard pipes	Caulk between pipe and wood	
	Draft through floor	Insulate or lay plastic sheet beneath floor or install carpeting	
	Fireplace hearth-to-floor joint	Caulk	
	Sunken floor or stairs	Caulk around corners and seams	
	Electric junction boxes	Stuff fiberglass behind	

Table 2-3—*Continued*

Location	Problems	Remedies	Other Problem Locations
Walls	Electric outlets and light switches	Insulating gaskets	Walls in closets; cracks behind cabinets and cupboards; plumbing penetrations under sinks, behind toilets, in laundry rooms, in bathtubs and showers; wall-mounted light fixtures; places where dissimilar materials come together
	Floor-to-wall seam (beneath baseboards)	Caulk	
	Cracked plaster or paneling	Clean and repair	
	Plumbing and wiring penetrations	Caulk or fill with fiberglass	
	Fireplace mantel-to-wall seam	Caulk	
	Medicine and other cabinets	Stuff fiberglass behind	
	Built-in cabinets, bookcases, appliances	Stuff fiberglass behind	
	Furnace registers	Caulk between duct and lath	
	Drafty fireplace	Install, repair or close damper	
Ceilings	Furnace registers	Caulk between duct and lath	Do not use flammable materials around or behind recessed light fixtures or junction boxes; leave 3-4'' of breathing space
	Conventional light fixtures	Caulk around wiring penetrations	
	Recessed lighting	Caulk around edges and seams	
	Wall-to-ceiling joint	Caulk if necessary	
	Stairways	Caulk where they meet ceilings	
	Fans and vents	Caulk between duct and lath or stuff fiberglass around duct	
	Cathedral ceiling beams	Caulk at wood-to-plaster seam	
	Exposed brick chimney	Caulk at wood-to-plaster seam	
	Dropped ceiling	Plastic sheet behind ceiling	

[*Continued on next page*]

Table 2-3—*Continued*

Location	Problems	Remedies	Other Problem Locations
Basement	Cracked foundation or walls	Clean and patch or caulk	Don't forget to weather-strip and caulk basement doors, bulkheads and windows
	Flue or vent pipe shafts to attic and/or roof	Stuff with fiberglass	
	Bypass through cinder block wall	Stuff with fiberglass	
	Fireplace hearth and chimney	Caulk	
	Plumbing, sewage and electrical penetrations	Caulk or stuff with fiberglass	
	Air ducts or hot pipe shafts	Caulk or stuff with fiberglass	
	Stairway and wall stud bypasses	Fill with fiberglass	
	Sill plate, band joists	Caulk, cover with fiberglass	
Attic	Fireplace chimney joints	Caulk or fill with fiberglass	
	Plumbing, vent, duct, electrical penetrations	Caulk or fill with fiberglass	
	Ceiling fans	Caulk around edges	
	Ceiling/wall-floor joints	Caulk	
	Heating and cooling ducts	Tape and/or insulate	
	Wall or roof cavities	Fill with fiberglass	
	Attic hatch	Weather-strip and insulate	

Insulation

The value of insulation lies in its ability to reduce the rate of conductive heat loss. Insulation in exterior walls and attics reduces the demand on your furnace or boiler to replace lost heat, thereby saving fuel. During the summer, insulation slows the transfer of heat from outdoors to indoors, thereby allowing your apartment or house to stay cooler. Insulation also reduces heat loss from domestic water heaters and the hot water pipes connected to them and from the hot water pipes and ducts in space-heating systems.

Most of the insulating materials on the market are good ones, but it is important to understand the characteristics and limitations of each type of material, because some can only be used in certain applications, while others can be used in many ways.

Materials used as insulation are available in five forms:

Loose or fibrous fill is sold in bags.

Batts or blankets are sold in bundles or rolls.

Foam is sold as a liquid and can only be contractor-applied (urea-formaldehyde foam is currently

Table 2-4
R-Values for Insulation Commonly Used in Residential Buildings

Material	Physical Form	Typical R-Value per Inch	Thickness for R-11 (ins)	Thickness for R-19 (ins)	Thickness for R-30 (ins)
Cellulose	Loose fill	3.7	3	5	9
Fiberglass	Loose fill	2.5	4½	7½	12
	Batt/Blanket	3.1	3½	6	9½
	Board	4.5	2½	4	6½
Polystyrene Extruded	Rigid foam board	5.4	2½	4	6½
Expanded	Rigid foam board	3.5	3	5	9
Vermiculite	Loose fill	1.8	6	10½	16½

NOTE: Other insulating materials not commonly used or readily available for residential construction include: rock wool (R-3.3/inch), perlite (R-2.8/inch), polyurethane board (R-7/inch), isocyanurate board (R-8/inch) and urea-formaldehyde foam (R-2.5 to 4/inch after curing).

banned in Massachusetts and Connecticut for health reasons, but it is available elsewhere).

Rigid foam board is commonly sold in 2 by 4-, 2 by 8- and 4 by 8-foot sheets.

Sprayed foam (urethane) is sometimes used in small buildings in exterior and interior applications.

Each of these materials has a different resistance to the conduction of heat which is expressed as its R-value. The higher the R-value of a material, the better its ability to resist heat flow. Table 2-4 shows the R-values for various kinds and thicknesses of insulation used in residential buildings. Table 2-5 describes these insulating materials in more detail.

Vapor Barriers and Ventilation

Water vapor can be a major problem with some types of insulation. Air inside a building always contains some water vapor, which can pass through cracks and joints in insulated walls and ceilings. If the outdoor temperature is cold enough, water vapor will condense and be retained in fiberglass and cellulose insulation. This not only renders it less effective because it is wet, but can make it permanently useless. Vapor that condenses in walls can also cause paint to peel and wooden framing materials to rot. Using vapor barriers and adequate ventilation are the two primary ways of guarding against moisture problems.

Vapor barriers are usually sheets of coated paper, foil or plastic that serve to impede the flow of water vapor through insulation. These barriers should be used with all loose and batt or blanket insulation and with fiberglass board. Most plastic boards act as their own vapor barrier but must be well sealed at joints and penetrations. The barrier should be placed on the warm or living space side of insulation. That

[*Continued on page 58*]

Table 2-5
Types of Insulating Materials

Type and Description	R-Value	Material Cost/Ft²/ Inch Thickness ($)	Installation Cost/ Ft²* ($)	How Marketed	How Installed
Cellulose—usually recycled paper, but may be made from wood or other plant fibers	R-3.7/ inch	0.04	0.06	Usually in 30-pound bags; one bag covers 28 ft² × 6'' deep or 56 ft² × 3'' deep	Poured or blown in; blowing is better because process fluffs up material, increasing its insulative value and extending coverage
Fiberglass—mineral fiber as loose fill, batts or blankets and rigid board	R-2.5/ inch (loose fill poured)	0.46	0.56	Available as "pouring" wool sold in bags, or as "blowing" wool sold only through insulating contractors; batts are similar to blankets, except that they are cut to specific lengths, while blankets come in long rolls; rigid board available in several thicknesses	Can be poured, blown in or placed in cut lengths, depending on form; refer to table 2-6 for various applications Often used in insulating flat or built-up roofs
	R-11 (batts)	0.15	0.08		
	R-19 (batts)	0.24	0.09		
	R-30 (batts)	0.32	0.10		
	R-38 (batts)	0.50	0.11		
	R-4.5/inch (board)	0.24	0.15		
	R-5.5/inch (board)	0.41	0.15		
Vermiculite—loose fill; a type of mica expanded by heat and treated with asphalt for moisture resistance; used primarily in masonry block construction	R-1.8/ inch	0.07	0.07	In bags; one bag will cover 6½ ft² × 6'' deep or 13 ft² × 3'' deep	Poured

Precautions	Advantages	Disadvantages	How to Correct Problems
Very important to use vapor barriers and good ventilation because condensation greatly reduces insulative value	Highest R-value for material cost "Breathes" and is therefore good in frame walls and attics Small tufts of cellulose get into tight spaces more reliably than other types of insulation	Must be treated for fire resistance; quality control of this treatment is of primary importance; according to American Society for Testing Materials (ASTM), borax chemicals are only ones acceptable for assuring permanent fire resistance; ammonia sulfate, which separates from insulation in a warm attic, or aluminum sulfate, which is corrosive to metals, are both unacceptable Becomes ineffective if soaked Can be prone to vermin if not treated Cannot be used below grade or with masonry or metal construction Can settle, which reduces insulative value	If cellulose becomes soaked and remains wet, must be removed; source of moisture must be eliminated and new insulation installed If soaked can dry out, but make sure it is dry before adding new insulation If vermin-infested, cellulose must be removed and replaced with new, treated insulation If settling occurs, more insulation can be added If improperly fireproofed, should be replaced by properly treated material
When handled, fiberglass is irritating to skin, eyes and lungs; mask, goggles, gloves and full body clothing should be worn	High insulative value, especially in batt or blanket form or high pressure blown Excellent fire resistance If soaked, will regain insulative value upon drying "Breathes" and is therefore good in frame walls and attics As a rigid board, does not warp or buckle, yet is quite flexible	Should not be placed in contact with moisture; acts as wick, drawing moisture into itself and reduces insulative effectiveness until it dries out Board has little strength when compressed, compared to other insulative boards Settles (loose fill) Can dam up on obstructions if blown in	If soaked and remains wet, fiberglass must be removed; source of moisture must be eliminated and new insulation installed If settling occurs, more insulation can be added To reduce problem of fill damming on obstructions, hose blowing insulation should be pushed as far as possible into space to be insulated, and then slowly raised as insulation is blown in
None	Nonflammable Handles easily If soaked, will regain insulative value upon drying "Breathes"	Low R-value	. . .

[*Continued on next page*]

Table 2-5—*Continued*

Type and Description	R-Value	Material Cost/Ft²/Inch Thickness ($)	Installation Cost/Ft²* ($)	How Marketed	How Installed
Perlite—loose fill; type of volcanic rock expanded by heat into glasslike pellets; not naturally water-resistant and must be treated with silicone; used primarily in masonry block construction	R-2.8/inch	0.07	0.07	In bags	Poured
Polystyrene—rigid foam boards, either expanded or extruded; expanded polystyrene typically has a lower R-value and breaks up more easily than extruded polystyrene	R-3.5/inch (expanded) R-5/inch (extruded)	0.11 0.26	0.15 0.15	In boards, varying in thickness from ½ to 3''; most boards 2' × 8'	By placing in space to be insulated; kept in place with adhesives, nails, or other fasteners
Polyurethane and polyisocyanurate—rigid foam boards	R-7/inch	0.39	0.15	In boards of varying thicknesses	By placing in space to be insulated; often used between building sheathing and exterior siding; kept in place with adhesives, nails or other fasteners
Urea-formaldehyde—mixture of five chemicals combined in liquid form which is foamed into wall cavities	R-2.5-4/inch (after curing)	0.08	0.13	In liquid form, but only available wholesale, (Because of serious questions regarding its health effects, currently not marketed in Massachusetts and Connecticut, and it may be banned by the federal Consumer Product Safety Commission)	Installed professionally by foaming into wall cavities; important that ingredients are combined in exact proportions and within narrow temperature range; temperature of insulated wall cavity should be in range of 23 to 86°F for 4 days after installation

SOURCE: Material and installation costs are from Robert Snow Means Company, Inc., *Means® Building Construction Cost Data*, 38th ed. (Kingston, Mass., 1979).
*The installation cost of each additional inch of insulation is usually much less than the cost of installing the first inch.

is, batts placed against walls or attic ceilings should have the vapor barrier facing into the room.

Most batt or blanket insulation comes with a vapor barrier as a backing, while loose insulation or fiberglass board requires installation of a separate barrier, usually a 6-mil-thick polyethylene film. Sometimes a vapor barrier cannot be easily installed, such as in finished wall cavities where insulation is blown in. Painting the wall with a vapor barrier

Precautions	Advantages	Disadvantages	How to Correct Problems
Can prove irritating to lungs when poured; mask should be worn	Nonflammable Effective over wide temperature range, up to 1,800°F If soaked, will regain insulative value upon drying "Breathes"	Low R-value	. . .
Expanded polystyrene requires very careful handling to prevent breakage	Good for high moisture conditions because it is its own vapor barrier Can be used where very little space is available for insulating	Flammable and must be protected by fire-resistant material	. . .
None	Good for high moisture conditions because they are their own vapor barriers Can be used where very little space is available	Flammable and must be protected by fire-resistant material	. . .
While not recommended for wood frame construction, urea-formaldehyde may have useful application in masonry construction	Flows easily around piping and other obstructions Has relatively high insulative value	Can emit noxious odor; can cause health problems May stain walls Can crack interior wall surface if installed with too much pressure May shrink up to 40 percent if improperly installed Somewhat flammable	Good oil-based paint or plastic-coated wall covering can limit release of noxious odors Stained walls should be sealed and repainted If fumes cannot be reduced, or if insulation shrinks greatly, urea-formaldehyde foam must be removed; this involves removing wall

primer or with two coats of a good oil-based paint will adequately hinder the transmission of water vapor. Other possibilities are to cover the wall with ceramic tile, plastic-coated paneling or paper. These measures must be very carefully pursued in high-vapor areas such as bathrooms and kitchens.

Ventilation, or the circulation of air, is also important for controlling moisture problems because it provides a way to remove water vapor from insulation and structural materials that may be harmed by water (see figure 2-5). In wood-frame buildings, the outer surface of exterior walls should be made of materials that allow ventilation of air through them so that moisture can

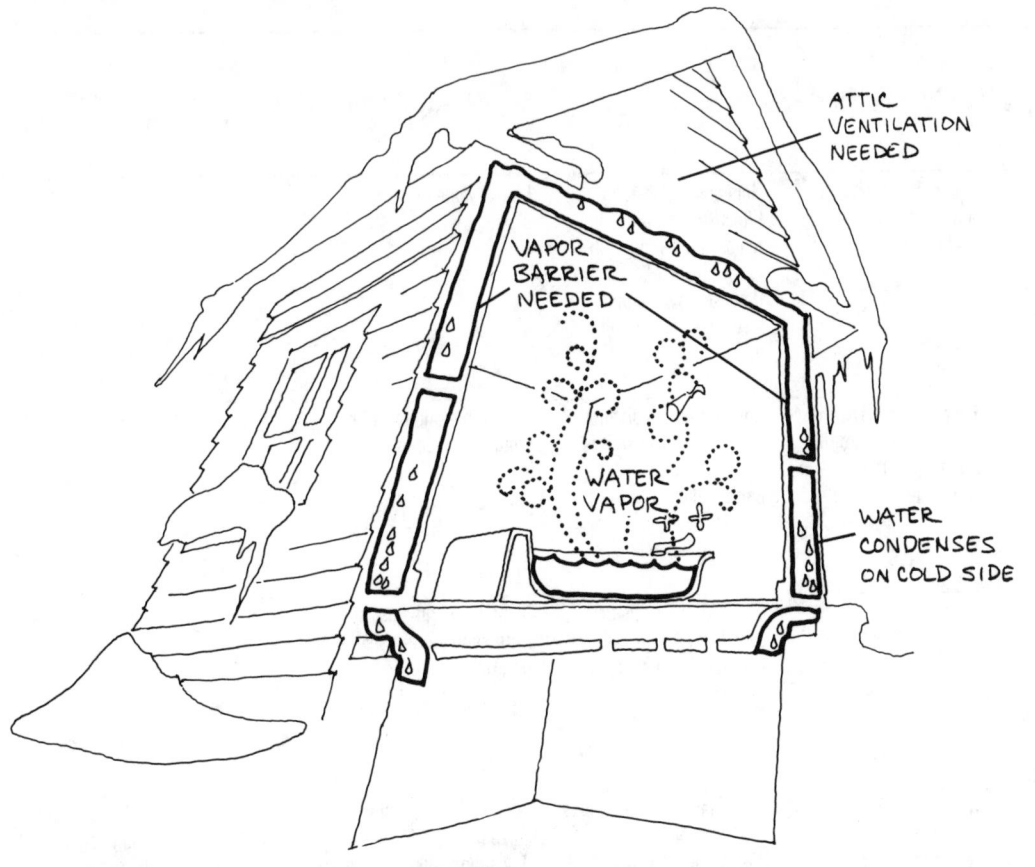

ATTIC VENTILATION NEEDED

VAPOR BARRIER NEEDED

WATER VAPOR

WATER CONDENSES ON COLD SIDE

FIGURE 2-5: AREAS WHERE CONDENSATION CAN OCCUR IN INSULATION

escape. It is important to make sure that all sheathing and siding materials added to these structures allow for this ventilation. In addition, insulated attics must have good cross ventilation. A rule of thumb for attics with a vapor barrier and insulation is that one square foot of ventilated opening must be provided for every 300 square feet of insulated attic floor. Double this number if there is no barrier. The most common places to ventilate attics are at the eaves, the ridge or the walls (see figure 2-6).

Adding Insulation

Some buildings, particularly those built since 1940, already have some insulation in their walls or attic, although it may be inadequate. In some cases, additional insulation can be installed. It should be either cellulose or fiberglass, in batts for attics or loose fill for attics and walls, and should have no vapor barrier or a perforated barrier. It is not generally possible to add insulation to wall cavities previously filled with batts or blankets, although if the old material was loose fill that has settled, more can be added.

For those buildings where wall insulation is either not feasible or too expensive, walls can be insulated temporarily by hanging or attaching such materials as quilts, pictures with insulative backing or tapestry wall hangings. Also, some pieces of furniture, such as bookshelves (with books!), can reduce heat loss when placed along outside walls.

Table 2-6 shows where the various insulating materials are applied in residential buildings, and following the table is a sample insulation contract that is recommended by the Commonwealth of Massachusetts. Installation of vapor barriers, ventilation and type of insulation used are among the points covered. All property owners who are planning to insulate their buildings should review this contract carefully and model their own agreements upon it. Before undertaking installation of any type of insulation, it is a good idea to consult either an expert in the field or one of the more comprehensive references listed at the end of this chapter. Such a precaution will minimize the chances of problems arising after installation.

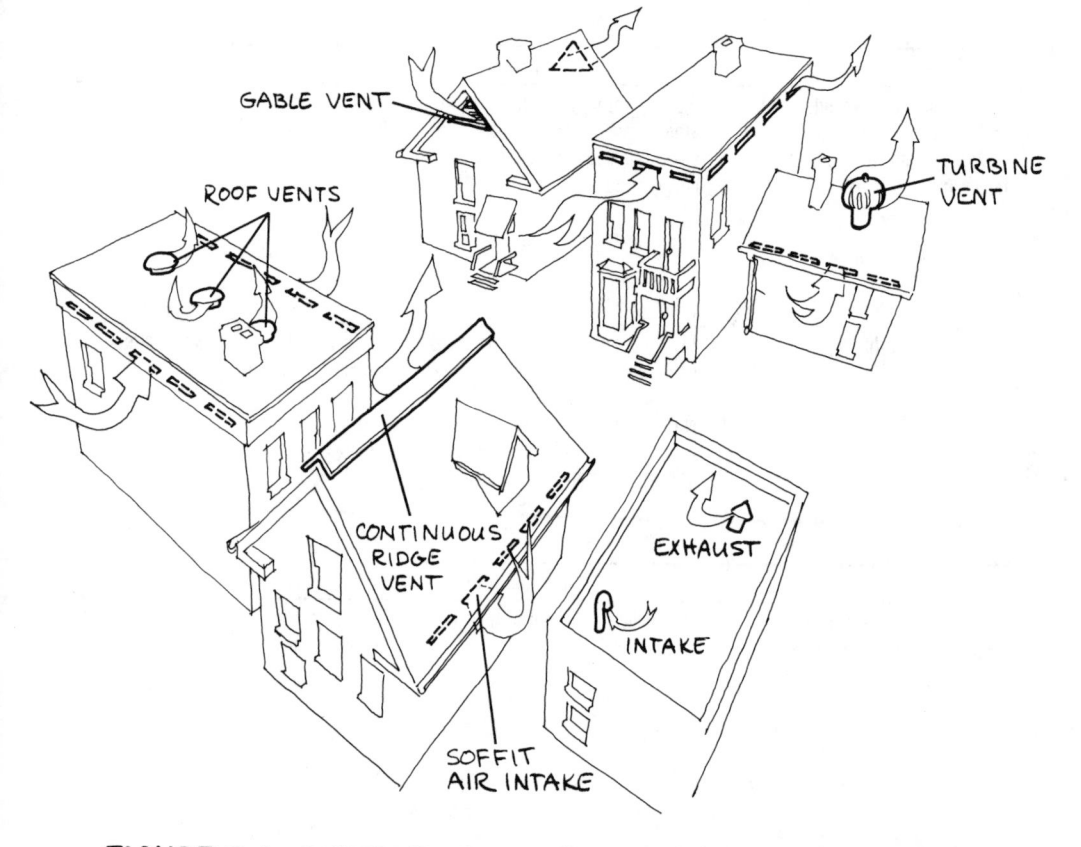

GABLE VENT

ROOF VENTS

TURBINE VENT

CONTINUOUS RIDGE VENT

EXHAUST

INTAKE

SOFFIT AIR INTAKE

FIGURE 2-6: DIFFERENT ROOF VENTILATION SCHEMES

Table 2-6
How and Where to Install Insulation

Type	Where Appropriate	Involves
Attic Floor Insulation		
Loose fill (cellulose, mineral wool or fiberglass)	Used where a person can work in the attic space or get a blower nozzle in	Emptying bags of loose material into the spaces between the attic floor joists, or blowing the material in with a hose and nozzle
Batt or blanket (fiberglass)	Used where a person can work in the attic space	Placing the insulation in the spaces between the joists of the attic floor
Roof Insulation		
Rigid foam board (extruded and expanded polystyrene, polyiso-cyanurate, polyurethane, fiberglass, fiberboard)	Placed over roof sheathing when attic cannot be insulated or when exposed joist "cathedral" ceiling is desirable	Roofing is removed; insulation is then attached to sheathing and new roofing installed over it
Batt or blanket (fiberglass)	Placed between rafters when attic floor cannot be insulated; used with a cathedral ceiling that has finished surface attached to the bottom of the roof rafters; or when house has no attic or the attic is to be heated	Installing insulation between rafters, then installing new ceiling, if desired
Ceiling Insulation		
Rigid ceiling tile (extruded or expanded polystyrene or fiberglass)	Used when attic floor or roof pitches cannot be insulated	Installing suspended ceiling with insulated panels set in

Sketch	Precautions	For Building Types*
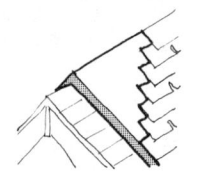	Keep insulation 4'' away from chimneys and lighting fixtures; cellulose must be factory-treated with borax chemicals to be nonflammable; it is best to blow in cellulose so that it puffs out more, thereby providing better insulative value and extending the coverage of the material; vapor barriers and ventilation very important	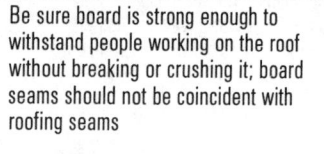
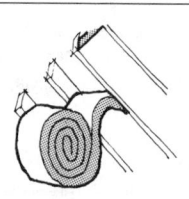	If insulation has foil or paper backing which creates a vapor barrier, foil or paper must face down toward living space (warm side); edges should be stapled to joists to make good seal; wear protective clothing (mask, goggles, gloves, all-body clothing); provide adequate ventilation	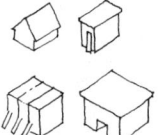
	Be sure board is strong enough to withstand people working on the roof without breaking or crushing it; board seams should not be coincident with roofing seams	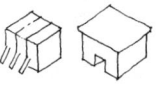
	Space must be left between top of the insulation and bottom of the roof sheathing for ventilation (so roof structure doesn't become wet and rot); this space must be vented to outside air at eaves and, if possible, at ridge; wear protective clothing when installing	
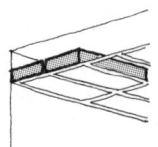	Insulation lowers ceilings 6'' or more; walls that pass through new ceiling leave gaps in insulation; material must have vapor barrier facing living space	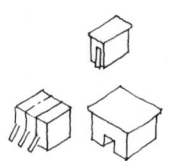

[*Continued on next page*]

Table 2-6—*Continued*

Type	Where Appropriate	Involves
Ceiling Insulation—*Continued*		
Batt or blanket laid on suspended ceiling (fiberglass) or Loose fill blown in on suspended ceiling	Used where attic floor or roof pitches cannot be insulated	Installing suspended ceiling with batts laid on top or Installing ceiling joists and sheetrock and then blowing in cellulose or fiberglass insulation
Frame Wall Insulation Above Ground		
Loose fill (cellulose, mineral wool or fiberglass)	Used when not making major alterations or replacing wall finishes	Blowing material into wall cavities through small holes in each stud space—either from inside through plaster, or outside through siding and sheathing; walls must be patched afterwards
Batt or blanket (fiberglass)	Used when interior finish is being removed or when wall cavities are being widened with additional studs	Removing interior wall finish and placing fiberglass batts or blankets between studs; then installing new wall finish
Masonry Wall Insulation Above Ground		
Rigid foam board (extruded or expanded polystyrene, polyurethane, polyisocyanurate)	Against inside or outside face of wall	Affixing boards with adhesive or nails to inside face of existing plaster wall, then applying fire-retardant finish such as sheetrock; affixing boards with adhesive or nails to exterior wall, then covering with siding (this is preferred application)
Batts or blankets (fiberglass)	Against inside face of original finished wall	Building a new stud wall against the inside face of the masonry wall, placing fiberglass between studs and installing new finished wall

Sketch	Precautions	For Building Types*
	Insulation lowers ceilings about 6''; walls that pass through new ceiling leave gaps in insulation; good vapor barriers are difficult to assure; check window and door heights to see if lowering ceiling is feasible	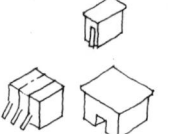
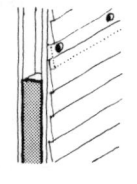	Be careful to assure that all spaces are filled—allow for diagonal corner bracing, blocking, etc.; since it is impossible to put in vapor barrier, use good oil base or vapor barrier paint on interior wall surface; fiberglass has greater tendency than cellulose to catch on nails or wires and dam up, leaving voids	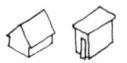
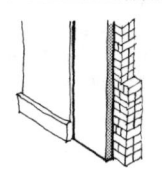	Vapor barriers must be tightly stapled to achieve good seal; plastic barriers should be joined by tape; wear protective clothing	
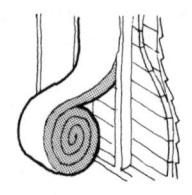	Rigid foam is flammable and must be covered by a fire-resistant finish such as plaster or drywall; window and door trim may need to be modified; baseboard trim may need to be modified; vapor barrier should be added to warm side if fiberglass board is used; foam must be covered with cement stucco or siding; windows, door frames and other trim must be modified	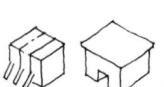
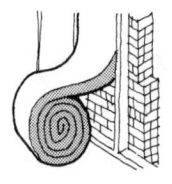	Room dimensions reduced approximately 4 to 6''; window and door trim must be modified, as may baseboard trim; could be conflict with radiators or baseboard heaters	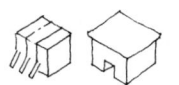

[*Continued on next page*]

Table 2-6—*Continued*

Type	Where Appropriate	Involves
Wall Insulation Below Ground		
Rigid foam board (expanded or extruded polystyrene, polyurethane, polyisocyanurate)	Against face of inside masonry walls—where accessible; as perimeter insulation, against outside face of masonry wall below grade	Affixing boards to inside of wall, then applying fire-retardant material such as sheetrock; affixing boards to outside of wall, then covering with a waterproof finish and providing for exterior perimeter drainage
Batt or blanket (fiberglass)	Against inside face of masonry walls, where accessible	Building a new stud wall against the masonry, placing fiberglass batts between studs and installing new finish wall
Basement Ceiling Insulation		
Batts or blankets (fiberglass)	Over unheated crawl space or basement	Fitting batts between floor joists with vapor barrier up (toward warm side) and supporting them with chicken wire or other wires
Duct & Pipe Insulation		
Flexible blankets with facing (fiberglass)	Around ducts	Installing insulation cut to length and taping at seams
Semirigid with jacket (fiberglass)	Around pipes	Installing over pipe and taping at seams
Flexible (closed-cell neoprene)	Around pipes	Installing over pipe and taping at seams
Rigid (closed-cell urethane)	Around pipes	Slipping over pipe and taping at seams
Water Heater Insulation		
Batt or blanket with jacket (fiberglass)	Around tank	Wrapping blanket around tank and taping at seams

*Single-family house 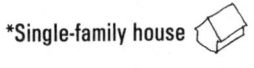   Triplex    Row house    Multi-unit apartment building

Sketch	Precautions	For Building Types*
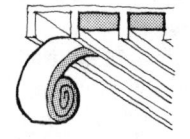	When installing perimeter insulation, exterior face of the masonry walls must be waterproofed prior to installation; extruded polystyrene is normally used for this application	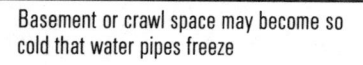
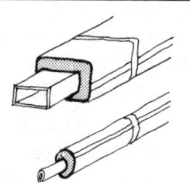	Fiberglass must not come in contact with moisture as it will act as wick, drawing moisture into itself; be sure foundation wall is well drained and otherwise treated to prevent moisture from coming through	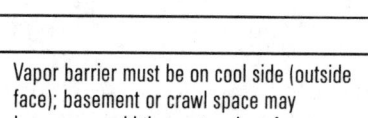
	Basement or crawl space may become so cold that water pipes freeze	
	Vapor barrier must be on cool side (outside face); basement or crawl space may become so cold that water pipes freeze	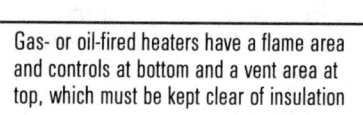
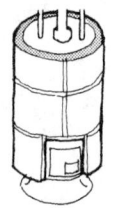	Gas- or oil-fired heaters have a flame area and controls at bottom and a vent area at top, which must be kept clear of insulation	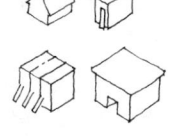

A Model Insulation Contract

This is a sample insulation contract used in Massachusetts. Review it carefully before hiring an insulation contractor. Any agreements that you may enter into with a contractor should incorporate the sections of the contract which pertain to your particular situation.

GENERALLY ACCEPTED PRACTICES FOR INSTALLING INSULATION IN EXISTING HOMES

1. Vapor barriers, if installed, will be placed on the warm (interior) side of the area to be insulated.
2. Before installing insulation in the ceiling/attic area, the Contractor will check to determine whether there is adequate ventilation. If there is inadequate ventilation, the Contractor will inform the Homeowner. Inadequate ventilation may result in moisture damage and may impair the effectiveness of the insulation (at least 1 square foot of ventilation for every 300 square feet of insulation with vapor barrier).
3. After an attic has been completely insulated, the Contractor will check to insure that the vent areas have not been covered with insulation during the work process.
4. The Contractor will use or construct a protective barrier to keep combustible insulation a minimum of 3 inches away from heat sources (such as recessed light fixtures, fans and chimneys).

5. If additional attic insulation is being installed to already existing insulation, a vapor barrier will not be used. If the new insulation is a batt/blanket type with an attached vapor barrier, this vapor barrier will be extensively perforated before installation to avoid the problem of trapped moisture and condensation.
6. The Contractor will locate obstructions in the wall cavities before installing insulation in sidewalls. The Contractor will blow insulation in through two holes in each stud bay. If the blow is difficult, additional holes will be used. Blow holes will be left unplugged and may be screened before siding is replaced.
7. If foam insulation is accidentally sprayed on bordering glass or aluminum areas, it will be removed immediately and these surfaces rinsed thoroughly with water. The Contractor will always first test and weigh a sample of the material (to determine the correctness of its density) before beginning the real application process and will provide a written record to the owner of this test. Where urea-formaldehyde foam insulation is to be used, it will not be installed in any attic and will only be installed when the temperature of the exterior surface of the cavity in which foams are to be applied is within the rate of -5 to 30°C (23 to 86°F), as specified by the Department of Housing and Urban Development. (Editor's note: Complaints of health problems

have been reported by some occupants of houses with urea-formaldehyde insulation. There is currently a moratorium in Massachusetts and Connecticut on installation of this material and a national ban on urea-formaldehyde foam insulation is under consideration by the federal Consumer Product Safety Commission.)

8. If loose fill insulation is being installed, the Contractor will keep an accurate count of how many bags are used in order to insure that the correct amount of loose fill insulation is installed.

9. The safety of on-site manufactured cellulose is extremely questionable. There is a definite problem with the installer's ability to ensure that the insulation is properly treated for flame resistance under these specific conditions. Where cellulose insulation is installed, the Contractor will display the flame retardancy and Class I label on all bags to be used in the installation. The Contractor will verify that borax chemicals are used as flame retardants.

10. Except as otherwise specified above, the Contractor agrees to perform the work described in Part I of the following agreement according to the manufacturer's installation instructions.

AGREEMENT

READ THIS AGREEMENT AND MAKE SURE YOU UNDERSTAND IT BEFORE SIGNING IT. MAKE SURE ALL BLANKS ARE COMPLETED AND ALL PROVISIONS WHICH DO NOT APPLY ARE CROSSED OUT. THIS AGREEMENT HAS LEGAL FORCE AND EFFECT AND BINDS THOSE WHO SIGN IT.

This Agreement is made on_____ between

_____ of _____
(Contractor's Name)   (Contractor's Address)

_____
(Contractor's Phone No.)

hereinafter called "Contractor," and _____
(Homeowner's Name)

of _____ hereinafter called "Owner."
(Homeowner's Address)

[Continued on next page]

I. MATERIALS AND WORK

Contractor agrees to make energy conservation improvements on the Owner's premises to the extent and under the conditions specified below:

A. Contractor agrees to furnish all materials and labor for the following energy conservation projects (e.g., insulate attic, insulate exterior walls, install storm windows) and to state for each project:

1. any major alterations to the premises required;
2. the type of materials for the project, including a statement of total R-value to be achieved by each insulation material;
3. the type of vapor barrier, if any, Contractor will install;
4. the type of fire barrier, if any, Contractor will install;
5. the type of ventilation, if any, Contractor will install;
6. the cost of each separate project.

B. Contractor agrees to fill all cavities in each project area, except as it may interfere with proper ventilation and fire safety considerations.

C. Contractor agrees to properly protect the property of the owner at each project work site and adjacent areas, and to restore the premises to a condition similar to that prior to commencement of work. Contractor further agrees to leave the work area in a neat and orderly condition upon completion of work.

D. Contractor agrees to secure, at his own expense, if required, all city, town or state permits necessary to do the work.

E. Contractor agrees to use only materials whose labels indicate conformance with these federal specifications:

INSULATION MATERIALS	FEDERAL SPECIFICATION NUMBERS (unless superseded)	
• **Loose Fill**		The Federal Government requires
Fiberglass	HH-I-1030A	that a Federal Specification
Cellulose	HH-I-515C	number for those insulations
Vermiculite	HH-I-585B	that meet its standards appear on
		the packages of insulation. These
• **Batts/Blankets**		specifications cover such
Fiberglass	HH-I-521E	characteristics as fire retardancy,
		R-value and corrosiveness.
• **Rigid Board**		
Polystyrene (extruded & expanded)	HH-I-524B	
Urethane	HH-I-530A	
Fiberglass	HH-I-526C	
• **Foam** (See Generally Accepted Business Practices No. 7)		

II. PRICE

Contractor agrees to do all work described in Part I for no more than the price

of _____ .
   (amount)

III. PAYMENT

Contractor shall be paid by Owner according to the following schedule:

IV. COMMENCEMENT AND COMPLETION OF WORK

Unless agreed to in writing, Contractor will not order materials or begin work before the third day following the signing of this Agreement.

Contractor will begin the work on or about _____
                                                    (date)

and shall work each day thereafter until work is completed, barring delay caused by circumstances beyond Contractor's control.

V. INSURANCE

Contractor will be responsible to Owner or any third party for any property damage or bodily injury caused by himself, his employees, or his subcontractors in the performance of, or as a result of, the work under this Agreement. Contractor agrees to carry insurance to cover such risks and, if requested to do so by the Owner, will provide an insurance binder detailing the extent of his coverage.

VI. SUBCONTRACTING

Contractor agrees that, notwithstanding any agreement for materials and/or labor between Contractor and a third party, Contractor is responsible to Owner for completion of all work described in Part I in a timely and workmanlike manner.

VII. OTHER CONDITIONS

Contractor agrees to follow the Generally Accepted Practices for Installing Insulation in Existing Homes, attached to, and incorporated by reference in, this Agreement.

VIII. MODIFICATION

This Agreement, including the provisions relating to price (II) and time (IV), cannot be changed except by a written statement signed by both Contractor and Owner. However, cancellation by Owner is allowed in Accordance with paragraph X.

[*Continued on next page*]

IX. ADDITIONAL TERMS
(Insert any additional terms agreed to by the Contractor and Owner.)

X. NOTICE OF CANCELLATION
Owner may cancel this Agreement if signed at a place other than Contractor's address, if Owner notifies Contractor in writing of his intention to do so no later than midnight of the third business day following the signing of this Agreement. The following language addressed to Owner regarding notice of cancellation is required by statute:

YOU MAY CANCEL THIS AGREEMENT IF IT HAS BEEN CONSUMMATED BY A PARTY THERETO AT A PLACE OTHER THAN AN ADDRESS OF THE SELLER, WHICH MAY BE HIS MAIN OFFICE OR BRANCH THEREOF, PROVIDED YOU NOTIFY THE SELLER IN WRITING AT HIS MAIN OFFICE OR BRANCH BY ORDINARY MAIL POSTED, BY TELEGRAM SENT OR BY DELIVERY, NOT LATER THAN MIDNIGHT OF THE THIRD BUSINESS DAY FOLLOWING THE SIGNING OF THIS AGREEMENT.

SEE THE ATTACHED NOTICE OF CANCELLATION FORM FOR AN EXPLANATION OF THIS RIGHT.

Contractor and Owner hereby agree to the above terms.

_____ (Owner's signature)

_____ (Contractor's signature)

LIMITED WARRANTY
This WARRANTY covers all insulation materials installed by

_____
(Contractor's Name)

_____
(Contractor's Address)

_____ , at the home of _____
(Contractor's Phone No.)   (Homeowner's Name)

_____
(Homeowner's Address)
as described in the attached Agreement.

1. Contractor warrants to Owner that all work will be done in a workmanlike manner, free from defects, and in conformance with all specifications mentioned in Part I of the Agreement.

2. Contractor agrees that, if any defect in materials or workmanship arises within _____ , he will repair such defects and bring the work up to the standards required under the Agreement at no additional expense to Owner.

3. Any claims made by Contractor to Owner regarding cost or fuel savings shall be stated here:

The Contractor makes no other claims regarding cost or fuel savings.

This warranty gives you specific legal rights, and you may also have other rights which vary from state to state. Under Massachusetts law, sales of goods carry an implied warranty of merchantability and fitness for a particular purpose.*

In order to obtain performance of these Warranty obligations, the person to contact is the Contractor, unless otherwise specified here:

_____
(Name)

_____
(Address)

_____
(Contractor's Signature)

*Contact your state Attorney General's office for specific information on such warranties.

PRIME WINDOWS AND DOORS

Prime windows and doors are not generally included in discussions of weatherization, but they are an exceptionally critical part of the building envelope since they constitute a primary barrier to the escape of heat. Constant use over many years causes wear and tear and makes their fit less tight. Frames may rot, sag or crack and pieces of sash may fall out, leaving holes through the wall.

Joints and cracks around windows and doors are major areas of infiltration through the envelope. Indeed, a ¼-inch space between the bottom of a typical door and its threshold is the equivalent of a 3-inch-diameter hole straight through the door. Before doing any thinking about storm windows, you should repair or replace defective prime windows and doors.

Prime windows and doors should be inspected every year for problems and repaired, if necessary, to ensure their continuing effectiveness. Broken windowpanes are candidates for immediate replacement, but even whole windowpanes may require some work. Window glazing putty tends to dry, crack and fall out after many years, leaving cracks around the windowpanes through which air may enter. If much putty is cracked or missing, remove the panes, clean out the old putty and reglaze the window.

In double-hung windows and sliders, the window sash should fit tightly into the channels provided for this purpose. If the sash rattles in the frame, installation of metal channel guides or spring metal weather stripping can provide a tighter fit. This problem may be more difficult to fix if your windows have metal frames. Sash cords should be repaired, if only because a window that is not counterbalanced can be dangerous and cause injury. All double-hung windows should have sash locks. These are not only needed to keep out burglars but also to force a tight fit between the upper and lower sashes. Cracked framing should be patched or replaced. If your windows and doors are in particularly bad shape, no amount of repair will make them much tighter than they already are. You should consider replacing them.

Cracks between the plaster and wood or metal window and door frames should be patched with caulk. If they are very large, they should be filled with foam caulk and new plaster should be applied. Joints between the moving and stationary parts of windows and doors should be weather-stripped.

STORM WINDOWS AND DOORS

Storm windows and doors also reduce heat loss. The purpose of these devices is to create a pocket of "dead" air between the window or door and the storm. This air pocket reduces the transfer of heat from indoors to outdoors, thereby acting as insulation. If you live in an area where storm windows are not cost-effective but you find it necessary to replace your prime windows, you should install double-pane windows. The optimum combination is a double-pane prime with a storm as a third glazing layer.

In order for storm windows to work, both the prime and the storm windows must be sealed tightly, except at the storm window weep-holes, or else their combined effectiveness to reduce the flow of heat is diminished. Storm doors work the same way as storm windows, although with their higher cost the payback period is longer. They will certainly increase the comfort of the rooms in which they are located. A vestibule, which has the same purpose as a storm door but includes a small lobby between the two doors, can be a desirable addition to a building, particularly an apartment building.

Maintenance

Storm windows and doors should also be inspected and, if necessary, repaired every year in order to maintain their effectiveness. They should be cleaned to ensure that the sashes will fit tightly in their frames. In addition, the sills should be cleaned so as to keep the weep-holes clear. If these holes are plugged, water vapor condensate cannot escape and the windows will steam up. Over a long period of time, such condensation can cause rotting of wooden prime window frames. Caulking between the storm window and frame of the prime window should be checked for openings or cracks. If the caulk is ineffective it should be replaced.

Finally, all parts of the frame and sash should be checked to make sure they have not bent or warped out of shape. If you find that you cannot make your storm windows fit tightly, then you should consider installing new ones.

Shopping for Storm Windows

Storm windows are usually available with aluminum frames. Wood-framed storms are available but generally only when buying an entire window assembly. It may be possible, however, to order such frames from a millwork. The manufacturers of storm windows (and storm doors) usually produce two or three lines of different quality. While the least expensive lines do not hold up very long, regardless of the manufacturer, moderately priced storm windows (and doors) are normally the best buy.

When shopping for storm windows, you should ask to see cut cross sections of the various quality frames and to have the merits and demerits of each sample explained. Look carefully at the thickness of the metal and how it is designed for stiffness and operation. Also, check how the corners of the frames and sash are constructed. The least expensive storm windows tend to have thin frames and weak corner joints which will warp out of shape quickly, thereby reducing the insulating value of the window.

Two final features to look for in metal storm windows are the gasket and the thermal break. The first is a rubber or plasticlike part of the window that prevents the glass from coming in contact with the metal; the second prevents the interior surface of the metal from coming in contact with the exterior surface. Both reduce conductive heat losses by conduction through the metal.

Alternatives to Storm Windows

If storm windows are too expensive to install in your home, there are other ways to reduce heat loss through windows. These are usually not as effective as permanent storms. Temporary storm windows can be made of plastic sheets or film. They should be installed inside the window rather than outside because both wind and sun tend to reduce the lifetime of plastic. It is essential that the plastic sheet be sealed tightly all around the window frame. Use tape to provide this seal.

Another way of reducing window heat loss is to install thick curtains or drapes that close tightly. A valance may be needed to prevent airflow at the top. Thermal curtains, especially designed to provide insulative value, are now available. Usually these curtains are made of layers of fabric or plastic film with insulating pockets of air between these layers. Thermal curtains often are set into a frame so that when closed they form a tight seal. Another alternative, thermal shutters, are made of insulating materials and fit very tightly when they are closed. (Movable insulation is discussed in greater detail in "Passive Solar Heating and Natural Cooling," chapter 5.)

For Further Reference

American Ventilation Association. *Handbook of Moving Air.* Houston, Tex., 1977.

Burch, D. M., and Hunt, C. M. *Retrofitting an Existing Wood-Frame Residence for Energy Conservation.* Washington, D.C.: National Bureau of Standards, 1978. Order no. NBS B55 105.

Byalin, J. *Women's Energy Tool Kit.* New York: Consumer Action Now, 1980.

Central Mortgage and Housing Corporation of Canada. *The Conservation of Energy in Housing.* Ottawa, Ontario, 1977. Order no. NHA 5149.

Collins, B. L. et al. *Energy Effective Windows.* Washington, D.C.: National Bureau of Standards, 1978. Order no. NBS SP 512.

David Conover and Associates. "A Guide to Thermal Insulation." Boston, 1977-78.

Farallones Institute. *The Integral Urban House.* San Francisco: Sierra Club Books, 1979.

Harrje, D. T.; Dutt, G. S.; and Beyea, J. "Locating and Eliminating Obscure but Major Energy Losses in Residential Housing." *ASHRAE Transactions* 85 (1979): Part II.

Harrje, D. T.; Dutt, G. S.; and Gadsby, K. J. "Isolating the Building Thermal Envelope." Princeton, N.J.: Center for Energy and Environmental Studies, Princeton University, n.d.

Hastings, S. F., and Crenshaw, R. W. *Window Design Strategies to Conserve Energy.* Washington, D.C.: National Bureau of Standards, 1977. NBS Building Science Series 104.

Höglund, I., and Wänggren, B. *Studies of the Performance of Weatherstrips for Windows and Doors.* Stockholm: Swedish Council for Building Research, 1980.

Institute for Local Self-Reliance. *Weatherization Materials Handbook.* Washington, D.C.: Community Services Administration, 1979.

Knight, P. A. *Home Retrofitting for Energy Savings.* New York: McGraw-Hill Book Co., 1981.

Lawrence Berkeley Laboratory. *Windows for Energy Efficient Buildings,* nos. 1 & 2. Berkeley, Calif., 1979 and 1980.

Leckie, J. et al. *More Other Homes and Garbage.* San Francisco: Sierra Club Books, 1981.

Massachusetts Executive Office of Energy Resources. *Understanding Energy Efficiency in the Home.* Boston, 1978.

NAHB Research Foundation. *Insulation Manual: Homes, Apartments.* 2d ed. Rockville, Md.: National Association of Home Builders, 1979.

Ohio Energy and Resource Development Agency. *Manual for Cost-Effective Weatherization of Low Income Housing.* Columbus, Ohio: Battelle Columbus Laboratories, 1977.

Socolow, R. H., ed. *Saving Energy in the Home.* Cambridge, Mass.: Ballinger Publishing Co., 1978.

U.S. Department of Energy. *Find and Fix the Leaks.* Washington, D.C.: U.S. Government Printing Office, 1981. Order no. PUB 384/500.

_____. *Residential Conservation Service Auditor Training Manual.* Washington, D.C.: U.S. Government Printing Office, 1980.

U.S. Department of Housing and Urban Development. *In the Bank. . . Or Up the Chimney?* Washington, D.C.: U.S. Government Printing Office, 1977. (Also published as: *How to Keep Your House Warm in Winter, Cool in Summer.* New York: Cornerstone Library, 1977.)

Wilson, T., ed. *Home Remedies.* Philadelphia: Mid-Atlantic Solar Energy Association, 1981.

Wing, C. *From the Walls In.* Boston: Little, Brown and Company, 1979.

3 Heating and Cooling Systems

Much attention has been given to the value of energy savings from improvements to the building envelope, but there is also a great amount of energy that can be saved by improving and modifying the existing heating system. In New England, for example, in buildings heated by oil or natural gas, an average of only 55 percent of the energy content of these fuels is actually delivered as usable heat. There is little reason to believe that this percentage is significantly different in other parts of the United States, but with more efficient heating systems this figure could rise to 75 percent and higher. A simple furnace or boiler tune-up, for example, can reduce fuel consumption by up to 15 percent. After that there are a number of modifications that can be done to increase efficiency. In recent years the high price of heating fuels has spawned a great deal of research and development of new components and complete high-efficiency heating systems. Existing systems that aren't too old or worn out can be retrofitted with high-efficiency components and controls. In other cases, total replacement with a new system is the best move in terms of cost-effectiveness.

HOW HEATING SYSTEMS HEAT

Before discussing how the efficiency of heating systems can be improved, it is necessary to describe how these systems work. Heating systems that burn a fuel (for example, oil, gas, wood, coal) need a *combustion chamber* where the fuel is burned to produce heat, and a *flue* to exhaust the resulting smoke and hot gases. With oil or gas systems, the fuel is fed automatically into the combustion chamber and then ignited by a *burner*. With wood, the fuel must be regularly replenished by hand (except for some large-capacity systems), and with coal, manual or automatic loading systems are available. The heat produced in the combustion chamber is delivered to the living space via pipes (hot water or steam system) or ducts (warm air system). Woodstoves are generally used to heat the living space directly, although there are wood-fired furnaces and boilers available.

In central heating systems, hot water, warm air and steam are distributed by pumps, fans or natural convection throughout the building to provide warmth. Steam delivers warmth by passing through cast-iron radiators. Hot water delivers warmth the same way, with baseboard heaters substituting for radiators in modern systems. Warm air delivers heat by being blown directly into the room.

In gas or oil systems the parts of the heating system include a burner, a central combustion chamber located within either a *boiler* if the distribution system is steam or hot water, or a *furnace* if the distribution is air, pumps for hot water or fans for air, connecting pipes for steam and hot water, ducts for air, and either radiators, baseboard heaters, registers or grilles. Wood and coal boiler and furnace systems have the same parts, except they do not have a burner. Some buildings are heated by electricity, most commonly by baseboard resistance heaters that resemble hot water baseboard heaters. Electricity passes through metal coils in these heaters, causing them to become hot; the heat is then delivered to the room by natural convection. Newer buildings may have electric resistance furnaces or heat pump forced air systems.

Regardless of the type of heating system, each building must have a control that turns the system on or off as needed. Usually this control is a *thermostat,* which is set at the desired temperature for the building. If the room temperature drops below the thermostat setting, the heating system turns on automatically. When the desired room temperature is reached, the thermostat turns the system off. The number of thermostats in the building depends upon the number of *zones* or *distribution loops.* There should be one thermostat for each zone. To determine how many zones are in your building, look for the number of distribution pumps or fans; usually only one pump or fan is needed for each zone.

Once you understand how your heating system operates, you can begin to assess what must be done to improve its performance so that more heat is produced for every unit of fuel burned. Sometimes these improvements involve just a basic tune-up of the boiler or furnace. In other cases, improvements will include replacement of parts or, in some cases, replacement of the entire heating system.

As you read about the options for improving the performance of heating

[*Continued on page 82*]

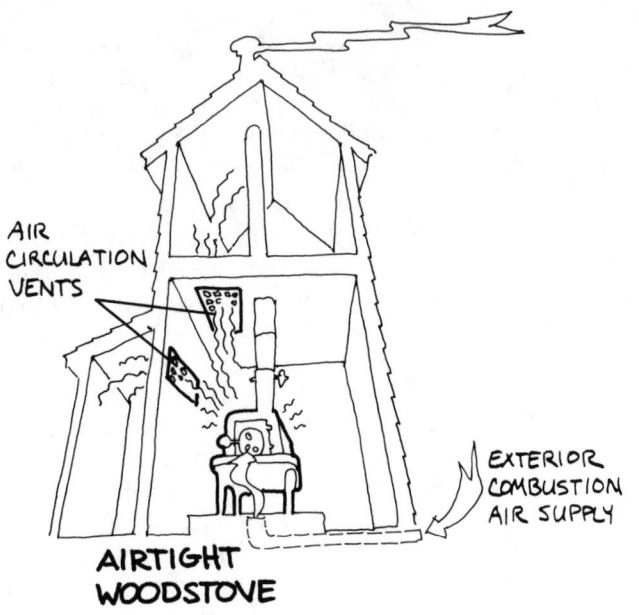

AIR
CIRCULATION
VENTS

EXTERIOR
COMBUSTION
AIR SUPPLY

AIRTIGHT
WOODSTOVE

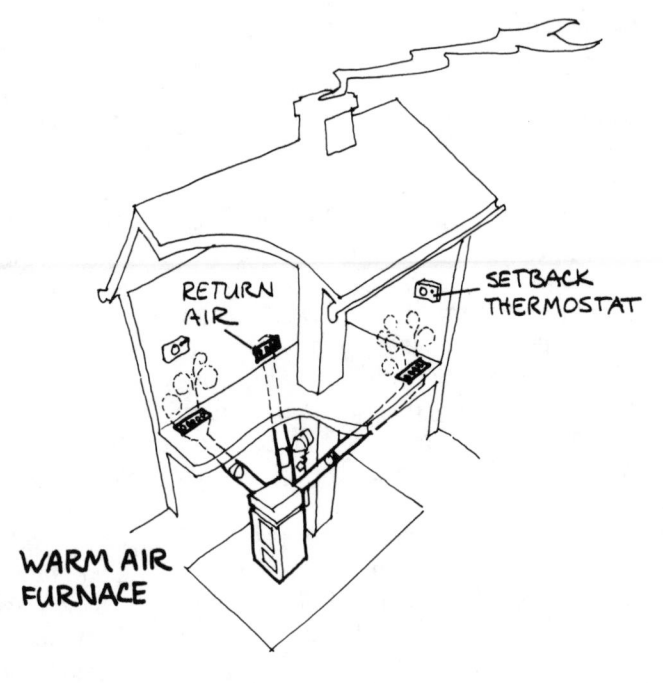

RETURN
AIR

SETBACK
THERMOSTAT

WARM AIR
FURNACE

FIGURE 3-1: HEATING SYSTEMS

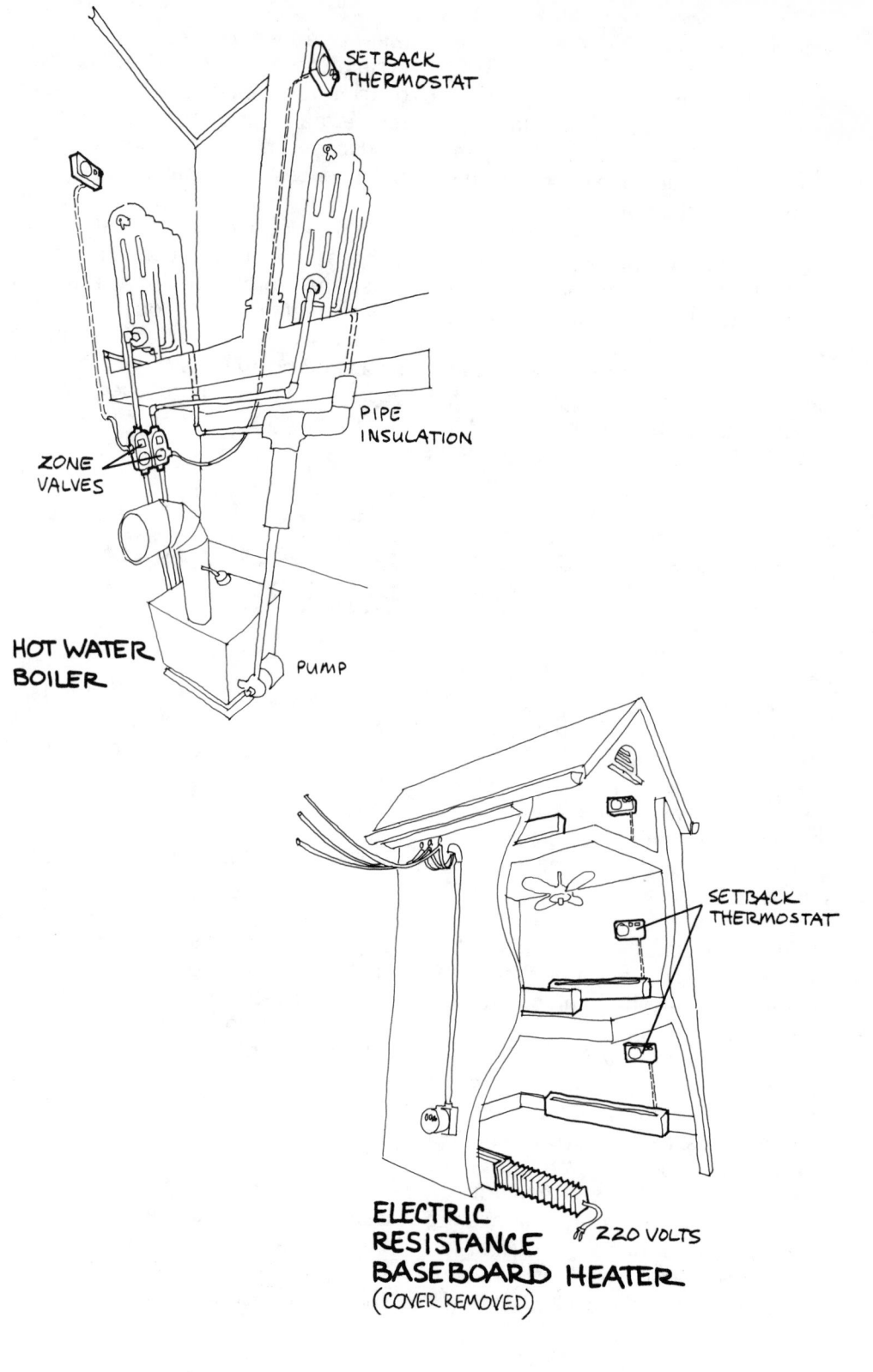

SETBACK
THERMOSTAT

PIPE
INSULATION

ZONE
VALVES

HOT WATER
BOILER

PUMP

SETBACK
THERMOSTAT

ELECTRIC
RESISTANCE
BASEBOARD HEATER
(COVER REMOVED)

220 VOLTS

systems, keep in mind the following points:

Improving the efficiency of your heating system will save fuel. In some cases, improving your heating system can save more fuel than improving the building envelope. Thus, when you set out with a certain sum of money to invest in energy conservation, don't just look at improving the envelope. Consider the building *and* its heating system.

Weatherization improvements affect the performance of your heating system. Weatherization reduces the amount of heat lost from a building, which consequently reduces the amount of heat input it needs. As a result, the heating system may become oversized for the building's reduced load. With an upgraded envelope, the heating system may only have to be on for short intervals to maintain a comfortable temperature. Frequent on-off "cycling" decreases efficiency. During the off part of the cycle, heat from the still-hot combustion chamber is lost up the flue. During the on part of the cycle, heat is lost while the combustion chamber is warming up, because the efficiency of the burn is less in a cool chamber. Therefore, when you weatherize a building extensively, you should also consider downsizing the heating system to make it more compatible with the building's reduced heating load.

Any changes in your heating system should allow for possible conversions to a renewable fuel, such as solar, in the future. As the price of conventional fuels rises, the price of renewable fuels becomes more competitive. If you are making any changes in your heating system because of age or oversizing, you should try to make those changes compatible with the requirements of a renewable energy system. Solar heating systems, for example, cannot use steam distribution because sun-generated temperatures are not high enough to produce steam. Also, additional radiators or longer baseboard heaters are usually needed to heat a room with solar-heated water because the water is not as hot as that produced by a boiler and must therefore have more radiating area to heat the room. You may wish to keep this in mind if you change from a steam to a hot water system in order to accommodate future conversion to a solar heating system.

OIL AND GAS FURNACES AND BOILERS

Maintenance

Furnace and boiler efficiency tend to deteriorate with time, which means that less heat is delivered for each gallon of oil or therm of gas burned. The causes of reduced efficiency include the following:

Too little or too much combustion air. For oil or gas to burn efficiently, exactly the right amount of air must be present. If there is too much or too little, the combustion efficiency drops.

Accumulation of dirt on the nozzle. The nozzle, through which gas or oil is released to the combustion chamber, can accumulate dirt that will hinder the fuel flow. As the flow is impeded, the delivery of heat to the distribution system is reduced.

Deposit of minerals from water on boiler walls. These minerals act as an insulator and can impede the flow of heat from the combustion chamber to the distribution system.

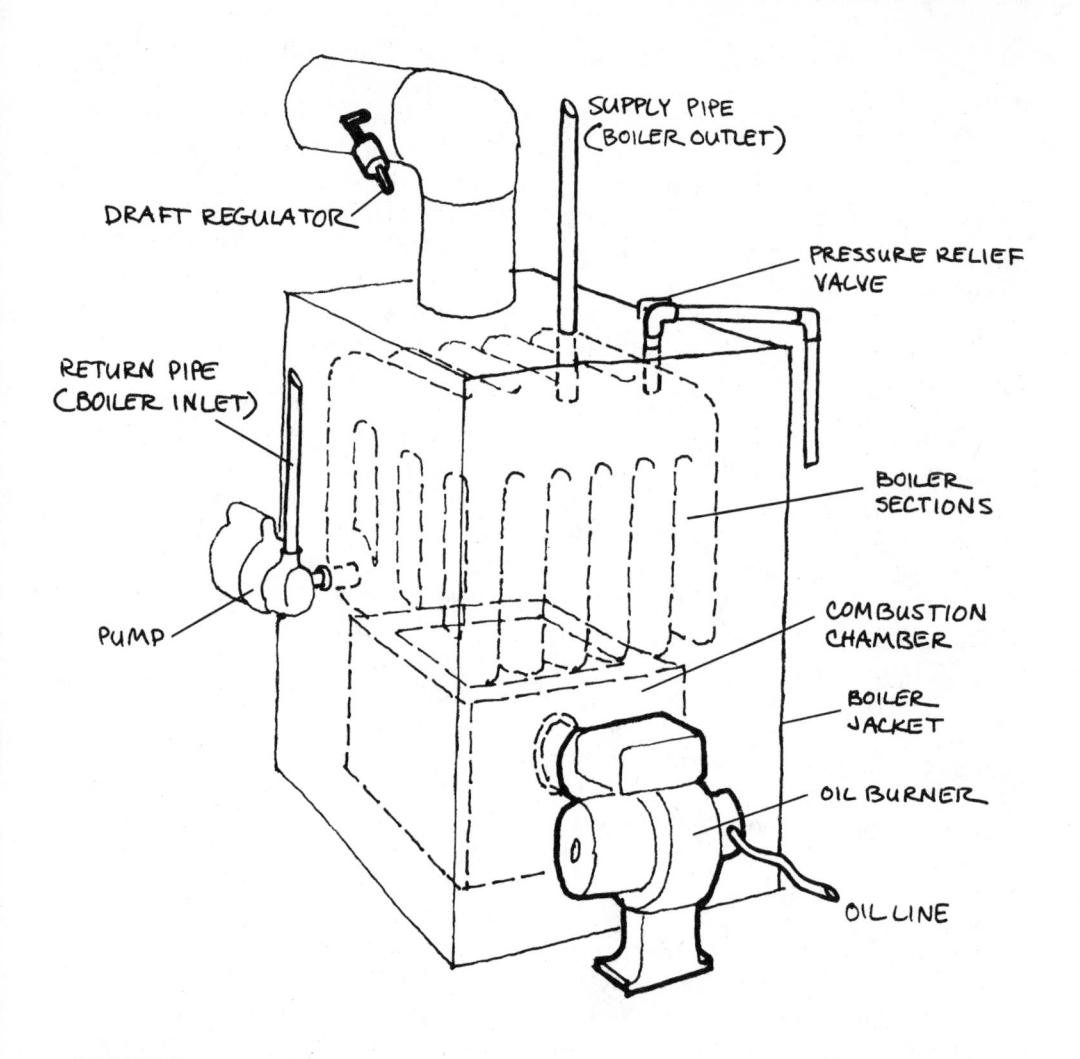

FIGURE 3-2: HOT WATER BOILER (OIL SHOWN-CAN ALSO BE GAS)

Buildup of soot in the combustion chamber of an oil burner. Soot is a good insulator—⅛ inch of soot has the insulative value of 1 inch of fiberglass. As the oil combustion chamber becomes coated with soot, the rate of heat transfer to the distribution system is reduced. Natural gas burns cleanly and does not produce soot.

To determine whether the performance of your oil or gas boiler or furnace is deteriorating, you should have its efficiency tested regularly—once a year for oil and every other year for gas. Unless you have a regular maintenance contract with your oil or gas dealer, you can assume that your system has not been regularly tuned. Many oil delivery companies, gas utilities and independent gas-fitting companies have employees capable of performing the test.

The efficiency of a furnace or boiler is determined by measuring the temperature of the flue gases, the concentrations of carbon dioxide and smoke in the flue gases of an oil-fired system, or the concentrations of carbon dioxide and carbon monoxide in the flue gases of a gas-fired system. The temperature of the flue gases indicates how much waste heat is going up the flue and out of the building. The higher the flue temperature, the lower the efficiency of the system. Normal flue temperatures are 400 to 600°F for gas-fired systems, 500 to 700°F for oil-fired systems and 350 to 400°F for oil systems with a flame retention burner.

The concentration of carbon dioxide (a by-product of combustion) in the flue gases indicates how completely the gas or oil is burned. Typical carbon dioxide concentrations are 8 to 10 percent for gas systems and 8 to 12 percent for oil systems.

The concentration of carbon monoxide in the flue gases of a gas system also indicates how complete the burn is. Carbon monoxide is toxic. Therefore, levels of this gas greater than one-tenth part per million (0.1 ppm) are undesirable. The concentration of smoke in the flue gases of an oil system indicates how clean the burn is. As the smoke concentration increases, system efficiency decreases. Flue temperatures and the concentrations of carbon dioxide and carbon monoxide are measured through a ¼-inch hole in the flue pipe, about 6 inches above its point of emergence from the furnace or boiler.

Although a professional should make a complete efficiency test, you can monitor system efficiency yourself by placing a flue thermometer (about $20) in the test hole and keeping a record of flue temperatures. The temperature should be taken 10 to 15 minutes after the burner comes on. If the temperature rises more quickly or to a higher point than during previous readings, it is likely that your furnace or boiler requires some maintenance. (Note that such a test cannot be done until your system has been tuned at least once.)

If an efficiency test indicates a low efficiency the unit must be tuned. Old units have efficiencies of 50 to 75 percent while new ones can exceed 80 percent efficiency. High flue temperatures in an oil-fired boiler or furnace indicate a buildup of soot that must be cleaned out of the combustion chamber. Tests that show abnormal carbon dioxide concentrations often indicate that too little or too much air is entering the combustion chamber. This can be caused by overly tight or leaky furnace seals, an improperly set air intake on the burner, barometric dampers or draft regulators that stick open or shut, a long fuel run or a blocked or improperly sized flue. The person conducting the test should check all of these items as part of the tune-up. In a gas system it's important to make sure that the nozzle is clean. In an oil system be sure to check that the fuel lines feeding the burner are clean and that the burner is in good shape. Boiler water in small units should be checked for cleanliness and changed as needed. On large boilers (where the consumption rate is greater than about 10 gallons of oil or 10 therms of gas per hour), boiler water should be changed regularly and may require treatment in order to maintain cleanliness. If there are filters attached to the furnace, they should be checked and cleaned every few weeks and replaced when necessary. A tune-up of a furnace can increase its efficiency by 5 to 10 percent; of a boiler, by 5 to 15 percent.

In addition to the problems just mentioned, there may be more dramatic problems that you can easily detect. These include corrosion of the flue pipe, pipe seals or joints; failure of pumps or fans; the release of strong odors from an air distribution system or boiler; repeated loss of boiler water; repeated ejection of hot water or steam through relief valves; and in the case of an oil-fired system, discharge from the flue with a high soot content. These problems indicate that repair or replacement of parts may be necessary.

Modifications for Higher Efficiency

While regular maintenance of gas or oil systems can prevent deterioration of efficiency, certain modifications can often increase efficiency. For furnaces, these include the following:

Altering the circulating fan so that it remains on after the burner has turned off. In many old furnaces, the fan that circulates heat to the building turns off when the burner turns off. The heat in the still-warm furnace is lost up the flue (and also to the basement). If the fan remains on after the burner has turned off, some of the residual heat can be directed usefully to the building, saving about 5 percent of the total fuel consumption.

Installing a two-speed circulating fan. Furnaces often have a high-speed circulating fan to facilitate delivery of heat on a cold day. However, the rapid delivery of heat is not as important on a warmer day, and slower delivery can adequately maintain the building temp-

erature. Since a high-speed fan consumes more electricity than a low-speed one, you can install a two-speed circulating fan attached to an outdoor temperature sensor. On a cold day, the high speed will operate, while on a warmer day the low speed will suffice.

Modifications to boilers include the following:

Installing aquastats that are controlled by outdoor temperatures (applicable only to hot water, not steam, boilers). Hot water heating systems usually deliver water to radiators or baseboard heaters at temperatures between 150 and 180°F. These high temperatures are necessary to heat a building adequately only on the coldest days, but are unnecessary on warmer days. During warmer periods, therefore, the boiler loses heat unnecessarily because the higher the boiler temperature, the higher the flue gas temperature and the greater the heat loss up the flue. To reduce this heat loss, equipment can be added to the boiler that senses the outdoor temperature and controls the aquastat, or water temperature regulator. When outdoor temperatures are at their lowest, the aquastat temperature is set high. When the reverse is the case, the aquastat setting is automatically reduced. Temperature-sensitive aquastats have been standard for many years on large boilers but have only recently become available for small ones. They cost between $120 and $200 and reduce fuel consumption by 10 to 15 percent, with payback ranging from one to two years.

Manually turning down aquastats (applicable only to hot water, not steam, boilers). Many heating systems are oversized for the heating loads of their buildings because system designers leave a large margin for error. Boilers designed to operate at temperatures

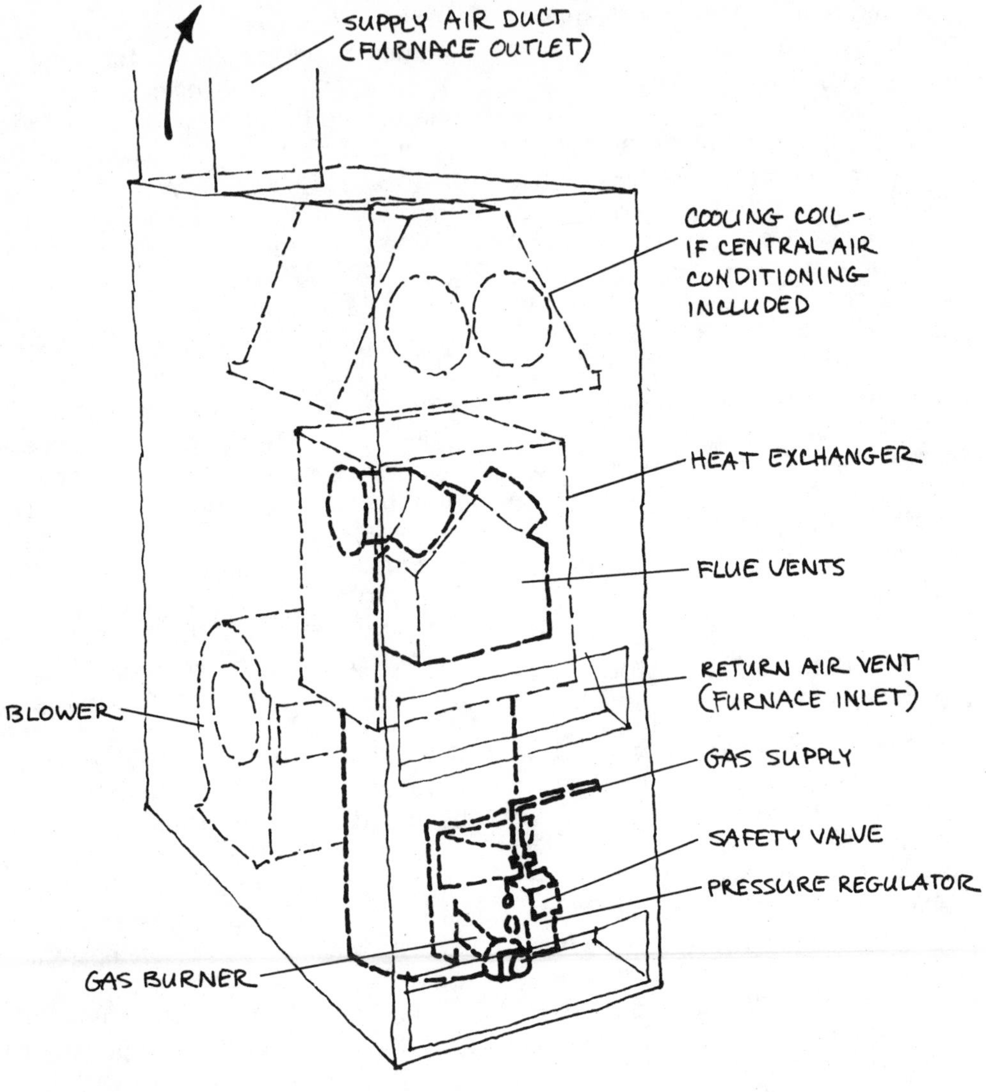

SUPPLY AIR DUCT
(FURNACE OUTLET)

COOLING COIL -
IF CENTRAL AIR
CONDITIONING
INCLUDED

HEAT EXCHANGER

FLUE VENTS

RETURN AIR VENT
(FURNACE INLET)

GAS SUPPLY

SAFETY VALVE

PRESSURE REGULATOR

BLOWER

GAS BURNER

FIGURE 3-3: WARM AIR FURNACE (GAS SHOWN - CAN ALSO BE OIL)

as high as 200°F can often be set back to 150°F or less and will still provide adequate heating. This setback can result in a fuel savings of 5 to 10 percent.

Modifications to both furnaces and boilers include the following:

Installing automatic flue dampers. After a burner has turned off, heat is lost up the flue of the still-warm boiler, even if it is properly sized for the building's heat demand. To reduce this heat loss, a damper can be installed in the flue that will automatically close when the burner is off. There are two types of automatic dampers: powered and passive. Powered units are wired to the burner ignition and open and close in response to burner operation (see figure 3-4). Passive units (used in gas systems) contain bimetallic blades that flex open as they heat up. Each opening cycle takes about three minutes, during which time combustion exhaust spills into the boiler room. These dampers are available in different sizes to fit a variety of flue diameters. To satisfy the fire codes of most states, flue dampers must be American Gas Association or Underwriters Laboratories approved. In addition, they

should be designed to fail in an open position so that combustion gases can be safely vented, and the actual damper blade should be of corrosion-resistant stainless steel. The damper should be installed between the barometric damper (which controls airflow while the burner is on) and the chimney.

Automatic flue dampers can save 10 to 15 percent of total fuel use and cost between $250 and $400, depending upon size. Paybacks range from one to two years and depend greatly on the type of system and its characteristics. Payback for a gas system with an unlimited open air-feed can be very rapid. A properly sized and tuned oil

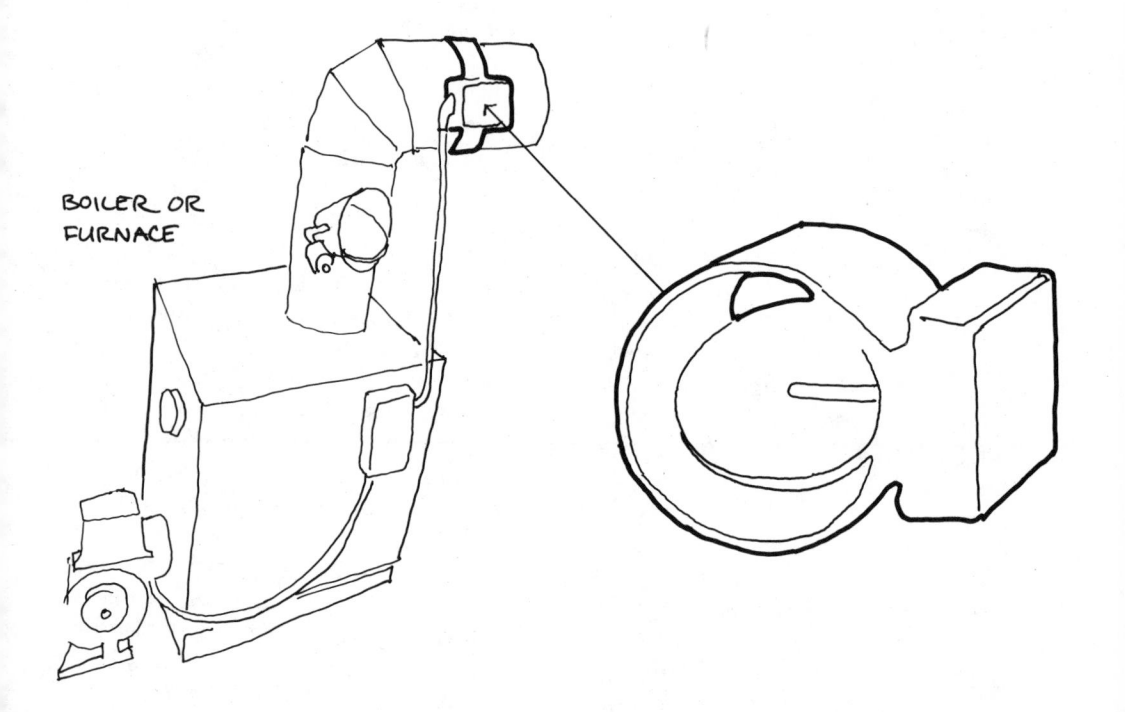

BOILER OR
FURNACE

FIGURE 3-4: AUTOMATIC FLUE DAMPER

system can make far less use of such a device. The savings provided by a flue damper installed on a system with a flame retention burner will only be in the range of 5 percent, because the burner itself acts as a sort of damper. Some burner manufacturers will void their warranties for installation of a damper. In order to determine whether such a device is appropriate for your system, you should consult a heating expert.

Total System Replacement

Often, regular maintenance or modification of the furnace or boiler will not bring efficiency close to that of a modern, properly sized system. In such a case, it is often cost-effective to replace all or part of the system. There are two criteria for determining if replacement is in order: if, after regular maintenance, the combustion efficiency of the furnace or boiler is still less than 65 percent, and if the system is somewhat oversized for the building's load and cycles frequently.

On the coldest day of winter, a properly sized unit should be on continuously. Check how many minutes your system is on in an hour and then read the outside temperature. Consult table 1-1 to determine the length of time a properly sized unit should be on. If the on-time of your system is much less than that listed in the chart, it may be oversized. In moderate weather, a constantly operating system can indicate inefficient operation, leaks or poor heat delivery, so it is generally best to repair and tune up the system before deciding whether or not to replace it.

Various parts of a furnace or boiler system can be replaced or adjusted under a wide range of circumstances. For gas systems, these include reduction of the firing rate and installation of a pilotless ignition burner. For oil-fired units, these include replacement and downsizing of the nozzle orifice and replacement of the burner. For both types of systems, it may also pay to replace the furnace or boiler with a smaller one.

Modifications for Gas-Fired Systems

Reduction of the firing rate. If a furnace or boiler system is oversized, the easiest way to modify it is to reduce the rate at which gas is released (and wasted) into the combustion chamber during the burn. This can be carried out by your gas utility or an independent gas-fitting company, either of which should be qualified to determine by how much the firing rate should be reduced. On a steam boiler, reduction of the firing rate may also require changes in other parts of the system. The firing rate should be reduced by no more than 25 to 30 percent. Beyond this point, there is such a small flame that the air or water is heated inefficiently, causing the heating system to respond too slowly to a thermostat's demand for heat. Reducing the firing rate can reduce fuel consumption by as much as 10 percent.

Installation of a pilotless ignition burner. Gas burners in most buildings have continuous pilot ignition. New gas burners are ignited by an electric spark, a feature that reduces fuel consumption by 20 to 30 percent in small heating systems. The firing rate can be reduced on a new gas burner (within the constraints mentioned previously) to realize further fuel savings. Pilotless ignition burners range in price from $100 to $250.

Replacements for Oil-Fired Systems

Replacing and downsizing the nozzle orifice. If a furnace or boiler is over-sized, the easiest way to modify it is to reduce the size of the burner nozzle. This means that less oil is sprayed into (and wasted in) the combustion chamber during the burn. This change is inexpensive and can be done by the building occupant, although it is advisable to check with a heating system professional to determine how much the nozzle size should be reduced. The nozzle should be downsized by no more than 25 to 30 percent of its original size. Beyond this point, there is such a small flame that the system is unable to respond rapidly enough to the thermostat's demand for heat. Downsizing the nozzle can reduce fuel consumption by as much as 10 percent and costs between $40 and $60 if performed by a professional.

Replacing the burner. Old burners have two problems. First, the temperature of their flame tends to be relatively low, which causes inefficient combustion. Second, the flame tends to linger after the burner has turned off, leading to needless waste of fuel. Modern burners, called *flame-retention burners*, reduce these problems: The flame is hotter, and it extinguishes more readily when the burner turns off. These two features can reduce fuel consumption by 5 to 20 percent. In addition, the nozzle size can be reduced on a new burner (within the constraints mentioned previously) to realize further fuel savings. Flame-retention burners cost between $250 and $400.

Replacements for Gas and Oil Systems

Downsizing the furnace or boiler. If a furnace or boiler is very inefficient or highly oversized, it should be replaced. However, a new unit is expensive, ranging in price from $1,500 to $2,500, installed. Replacement should be considered only if no other changes can significantly improve the performance of the heating system. A prime candidate for such replacement, for example, is an oil-fired pancake, a converted coal-fired unit.

A new boiler or furnace should be sized by a heating professional based upon a detailed heat-loss analysis of the building. State building codes generally limit a new unit's capacity to no more than 125 percent of the building's calculated design heat load, although the specifics may vary from state to state. (Your state building code commission can provide specific numbers in this regard.) Note that steam boilers must not be downsized more than 25 percent or they may not adequately serve the existing radiator system.

Downsizing can result in fuel savings of 20 percent or more. If, however, the unit has a flue damper, the overcapacity becomes less important and downsizing becomes less economical.

Table 3-1
Typical Oil or Gas Furnace Improvements

	Measure	Frequency	Savings Potential (%)	Done By	Cost ($)	Payback (yrs)	Comments
Maintenance	Tune-ups—clean furnace, burner; check valves, pipes, inlets	Gas: every 2 years Oil: yearly	5-10	Licensed technician, gas fitter or supplier	30-70	0.25	Should be done as part of service contract
	Change filter regularly	Every 1-2 months	5	Building occupant	3-5	0.10	Do-it-yourself task
	Efficiency tests	Twice each heating season	...	Licensed technician	...	...	Should be done as part of tune-up
	Monitor temperature of flue	Every other week	...	Building occupant	...	...	Can signal need for tune-up
Modifications	Install temperature-sensing dual speed fan	...	About 300 KWH	Electrician, building occupant, fuel supplier	120	4-6	Reduces electricity consumption of forced air system
	Install a delay so that fan shuts off after burner goes off	...	5	Electrician, gas supplier, fuel supplier	140	3-5	More heat from combustion is delivered to living space
	Install flue damper	...	10-15	Gas supplier, heating contractor, fuel supplier	250-400	1-2	Must fail open. Must be AGA or UL approved
Replacements — Oil	Replace burner with flame retention burner	...	5-20	Licensed technician, fuel supplier	250-400	1-4	...
Replacements — Gas	Replace burner with pilotless ignition burner	...	20-30	Licensed technician, gas supplier	100-250	1-2	...
Replacements — Both	Downsize burner nozzle or reduce firing rate	...	10	Licensed technician, perhaps building occupant	40-60	1-2	Should downsize no more than 25-30%
	Replace furnace with smaller, high-efficiency unit	...	10-40	Licensed technician, fuel supplier, gas supplier	1,500-2,500	3-7	...

Correction factors: To find fuel savings, multiply by the appropriate factor for your zone. To find payback, divide by the same factor. Zone I: 0.33; Zone II: 0.67; Zone III: 1.00; Zone IV: 1.33.

Table 3-2
Typical Oil or Gas Boiler Improvements

		Measure	Frequency	Savings Potential (%)	Done By	Cost ($)	Payback (yrs)	Comments
Maintenance		Tune-ups—clean boiler, burner; check valves, pipes, inlets	Gas: every 2 years Oil: yearly	5-15	Licensed technician, fuel supplier or gas fitter	30-70	0.25	Should be done as part of service contract
		Efficiency tests	Twice each heating season	...	Licensed technician	...	...	Should be done as part of tune-up
		Monitor temperature of flue	Every other week	...	Building occupant	...	...	Can signal need for tune-up
Modifications		Install outdoor temperature sensing aquastat	...	10-15	Electrician, fuel supplier or gas fitter	120-200	2-4	Will lower water temperature during mild part of heating season; used only on hot water systems
		Manually set back aquastat	...	5-10	Fuel supplier, building occupant	...	...	Used only on hot water systems
		Install flue damper	...	10-15	Fuel supplier, heating contractor, gas fitter or supplier	250-400	1-2	Must fail open. Must be UL or AGA approved
Replacements	Oil	Replace burner with flame retention burner	...	5-20	Licensed technician, fuel supplier	250-400	1-4	...
	Gas	Replace burner with pilotless ignition burner	...	20-30	Licensed technician, gas supplier	100-250	1-2	...
	Both	Reduce firing rate, downsize nozzle	...	10	Gas supplier	40-60	1-2	Only on hot water systems. Should downsize no more than 25-30%
		Replace boiler with smaller unit	...	10-40	Licensed technician, gas supplier, fuel supplier	1,500-2,500	3-7	Best candidate for replacement is old "pancake" oil boilers converted from coal

Correction factors: To find fuel savings, multiply by the appropriate factor for your zone. To find payback, divide by the same factor. Zone I: 0.33; Zone II: 0.67; Zone III: 1.00; Zone IV: 1.33.

Table 3-3
Oil or Gas Steam Distribution System Improvements

	Measure	Frequency	Savings Potential (%)	Done By	Cost ($)	Payback (yrs)	Comments
Maintenance	Check valve (on 1-pipe systems)	Monthly	5-20	Fuel supplier, building occupant	...	...	Should be part of service contract
	Check traps (on 2-pipe system)	Yearly	5-10	Fuel supplier, licensed technician	...	...	Should be part of service contract
	Balance system	As needed	5-20	Licensed technician, perhaps building occupant	Varies	Varies	Should be part of service contract
	Clean radiators	Monthly	1-5	Building occupant	...	...	Necessary for good heat delivery
	Remove obstructions from front or top of radiators	Regularly	5-15	Building occupant	...	...	Necessary for good heat delivery
	Check for leaks	Yearly	Varies	Building occupant, fuel supplier	...	...	Leaks can reduce heating efficiency
	Insulate pipes	...	5-10	Building occupant technician	50-200	1-3	...
Modifications	Install Hartford loop	...	Varies	Plumber, technician	Varies	Varies	Prevents blockage of return lines with sediment
	Place reflective shields behind radiators	...	5	Building occupant	0.50-2.50 per radiator	Varies, very short	Easily done with cardboard and aluminum foil
Replacements	Replace steam system with hot water system	...	Varies	Plumber, heating engineer	Varies	Varies	More compatible with possible future conversion to solar heating

Correction factors: To find fuel savings, multiply by the appropriate factor for your zone. To find payback, divide by the same factor. Zone I: 0.33; Zone II: 0.67; Zone III: 1.00; Zone IV: 1.33.

STEAM HEAT DISTRIBUTION SYSTEMS

Maintenance

Regular maintenance of the heat distribution system is important in order to achieve and maintain adequate comfort in a building. Proper maintenance of a steam distribution system includes the following:

The valves on all one-pipe steam systems (characteristic of small buildings) should be working properly. If they are not functioning properly, air can get into the line and block the passage of steam. It is always cost-effective to replace these valves.

The traps on all two-pipe steam systems (characteristic of apartment buildings) should be checked to determine if they are working properly. These traps operate automatically to close the return line when steam enters the radiator in order to prevent the returning steam from being directed back into the boiler with the condensate (water) it is carrying.

The heating system should be balanced. All radiators within a distribution loop should be at about the same temperature. You can accomplish this by attaching a piece of clay to each radiator in a loop and pressing a thermometer into each piece. The valves on the radiators should then be adjusted until every thermometer gives approximately the same reading. If the heating system is not balanced, one part of the building may become too cold while another becomes too warm. To cool off the warm rooms, windows often have to be opened, which wastes fuel.

Radiators should not be obstructed. Objects placed in front of or on radiators prevent their radiating heat into the room.

Pipes should be checked for leaks. Leaks can reduce the efficiency of the heating system.

Radiators should be cleaned regularly. Dust reduces the ability of radiators to heat a room.

All hot pipes in unheated spaces should be insulated. Otherwise, heat is lost to parts of the building where it is not needed.

Radiators on one-pipe steam systems should be tilted slightly toward the supply/drain pipe. This allows the steam condensate to be drawn from the radiator and returned to the boiler.

Modifications

A modification to a steam distribution system that can improve performance

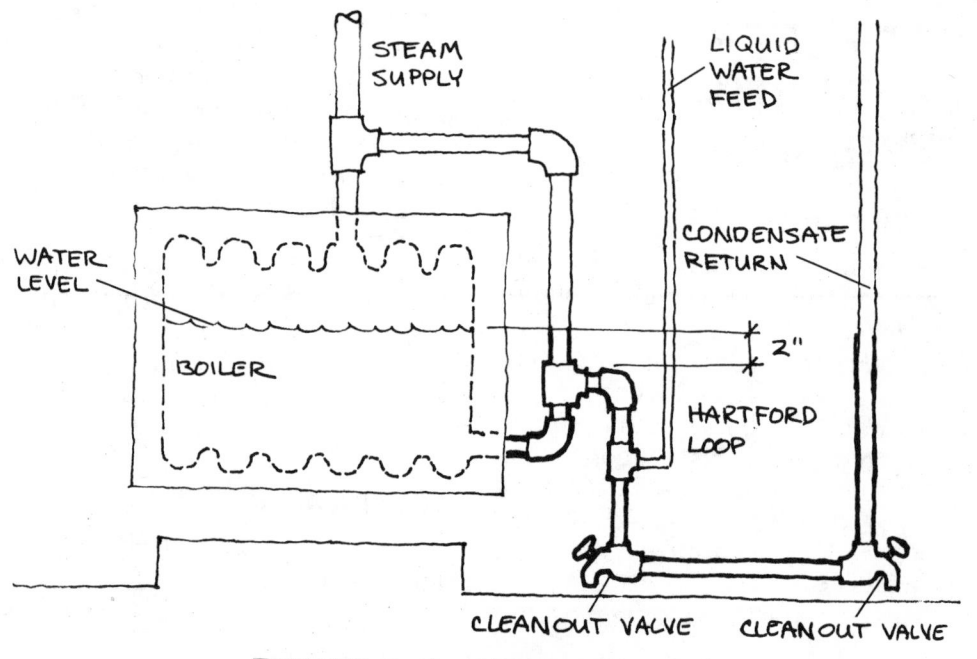

FIGURE 3-5: HARTFORD LOOP

is the installation of a Hartford loop near the boiler, on the condensate return line (see figure 3-5). This loop serves to collect and drain sediment that gathers in the condensate and eventually blocks its flow. The condensate sediment should be drained several times during the heating season, depending upon the steam system and water quality. Also, placing reflective shields between radiators and outside walls will reduce the amount of heat lost through those walls. You can do this easily using cardboard and aluminum foil.

Replacement

If the performance of a steam heating system is so poor that it has to be replaced, you might consider installing a hot water distribution system instead. This replacement would be compatible with future conversion to an active solar heating system.

HOT WATER DISTRIBUTION SYSTEM

Maintenance

Proper maintenance of a hot water distribution system includes the following:

All pumps must be serviced regularly. If they are not regularly checked for worn parts, the pumps can burn out, shutting down the distribution loops.

Radiators and/or baseboard heaters and pipes should be checked for leaks. Leaks reduce the efficiency of the heating system.

Radiators and/or baseboard heaters should be cleaned regularly. Dust on radiating surfaces, particularly on baseboard heaters, can act as an insulator and reduce the system's ability to heat a room.

Bleed valves on radiators and baseboard heaters should be working properly and used regularly. These valves are necessary to remove any air that might enter a distribution system. If air is not regularly removed from a hot water distribution system, it can prevent the flow of water, and hence heat, through the building. New systems often have automatic bleed valves, while older systems have manual ones. A radiator key for bleeding these valves costs about 25¢. If a radiator is cold, it may need bleeding.

The heating system should be balanced. All radiators and/or baseboard heaters within a distribution loop should be about the same temperature. You can accomplish this by attaching a piece of clay onto each radiator or baseboard heater in the loop and pressing a thermometer into each piece. The valves on each radiator and/or baseboard heater should then be adjusted until every thermometer gives approximately the same reading.

Radiators and/or baseboard heaters should not be obstructed. Objects placed in front of or on heaters prevent them from radiating their heat into the room.

All hot pipes in unheated spaces should be insulated. Otherwise, heat is lost to parts of the building where it is not needed.

Modifications

Placing reflective shields between radiators and outside walls will reduce the amount of heat lost through those walls. You can do this easily using cardboard and aluminum foil.

Replacement

If the performance of a hot water heating system is so poor that it has to be replaced, you might consider installing more baseboard heater length in the new system. With this additional radiating area, the circulating water can be less hot than before and still adequately heat the room. Such a change makes the distribution system compatible with a solar heating system because solar-heated hot water is not as warm as boiler-heated hot water.

FIGURE 3-6: STEAM AND HOT WATER RADIATORS

Table 3-4
Oil or Gas Hot Water Distribution System Improvements

	Measure	Frequency	Savings Potential (%)	Done By	Cost ($)	Payback (yrs)	Comments
Maintenance	Service pumps	Yearly	5	Fuel supplier, technician	...	...	Should be part of service contact
	Balance system	As needed	5-20	Licensed technician, perhaps building occupant	Varies	Varies	Should be part of service contract
	Check for leaks	Yearly	Varies	Building occupant, fuel supplier	...	...	Leaks can reduce heating efficiency
	Remove obstructions from front or top of radiators or baseboard heaters	Regularly	5-15	Building occupant	...	...	Necessary for good heat delivery
	Clean radiators or baseboard heaters	Monthly	1-5	Building occupant	...	...	Necessary for good heat delivery
	Check and use bleed valves	Monthly	1-15	Building occupant, fuel supplier	...	...	Necessary for good heat delivery
Modifications	Insulate pipes	...	5	Building occupant, technician	30-60	0.5-2	...
	Place reflective shields behind radiators	...	5	Building occupant	0.50 to 2.50 per radiator	Varies, but very short	Easily done with cardboard and aluminum foil
Replacements	Add more baseboard heaters	...	Varies	Plumber, heating engineer	Varies	Varies	More compatible with possible future conversion to solar heating

Correction factors: To find fuel savings, multiply by the appropriate factor for your zone. To find payback, divide by the same factor. Zone I: 0.33; Zone II: 0.67; Zone III: 1.00; Zone IV: 1.33.

HOT AIR DISTRIBUTION SYSTEM

Maintenance

Proper maintenance of a hot air distribution system includes the following:

All fans must be serviced regularly.

If they are not regularly checked for worn parts, the fans can burn out, shutting down distribution loops.

Air ducts should be checked for leaks. Leaks reduce the efficiency of the heating system. Look for any obvious holes or breaks and fix them with duct tape or caulk. The smoke test described in chapter 1 can be used for this purpose.

The dampers in air ducts and/or air

Table 3-5
Oil or Gas Hot Air Distribution System Improvements

	Measure	Frequency	Savings Potential (%)	Done By	Cost ($)	Payback (yrs)	Comments
Maintenance	Service fans	Yearly	5-10	Technician	...	...	Should be part of service contract
	Check for air leaks	Yearly	Varies	Technician, perhaps building occupant	...	...	Can use smoke test
	Check dampers and registers for seal	Yearly	Varies	Technician	...	...	Should be part of service contract
	Clean grilles or registers	Monthly	5-15	Building occupant	...	...	Necessary for good heat delivery
	Remove obstructions from front of grilles or registers	Regularly	5-15	Building occupant	...	...	Necessary for good heat delivery
	Balance system	As needed	5-20	Technician, perhaps building occupant	Varies	Varies	Should be part of service contract
	Insulate ducts	Once	5-10	Technician, building occupant	50-150	1-2	...
Modifications	Install dampers in ducts	...	Varies	Heating engineer or contractor	Varies	Varies	Allows rooms to be closed off to heat

Correction factors: To find fuel savings, multiply by the appropriate factor for your zone. To find payback, divide by the same factor. Zone I: 0.33; Zone II: 0.67; Zone III: 1.00; Zone IV: 1.33.

registers should be checked for proper seal and operation. These devices allow you to close off parts of the building from receiving heat.

All hot air ducts in unheated spaces should be insulated. Otherwise, heat is lost to parts of the building where it is not needed.

All grilles and/or air registers should be cleaned regularly. Dirt on these surfaces can impede the flow of hot air.

Air grilles and/or registers should not be obstructed. Objects placed in front of or on these openings prevent hot air from being blown into the room.

The heating system should be balanced. The air coming out of all the registers or grilles should be at about the same temperature within one distribution loop. This is difficult to achieve and usually requires the services of a heating system professional. Some balance can be achieved by controlling the degree to which air registers or dampers in air ducts are open, or by partially obstructing registers, but this may throw the system out of balance elsewhere.

Modifications

If you wish to be able to close off some rooms of your building from heat, it may be wise to install dampers in your air supply ducts. These tend to work more effectively than registers in stopping the flow of warm air.

CONTROLS FOR OIL- AND GAS-FIRED SYSTEMS

Heating system controls are extremely varied. They usually consist of thermostats which control indoor temperature. In order to carry out this function, system controls may also regulate valve and damper settings, pump and fan speeds, indoor humidity, exchange rates with outside air and distribution system temperatures. In addition, control systems may respond to changes in climate.

Maintenance

Thermostats generally should have thermometers mounted on or near them, and they should be checked occasionally with an ordinary thermometer to ensure accuracy. Thermostats should be cleaned frequently to remove dust and lubricated when necessary. Those with filters require periodic filter replacement. Wiring should be checked for electrical shorts.

Modifications

It is cost-effective to invest in setback thermostats to control your heating system. These instruments allow your

Table 3-6
Oil- or Gas-Fired Control Systems

	Measure	Frequency	Savings Potential (%)	Done By	Cost ($)	Payback (yrs)	Comments
Maintenance	Clean thermostats, and lubricate if applicable	Monthly	3-5	Building occupant	...	...	Dust can affect thermostat performance
	Check temperature reading and recalibrate if necessary	Yearly	Varies	Technician	20	0.16	Faulty thermometers can affect thermostat performance
	Check thermostat leads for shorts	Yearly	Varies	Building occupant	...	...	Shorts can limit thermostat performance
Modifications	Install setback thermostat	...	10-20	Technician, perhaps building occupant	80-100	1-2	...

Correction factors: To find fuel savings, multiply by the appropriate factor for your zone. To find payback, divide by the same factor. Zone I: 0.33; Zone II: 0.67; Zone III: 1.00; Zone IV: 1.33.

thermostat setting to be controlled by a clock that automatically sets the temperature up when the building is occupied and down when it is empty or when the occupants are asleep. Setback thermostats can easily be installed by the building occupant where a regular thermostat already exists. Most residential buildings need only a setback thermostat that can be set back twice during the day. In large apartment buildings, on the other hand, it might be useful to install a unit that can be set for each day of the week. Many urban health codes have different minimum temperature requirements for different times of the day as far as apartment buildings are concerned. Fuel savings from such a device range from 10 to 20 percent, and since the cost of a setback thermostat is between $80 and $100, the payback is generally less than one or two years (see "Setback Thermostats" in this chapter).

OIL- AND GAS-FIRED DOMESTIC HOT WATER SYSTEMS

The heating of domestic (tap) water is a major energy consumer in households, accounting for about 15 percent of total residential energy use. Generally, hot water is heated and stored in a tank that is separate from the main heating system. The tank should be well insulated; if it is warm to the touch, it probably lacks sufficient insulation. Unless your building has a dishwasher without a separate heating element, the temperature of the hot water tank should be set no higher than 120°F. If your dishwasher does not have such a

heating element, the tank temperature must be set at 140°F.

One way of saving energy is through more judicious use of hot water. This can be accomplished through the installation of flow control devices in showers and sinks. These controllers reduce the volume of water coming out of the tap or shower without reducing the pressure.

Many oil-heated and some gas-heated buildings heat hot water by running water through the boiler or furnace and then directing it to the tap. The use of this tankless system means that the burner must run all summer just to heat hot water, all the while suffering from heat losses up the flue. In such cases, it is cost-effective to invest in a separate and smaller domestic hot water tank system so that the space-heating system can be turned off during the summer. If the oil burner is turned off, the boiler or furnace should be cleaned so that the soot deposited on the walls of the combustion chamber does not cause corrosion. (A gas burner produces little or no soot.)

In general, a replacement water heater will use natural gas or electricity. (In some locations, it may be impossible to get a gas hookup.) Several alternative water-heating systems are available. For example, you may wish to modify your present water-heating system to allow for solar water heating to meet part or all of your needs. If you do not use very much hot water, a tankless or demand water heater may be suitable. These are small, wall-mounted units that use gas or electricity. They heat water almost instantaneously when you need it rather than keeping a large tank of water hot all of the time. Such systems are not very expensive, but their maximum flow rates may not be sufficient to serve households with high hot water consumption. In this case, they probably will save little or no energy.

If your replacement system must use electricity, you should look into the possibility of installing a heat pump water heater. These units are explained in more detail in "Heat Pumps" in this chapter. In very cold climates, such units are suitable only for nonheating season use, since they require air temperatures of 45°F or more in order to operate efficiently.

A gas-fired water heater should be drained every six months in order to remove the sediment that builds up at the bottom of the tank. If not removed, this sediment will act as an insulating layer between the water and the gas flame and reduce efficiency. If you find that your water heater is not adequately insulated, install an insulating jacket with 3 to 6 inches of fiberglass. This improvement costs $12 to $25, and the payback is a year or less (see "Insulation" in chapter 2).

NEW HIGH-EFFICIENCY OIL AND GAS SYSTEMS

Most new heating system technologies involve some sort of modification to conventional heat exchanger or burner design. In general, these changes result in steady-state efficiencies of as much as 85 to 90 percent, with seasonal efficiencies in the area of 75 to 85 percent. (This is compared to steady-state efficiencies of 78 to 80 percent for the most efficient new heating units of conventional design.)

Pulse Combustion/Condensing Flue Units

There is now available a heating system that uses spark ignition in the combustion chamber and that keeps exhaust gases down to a relatively cool 120°F instead of the usual 400°F or more. In this process almost all the useful heat is removed from the flue gases, which are cool enough to be exhausted through a plastic pipe. In this system seasonal and steady-state efficiencies can reach 90 percent or more. Air and natural gas are mixed in a small combustion chamber and ignited by a spark plug. The resulting pressure forces the hot exhaust gas through a heat exchanger, where water vapor condenses (releasing latent heat of vaporization). In subsequent cycles, the fuel mixture is ignited by residual heat. The flue gas condensate must be removed through a special drain. The condensate is a noxious liquid that, although likely to be full of tars (and therefore toxic), has yet to be fully analyzed (see figure 3-7).

Only one such system is currently available on the commercial market. It is manufactured by Hydrotherm, Inc., of Northvale, New Jersey, and has measured efficiencies of 91 to 94 percent. The cost of the system is between 50 and 100 percent higher than a conventional one, although the improved efficiency pays back the difference within five years. The Hydrotherm system is now only gas-fired, but an oil-fired pulse combustion boiler is manufactured in Europe and may be marketed in the United States in the near future.

Flue Economizer

A flue economizer is a small auxiliary air-to-water heat exchanger that is

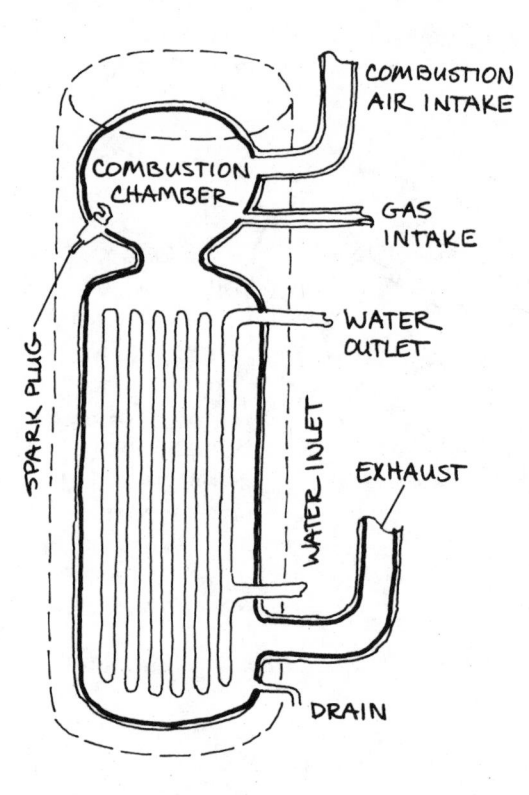

FIGURE 3-7: PULSE COMBUSTION BOILER (GAS)

collect soot and must be cleaned every two weeks or so during the heating season. Also, if the flue temperature is lowered by too much, moisture, corrosion and freezing may occur in the flue pipe. Air-to-air flue economizers are also available for about $100 to $200. These, however, save much less energy and are usually not tied into the central heating system. They are best for heating spaces near the flue.

Modified Flame Retention Burners

Conventional flame retention burners create a yellow flame, while modified flame retention burners create a blue flame in the combustion chamber. This is accomplished by recirculating unburned gases back through the flame zone, which allows more complete burning of the fuel. Another advantage of these systems is low soot formation. The best known blue flame system is manufactured by Blueray of Schuylkill Haven, Pennsylvania. It is available as a burner for retrofit to furnaces, or as a complete burner and boiler system for hot water distribution systems. Costs are roughly comparable to conventional flame retention units and new, high-efficiency boilers (see figure 3-8).

installed in the flue pipe. The unit captures and recycles usable heat ordinarily lost up the flue. This recaptured heat is used to prewarm water as it returns from the distribution system. Depending upon the age and design of the boiler and burner, a flue economizer can provide annual fuel savings of 10 to 20 percent. A flue economizer costs from $500 to $700 installed and has a payback on the order of two to five years. This device has a tendency to

Variable Fuel Flow Units

These burners throttle (or cut back) the fuel flow rate (and, hence, the flame size) as the system heating load varies. These burners offer conventional steady-state efficiencies but higher seasonal efficiencies. They are presently available for large apartment boilers and furnaces, but smaller, household-size units are not yet on the commercial market.

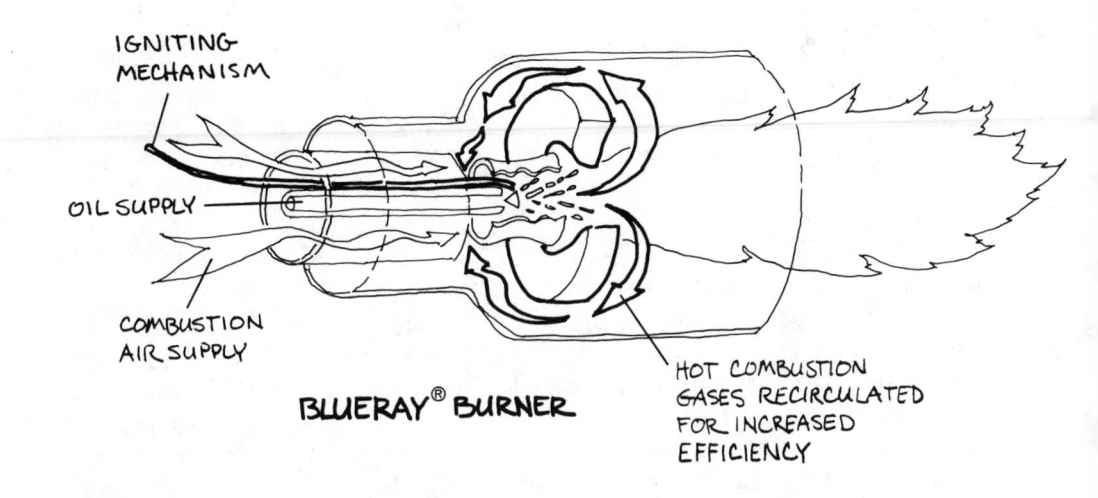

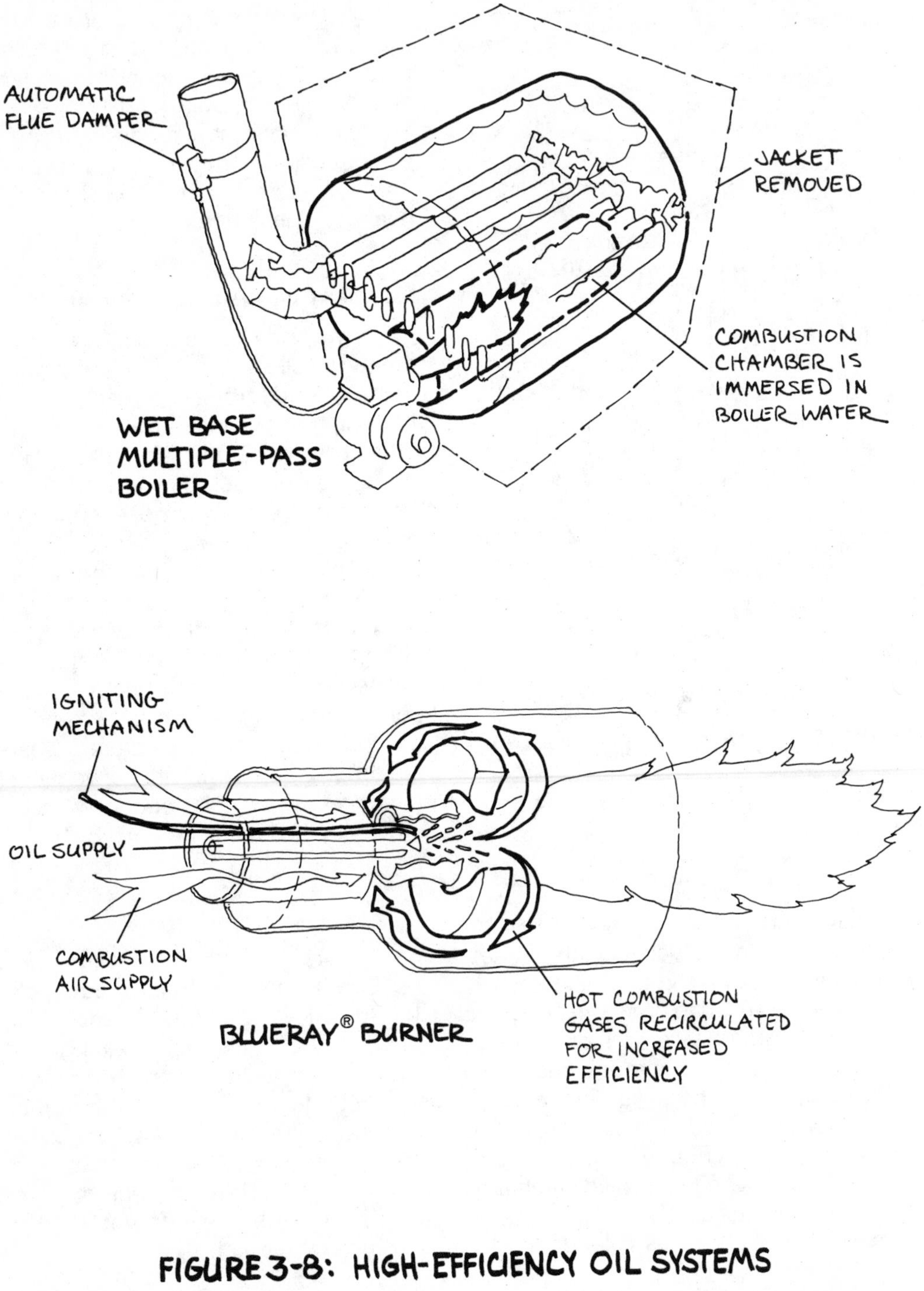

AUTOMATIC FLUE DAMPER

JACKET REMOVED

COMBUSTION CHAMBER IS IMMERSED IN BOILER WATER

WET BASE MULTIPLE-PASS BOILER

IGNITING MECHANISM

OIL SUPPLY

COMBUSTION AIR SUPPLY

BLUERAY® BURNER

HOT COMBUSTION GASES RECIRCULATED FOR INCREASED EFFICIENCY

FIGURE 3-8: HIGH-EFFICIENCY OIL SYSTEMS

Dual Oil/Gas Units

These are burners that can burn oil or gas as necessary. They offer no efficiency advantages, only the ability to switch fuels in the event of a shortage or a big price difference between one and the other. These burners are currently available as combination burner and boiler units for small systems, or as burners alone for large systems.

Small Internal Boiler Water Capacity (tankless)

These units offer slight advantages in seasonal efficiencies, as compared to conventional units, because there is less water to heat up and cool off. The same savings can probably be achieved with an automatic flue damper.

Most of the new systems described here and most of the conventional units on the market today are available only with heating capacities of 80,000 Btu per hour or more. There exists a major, unfilled need for efficient heating systems with outputs of 70,000 Btu per hour or less, capable of meeting the heating requirements of a well-weatherized house. For example, a set of weatherization improvements to the three smaller prototype buildings (described in chapter 1) can reduce design-day heating loads to 60,000 Btu per hour or less. Since one may wish to size a new heating system to meet smaller loads (for example, those occurring on a 10°F day, rather than a 0°F day, or for a warmer climate), even smaller heating systems are required. Unfortunately, such smaller systems require low fuel flow rates, and oil burners in particular tend to clog easily at these rates.

WOOD-FIRED HEATING SYSTEMS

Conventional wood-burning systems differ in several ways from oil- and gas-fired systems. There is, of course, no burner; the fuel is manually loaded into the combustion chamber. Wood fuel requires a large storage area relative to the size of an oil tank. The efficiency of wood systems varies more than that of conventional ones, depending upon the type of wood burned, the type of system and how the system is used. Finally, the gases produced during wood combustion contain considerably more particulate matter than those generated in conventional systems and may not be suitable for many areas with pollution problems. The basic reason for their current popularity is that for many people wood fuel is cheaper than other fuels. Another reason is largely psychological; wood burning fosters a sense of independence often lacking in today's world.

While the most common wood-burning systems are fireplaces and woodstoves, wood-burning furnaces and boilers with high efficiencies are also now available. These can be single or multi-fuel units that are capable of burning both wood and oil or gas. Recently new wood-burning systems have been developed that operate at efficiencies of 80 percent and higher.

The economics of burning wood in an urban location that is remote from readily available supplies are questionable. There is also the potentially serious problem that increased wood burning in densely populated urban locales will cause excessive air pollution. If you have access to a free or relatively inexpensive source of wood, and if your wood-burning system is a relatively

Table 3-7
Fireplace Improvements

	Measure	Frequency (yrs)	Savings Potential (%)	Done By	Cost ($)	Payback (yrs)	Comments
Maintenance	Clean chimney	3-5	...	Professional sweep, building occupant	60-200	Varies	Standard practice for safety
Modifications	Place sealing doors across face of fireplace	...	10-30	Building occupant	50-100	2-4	Prevents heat loss when fire dies; doors can be glass
Modifications	Install heating grates	...	5-10	Building occupant	50-200	3-8	Increases heat gain of room
Replacements	Install woodstove fireplace insert	...	10-15	Professional installer	Varies	Varies	Increases heat gain of room most of all

Correction factors: To find fuel savings, multiply by the appropriate factor for your zone. To find payback, divide by the same factor. Zone I: 0.33; Zone II: 0.67; Zone III: 1.00; Zone IV: 1.33.

efficient one, you can significantly reduce your oil or gas bills. But if you have to buy your wood from a dealer at premium prices and your system is operating at a low efficiency, you are likely to lose money on the deal.

Fireplaces: Maintenance

General maintenance of a fireplace chimney is important for safety reasons. Corrosive tars and liquids such as creosote are produced in wood combustion. If deposited in the chimney, they can be ignited and cause chimney fires. The fireplace and the chimney should be cleaned every three to five years to prevent this. In addition, during combustion, wood should be stacked to ensure a slow burn. Well-designed fireplaces can have combustion efficiencies of 30 to 40 percent; however, when the competing effects of increased infiltration are factored in, a fireplace will have an efficiency of no more than

about 5 percent and may, in fact, have a negative efficiency.

Modifications

The only truly viable modification for a fireplace is to replace it with an efficient woodstove, but building codes may not permit this. If you want to increase the efficiency of your fireplace, some helpful modifications include the following:

Installation of doors (which may be glass) across the face of the fireplace. These cut down on the heat lost after the fire dies, and they help regulate the burning rate.

Installation of heat grates or tubes. These grates, installed in the wall of the fireplace, direct some of the heat rising from the fire back into the room.

Some come with blowers to force convection. Even though these devices will improve efficiency, the average fireplace is still likely to be a net loser of energy. The most efficient device combines glass doors with a forced convection tube system.

Replacement

The best move to make with a fireplace is to install a woodstove insert. An insert is like one-half or two-thirds of a freestanding woodstove. It extends out from the fireplace opening, which is sealed to the insert. Because the stove projects into the room, it delivers much more radiant heat than a standard fireplace. And because it's a woodstove it also operates at much higher overall efficiency. The recent boom in the popularity of woodburning has led to the development of a number of commercial fireplace inserts. If you have more space in front of the fireplace, you can also consider installing a full-size woodstove, again by sealing off the fireplace around the flue pipe.

Woodstoves: Maintenance

A well-designed woodstove can have a combustion efficiency as high as 60 percent. The actual distribution efficiency will depend upon where the stove is placed, the source of combustion air and the amount of air infiltration into the house caused by the stove. Overall efficiency will, of course, be lower than the combustion efficiency. To maintain maximum efficiency and safety, the following steps should be taken:

The flue should be cleaned regularly and checked for corrosion. Tars and creosote are produced in wood combustion and can corrode metal flues and cause chimney fires.

The stove should be cleaned regularly to remove corrosive soot, and it should be checked for cracks and corrosion. Soot acts as an insulator and can impede heat transfer from the stove to the building.

Wood should be loaded in a manner to ensure a complete and steady burn. A complete burn is the most efficient.

The seals on the stove should be checked regularly. Poor seals can allow too much air intake, which reduces the efficiency of the burn.

The air intake should be set to ensure the most efficient burn. Too much or too little air can reduce the combustion efficiency.

Modifications

If you already have a woodstove and wish to improve its efficiency, the following steps can be taken:

The firebox can be lined with insulating firebrick. This will raise the temperature of the fire, improving efficiencies and reducing the quantities of combustion pollutants.

The length of the flue in the building can be increased. Longer flue runs allow for more of the heat produced in combustion to be transferred to the building; however, if the flue run is too long, creosote deposition will be increased.

The stove and flue can be painted a flat color. Metals painted a flat color are better radiators than those painted a metallic color.

If possible, the stove should be centrally located and near a masonry wall. Such a wall can store the heat produced by the stove and help to keep the building warm after the fire dies out.

Add an outside combustion air feed. Stoves normally consume already-warmed inside air, and much of this air is lost up the flue. An outside air feed allows combustion to be fed by cold outside air. This reduces cold air infiltration into the house.

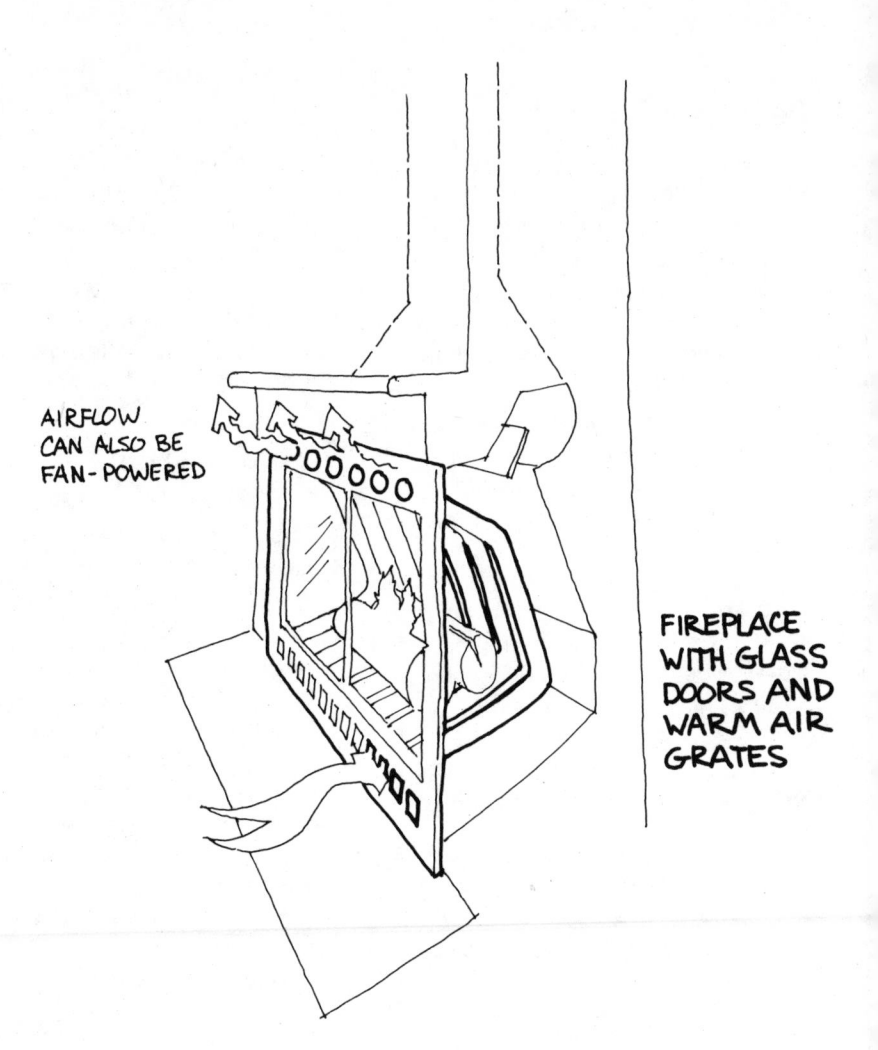

AIRFLOW
CAN ALSO BE
FAN-POWERED

FIREPLACE
WITH GLASS
DOORS AND
WARM AIR
GRATES

FIGURE 3-9: EFFICIENT FIREPLACE AND WOODSTOVE DESIGNS

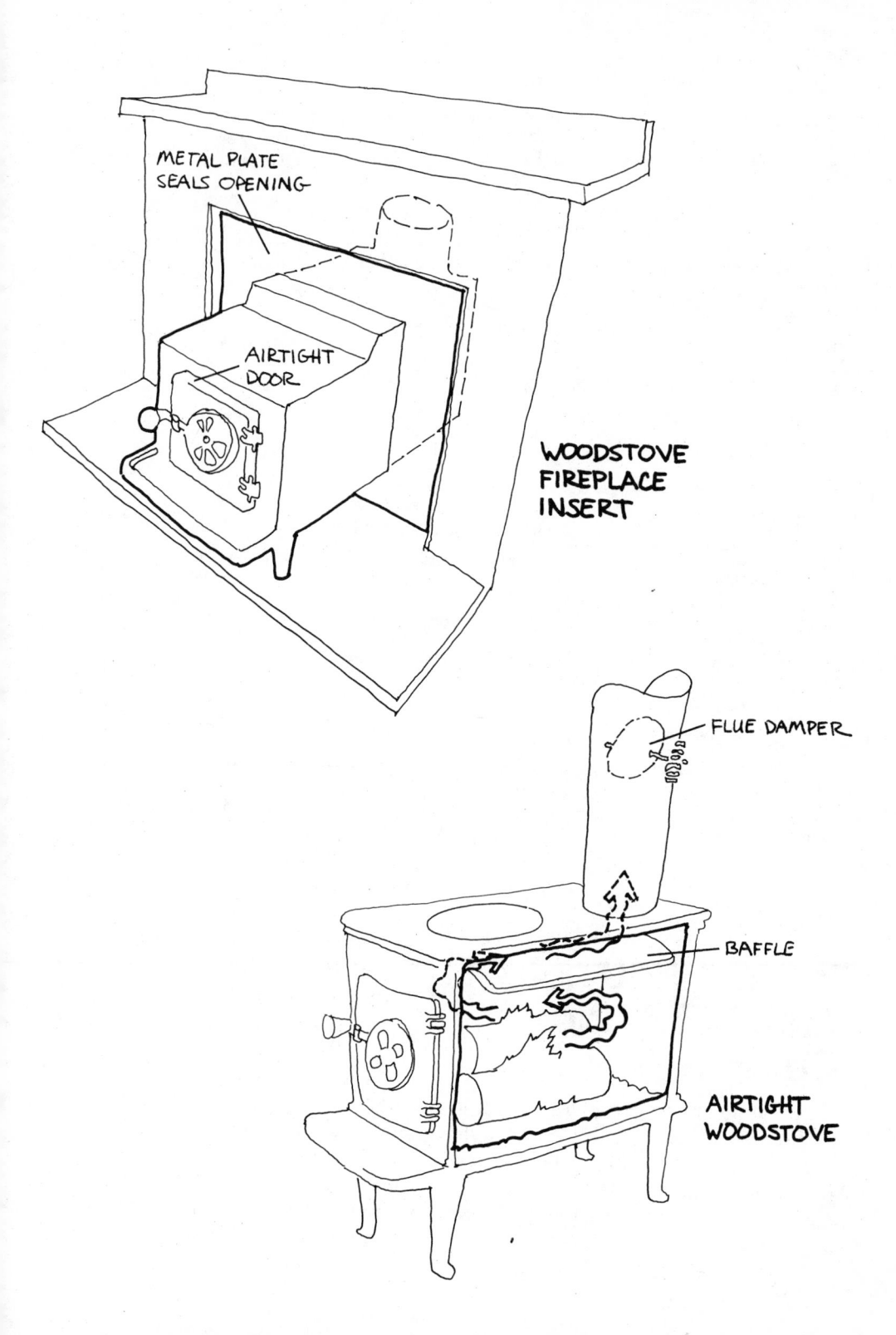

METAL PLATE
SEALS OPENING

AIRTIGHT
DOOR

WOODSTOVE
FIREPLACE
INSERT

FLUE DAMPER

BAFFLE

AIRTIGHT
WOODSTOVE

Table 3-8
Woodstoves

	Measure	Frequency	Savings Potential (%)	Done By	Cost ($)	Payback (yrs)	Comments
Maintenance	Clean flue and check for corrosion	Yearly	5-10	Professional sweep, building occupant	60-200	Varies	Important for safety
	Clean stove	Monthly	5-10	Building occupant	...	...	...
	Load wood properly	Regularly	10-15	Building occupant	...	...	Slow burn desired
	Check seals on stove	Regularly	Up to 5	Building occupant	...	...	Poor seals reduce efficiency of burn
	Check air intake	Regularly	5-10	Building occupant	...	...	Improper setting can reduce efficiency of burn
Modifications	Line firebox with insulating brick	...	5-15	Professional, or building occupant	Varies	Varies	Increases efficiency of burn
	Increase length of flue	...	5-15	Professional, or building occupant	Varies	Varies	Increases heat transfer to room
	Paint stove and flue flat color	...	5-10	Building occupant	About 20	Within 1 year	Flat paint a better radiator than metallic paint
	Relocate stove to more central location near masonry, if possible	...	10-30	Professional, or building occupant	Varies	Varies	Increases heat gain of room
	Add combustion air feed	...	20	Professional	Varies	Varies	Decreases infiltration into house

Correction factors: To find fuel savings, multiply by the appropriate factor for your zone. To find payback, divide by the same factor. Zone I: 0.33; Zone II: 0.67; Zone III: 1.00; Zone IV: 1.33.

WOOD-FIRED FURNACES AND BOILERS

As wood burning has increased in popularity in the United States, an increasing number of companies have begun to produce furnaces and boilers capable of burning wood logs. These systems function like conventional oil and gas systems and are intended for central heating of a building (unlike woodstoves and fireplaces which generally only heat one or a few rooms). Some systems are capable of burning two fuels, for example, wood and coal or wood and heating oil. These multi-fuel boilers and furnaces offer some added convenience in that, should the wood fire die out or be inadequate to heat the building, the back-up fuel can automatically take up the slack.

There are some disadvantages to wood and multi-fuel systems. The firebox of the furnace or boiler must be stoked, or loaded, with wood at periodic intervals. Depending upon system design and the heating requirements of your building, this can be as often as every 4 or 5 hours. Many wood system manufacturers claim that their system requires stoking only every 12 hours; it is best to be wary of such claims. Also, wood ash must be removed from the combustion chamber every so often in order to maintain system efficiency.

Some of the problems common to woodstoves and fireplaces are also characteristic of wood furnaces and boilers. Creosote buildup, with the accompanying hazard of chimney fires, can occur if the combustion temperature or air mixture in the firebox is incorrect. Wood systems also produce smoke and ash particulates, and this can be a disadvantage in areas with serious air quality problems. Finally, component lifetimes may possibly be less than those in conventional systems because of the greater variability in temperature and burning conditions characteristic of wood combustion.

Typically, a wood system has a combustion efficiency in the range of 65 to 70 percent, although several newer types of wood-burning central systems that promise higher combustion efficiencies are in the testing stage. These promise combustion efficiencies of as much as 80 percent through more complete burning of flue gases.

Wood furnaces and boilers do not cost much more than conventional oil or gas systems, and they do not require that major changes be made to the distribution system. At market prices, however, the cost of wood fuel may not be much lower than the cost of conventional fuels, so you should make cost and efficiency comparisons before installing a wood system. Furthermore,

the added maintenance problems of wood central systems may make them a less-than-ideal replacement for a conventional furnace or boiler as far as the urban homeowner is concerned.

ELECTRIC RESISTANCE SPACE HEATING

Some residential buildings are heated by electric resistance heaters. These usually run along the baseboards, but may also be in the form of wall or ceiling radiant units. They may be controlled by a room thermostat or have individual controls located on each heater.

Maintenance

Proper maintenance of electric resistance heaters includes the following:

The baseboard heater should not be obstructed. Objects placed in front of these heaters will prevent them from radiating their heat into the room.

The heaters should be cleaned regularly. Dust on radiating surfaces can act as an insulator and reduce the ability of the heater to warm a room.

If the heating system is controlled by a room thermostat, it should be checked regularly. The thermometer on the thermostat should be checked against the reading of an ordinary thermometer and recalibrated if necessary. In addition, the thermostat should be cleaned frequently to remove dust, lubricated if necessary and, if applicable, have its filters replaced periodically. Wiring should be checked for electrical shorts.

BASEBOARD
HEATER

WALL
HEATER
(WITH OR
WITHOUT
FAN)

FIGURE 3-10: ELECTRIC HEATERS

Modifications

Many electrical resistance heaters have individual thermostats without temperature controls. Usually the thermostat knob is graded from low to high or numerically from 1 to 5 or 10. Since it is difficult to assess room temperature and adjust the thermostat accordingly without some sort of thermometer, it may make sense either to replace the existing thermostats with ones graded by temperature or, better yet, to wire the baseboard heaters to a room thermostat control.

Replacement

Electric resistance heaters are expensive to operate. Furthermore, less than one-third of the energy contained in the fuels consumed at the power plant to generate the electricity is delivered to a building as usable heat. It may make sense to replace your electric resistance heating system with another, less expensive and more efficient system, such as natural gas or a heat pump.

Electric Resistance Hot Air Heating

In some parts of the United States (particularly where electricity is relatively inexpensive), many homes are heated by electric resistance hot air furnaces. In such a system, fans blow air across heated wire coils. The warmed air is then distributed through air ducts to registers or grilles in individual rooms. There is little that can be done to improve the efficiency of an electric resistance furnace, however, the distribution system suffers from the same ineffi-ciencies as gas and oil hot air systems. Therefore, all of the maintenance and modification recommendations listed in table 3-5 and in the text dealing with hot air distribution systems also apply here. In addition the filter in the furnace closet door or on the unit itself should be changed at least every six to eight weeks. (If your house has a heat pump with hot air distribution the same remarks apply.) Control systems for electric resistance furnaces are identical to those for oil and gas systems (see table 3-6).

Table 3-9
Electric Resistance Heaters

	Measure	Frequency	Savings Potential (%)	Done by	Cost ($)	Payback (yrs)	Comments
Maintenance	Remove obstruc-tions from front of heaters	Regularly	5-10	Building occupant	...	...	Heat delivery not effective if heaters blocked
	Clean heaters	Monthly	5-10	Building occupant	...	...	Heat delivery not effective if heaters dirty
	Check and calibrate thermostats, if applicable	Yearly	Up to 5	Building occupant, professional	Varies	Varies	...
Modifications	Replace thermostat with temperature-graded unit or room thermostat, if applicable	...	5-10	Electrician	20-100	Varies	Better control over temperatures achieved
Replace-ments	Replace heaters with heating system using another fuel	...	60 or more	Contractor	1,500-3,000 on average	3-5 on average	Electric heat is the most expensive energy

Correction factors: To find fuel savings, multiply by the appropriate factor for your zone. To find payback, divide by the same factor. Zone I: 0.33; Zone II: 0.67; Zone III: 1.00; Zone IV: 1.33.

HEAT PUMPS

A heat pump is a device, much like a refrigerator or an air conditioner, that removes heat from one area called the *source* and transfers it to another called the *sink*. Heat pumps can be used for both heating and cooling. Electric heat pumps can deliver up to three times more heat energy than is contained in the electricity used to power the device. Including efficiency losses at the electric power plant, from 75 to 100 percent of the heat energy in the original fuel can be delivered by an electric heat pump. As such, it is a highly efficient, energy-conserving technology and may be considered a desirable alternative to electric resistance space heating.

Operation

A heat pump works in the following way: When a substance changes from a liquid to a gas by boiling or vaporization, it absorbs heat; when it changes back from a gas to a liquid by condensation or liquefaction, it releases heat. A heat pump uses these properties to transfer heat. A substance called a *refrigerant* (usually Freon) with a boiling temperature far below room or outside temperatures is contained in a liquid state in a sealed coil (a heat exchanger) and exposed to a heat source (usually outside air, solar-heated water or groundwater). The refrigerant absorbs heat from the source and, as it warms up, boils. As a gas, the refrigerant is then pumped through a compressor and into another sealed coil, this one exposed to the sink. The compressor increases the pressure on the vapor, causing it to condense. As the refrigerant condenses, it releases heat which the sink (usually indoor air or a hot water heating system) absorbs. In this way, heat is removed from the source and transferred to the sink. The cycle can be repeated indefinitely. A heat pump operated in reverse (where the source is indoor air and the sink, outdoor air) has the effect of cooling, rather than heating. Cooling heat pumps are generally called air conditioners. Central heat pump systems are generally used for both heating and cooling.

In addition to the heat extracted directly from the source, heat pumps are also able to use waste heat produced by the compressor motor. Therefore, the amount of heat delivered by the heat pump to the sink equals the amount of heat extracted from the source plus the waste heat from the compressor motor.

The efficiency of any heat pump is given by the ratio of the heat delivered to the sink to the heat content of the electricity (or other fuel) consumed to power the compressor motor. This ratio is called the *Coefficient of Performance* (COP). Electric heat pumps typically have a maximum COP of 2 to 3 in comparison to a COP of 1 for electric resistance heating. The concept of COP is a tricky one. As a result of efficiency losses at the electric power plant, a heat pump using electricity produced by the burning of oil, gas or coal with a COP of 2.5 actually provides roughly the same amount of heat as would the direct, at-home burning of the fossil fuel with an efficiency of 75 percent. Once developed, heat pumps using natural gas may have COPs equal to an efficiency of as much as 150 percent.

It is important to realize that the COP of a heat pump decreases as the temperature of the source decreases because more work is required to supply the same amount of heat. As a consequence of this, more fuel is consumed. An electric heat pump with a COP of 2 at a source temperature of 45°F may only have a COP of 1.3 at 25°F. Therefore, electric heat pumps become more expensive than electric resistance heating at source temperatures below 20°F. This is a problem in northern climates for heat pumps that use outside air as a source. Most commercially available electric heat pumps are equipped with electric resistance auxiliary heaters that start up when outdoor temperatures drop below 20°F. This problem does not exist if the source is groundwater because, even under the coldest of conditions, groundwater temperature rarely drops below 40 to 50°F.

Another problem encountered by heat pumps operating under subfreezing conditions is icing of the outside heat exchanger. Most heat pumps have an electric resistance defrosting cycle which operates to melt ice before the heat pump can operate. This defrosting cycle further lowers the efficiency of an air source heat pump.

Heat Sources for Heat Pumps

Heat pumps can be designed to extract heat from a variety of sources. These include:

Air. If the cost of electricity is fairly low and winter temperatures remain in the 35 to 55°F range, electric heat pumps can supply heat at costs competitive with fossil fuels. In areas where January temperatures average 30°F or less and electricity is fairly expensive, air-source heat pumps are somewhat more expensive to run than oil or gas systems, although they remain less costly than ordinary electric resistance systems. Furthermore, operation of an air source heat pump over a wide temperature range tends to decrease seasonal operating efficiency and shorten system lifetime.

When they become commercially available, heat pumps with oil- or gas-fired compressor motors will be more efficient than conventional fossil fuel heating systems and could be backed up by conventional systems when air temperatures are too low. Gas-fired heat pumps may reach the commercial market within the next several years.

Groundwater. For the time being, the only way to increase the efficiency of a heat pump system in a colder climate is to ensure a moderately high and constant temperature heat source. This can be done with a groundwater source, which provides a more concentrated source of heat. While obtaining individual access to groundwater might pose problems, wells could be drilled to supply heat source water to several homes, a street or an entire neighborhood. A means to dispose of this groundwater is required for such a system.

Solar. A heat pump can also be installed and operated in tandem with an active solar heating system. The solar system provides warm source water for the heat pump. On clear days, the solar system can operate and either displace the heat pump or, by boosting the temperature of the storage tank, increase the efficiency of the pump and, therefore, the whole system. On mild, overcast days, the heat pump can compensate for low solar system efficiency and low tank temperature. Of course, on very cold, cloudy days neither system works very well and some sort of back-up heating system is necessary.

With the tandem arrangement, the solar system can operate at lower

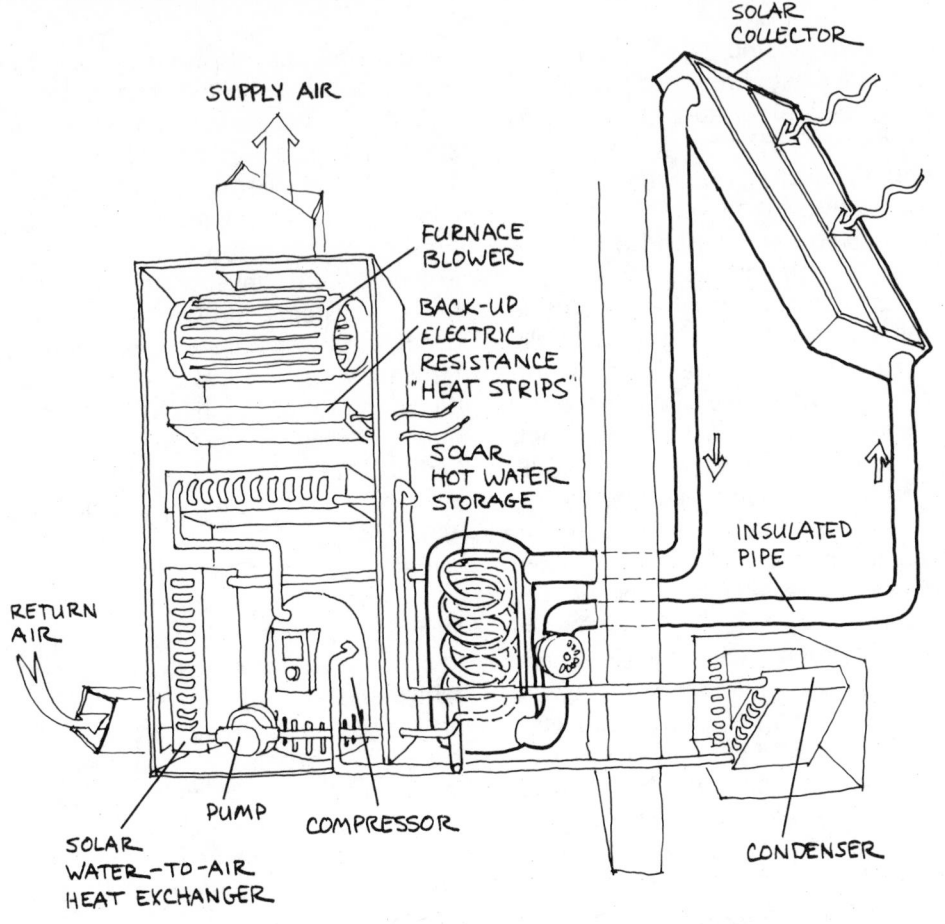

SUPPLY AIR

FURNACE
BLOWER

BACK-UP
ELECTRIC
RESISTANCE
"HEAT STRIPS"

SOLAR
HOT WATER
STORAGE

SOLAR
COLLECTOR

INSULATED
PIPE

RETURN
AIR

SOLAR
WATER-TO-AIR
HEAT EXCHANGER

PUMP   COMPRESSOR

CONDENSER

FIGURE 3-11: SOLAR-ASSISTED HEAT PUMP SYSTEM

temperatures, increasing the efficiency of the system, and can be constructed with single glazing, lowering the cost of the system. The cost of installing two complete systems is quite high (although tandem systems are available), so it might make sense to install a solar system with less collecting area, using it only to boost storage tank temperatures to the 70°F range for the benefit of the heat pump. In the summer, the solar system can supply domestic hot water. Figure 3-11 shows a heat pump/solar heating tandem system.

Ground. The ground is another potentially useful heat source. Ground temperatures below the frost line are generally the same as groundwater temperatures. Coils of pipe can be buried below this depth and the system can extract heat directly from the ground. About 1,000 square feet of yard space are required as a heat source for a single-family residence. The expense of digging up the yard, laying the pipe and relandscaping is likely to be high. J. A. Sumner describes such a system in *An Introduction to Heat Pumps.*

Waste heat. Heat pumps can also use waste heat from other energy-consuming systems as a source. Most conventional heating systems throw away a good deal of heat. Large buildings or complexes, such as hospitals, office buildings and shopping centers, often dump their waste heat into the air. A neighborhood or district heating system using heat pumps and a waste heat source could operate very efficiently. (Other uses of heat pumps are discussed in chapter 6.)

Hot water storage. In some areas, electric utilities offer rates that vary during the day. Often, these rates offer such low nighttime electricity costs that it is possible to use the heat pump to heat water at night, circulate the hot water through the distribution system during the day and save in energy costs over conventional fossil fuel systems (see "Load Management," chapter 4 and "Energy Pricing and Rate Structures," chapter 7).

Heat pump capacity is the rate of heating or cooling supplied at a specific temperature, usually around 45 or 50°F. It is expressed in MBh (thousands of Btu's per hour) or tons (1 ton = 12,000 Btu per hour). You should be careful when evaluating heat pumps on the basis of their advertised heating capacity. On a cold winter day (30°F), actual heating capacity is typically much lower than advertised capacity. In addition, you should also be careful when sizing a heat pump for a house. Heat pumps are customarily installed for air conditioning, especially in warmer climates, and are often sized to meet the cooling load of the house. In a cold climate, the cooling load is generally so small that heat pumps installed for air conditioning are inadequate for the total heating requirements of the house.

Heat pumps are somewhat more expensive than conventional heating units. For single or small multi-family residences, the installed cost of a heat pump runs around $1,200 per ton. Four to six tons is a common size for an uninsulated single family home. For larger systems, the installed cost drops to around $1,000 per ton.

Should You Install a Heat Pump?

Consider the following points before deciding whether or not to install a heat pump:

If solar heating is impractical or impossible and you presently have electric resistance heating, a heat pump makes economic sense.

If the choice is one between a conventional oil or gas system and a heat pump, the former is more economic at the present time; however, with time-of-use electric rates, heat pumps with short-term heat storage capacity may provide heat more cheaply than fossil fuel systems. Should gas-fired heat pumps become available, they will be more economical than conventional gas systems and much more so than electric resistance heat.

If you do have or will have access to a source of relatively inexpensive electricity, a heat pump is likely to be less expensive than oil, and perhaps gas, heat.

If you have access to a constant temperature source (groundwater, ground, solar energy or waste heat), a heat pump may be economical.

If a heat pump will necessitate installation of a completely new heat distribution system (for example, replacing electric baseboards with forced hot air), replacement of your present system may not be economical.

If you require air conditioning, heat pumps can be economically competitive with combined gas or oil heating and air conditioning. Alternatively, ice-making heat pumps may be a good choice. If time-of-use electricity rates are available, the economics can be even better because the machine can make ice for air conditioning when electricity costs are low. Such heat pump systems can also heat domestic hot water with reject heat.

ELECTRIC DOMESTIC WATER-HEATING SYSTEMS

Many buildings use electricity to heat domestic water. This is more expensive and less efficient than heating water with oil or gas, although an electric water heater is often easier to install.

Water is usually heated in a tank that should be well-insulated to minimize storage (or "standby") heat loss. Tank temperature should be set at 120°F. If the building has a dishwasher without a separate heating element, the tank temperature must be set at 140°F. You may also wish to install flow controls in showers and sinks in order to limit the consumption of hot water.

Maintenance and Replacement

A water heater should be drained every six months in order to clean out sediment that has built up at the bottom of the tank. If not removed, this sediment will act as an insulating layer and reduce the efficiency of the system. If your water heater is warm to the touch, it probably needs more insulation. An insulating jacket costs about $25, and the payback is about a year. (See "Insulation," chapter 2.)

If you are thinking about replacing your electric water heater, there are several alternatives to consider. First, you should consider installing a natural gas water heater, particularly if a gas connection is already present. Gas is much less expensive than electricity and is likely to remain so. If you cannot get a gas hookup, you might consider a heat pump water heater. Such a system is particularly appropriate if you live in an area with relatively mild winters. You might also consider installing a solar water-heating system to meet part of your needs. Both of these systems will require some type of back-up system, most probably a new electric one. In a warm climate, a heat pump water heater can perform as well or better than a solar water heater and at a lower cost. In colder climates, the heat pump can operate only during warmer months, while a solar system can operate year round.

Domestic Hot Water Heat Pumps

Heat pumps are available for heating domestic hot water. These units are available either for replacement of or retrofit to existing water heaters and can provide energy savings of as much as 50 percent over conventional electric water heaters. They can also provide some cooling and dehumidification on warm days by drawing heat out of the room. Domestic hot water heat pumps cost about $400 to $600 more than conventional units and possibly can save as much energy as solar domestic hot water at 20 to 25 percent of the

Table 3-10
Improvements for Electric, Oil or Gas Domestic Water-Heating Systems

	Measure	Frequency	Savings Potential (%)	Done By	Cost ($)	Payback (yrs)	Comments
Maintenance	Drain sediment from tank	Twice a year	Varies	Building occupant	...	...	Increases system efficiency
Modifications	Turn temperature control down to 120°F	...	5-10	Building occupant	...	...	A dishwasher with no separate heating element must have control set to 140°F; shut off electrical power to heater before adjusting thermostat
	Install low flow devices in showers and sinks	...	5-10	Building occupant, plumber	1-20	0.1-1	Reduces use of hot water
	Install insulating jacket (gas or electric)	...	10-20	Building occupant, technician	25	0.5-1	Necessary if tank is warm to touch; most tanks need more insulation
	Add solar heating component	...	50	Technician, plumber	1,500	10-15	Reduces use of oil or gas
	Add tankless demand-type heater	...	25-40	Technician, plumber	250-400	5-8	Best for houses with low hot water use
	Add heat pump water heater	...	50	Technician, plumber	800	5-7	Good only if gas hookup not available; may not be suitable for winter operation
	Install manual or timed switch	...	Varies	Technician, plumber	Varies	Varies	Can take advantage of off-peak electric rates; can reduce fuel consumption
	Replace tankless (oil) hot water system with tank system	...	Varies	Technician, plumber	200	4-5	Burner can be turned off during the summer
Replacements	If system is electric, install natural gas, heat pump, or solar water heating system	...	Varies	Technician, plumber	Varies	Varies	Lifetime cost is lower

Correction factors: To find fuel savings, multiply by the appropriate factor for your zone. To find payback, divide by the same factor. Zone I: 0.33; Zone II: 0.67; Zone III: 1.00; Zone IV: 1.33.

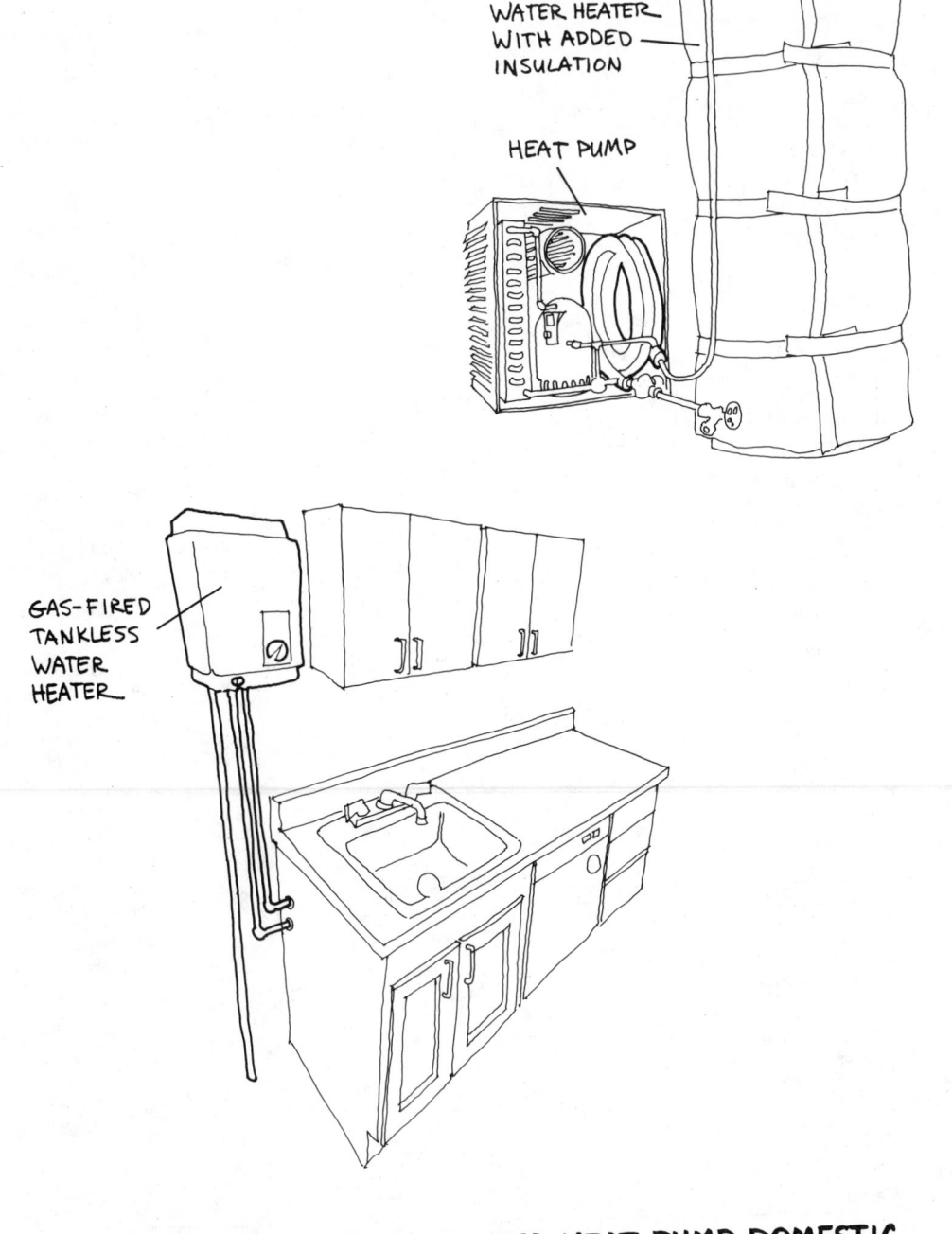

FIGURE 3-12: TANKLESS AND HEAT PUMP DOMESTIC WATER HEATERS

installed cost. There is, however, a larger continuing electricity cost associated with the heat pump system. Because available hot water heat pumps are air source systems, they are considered inappropriate for cold climates, unless backed up by electric resistance heating during the winter. Clearly, if placed in a heated space, a hot water heat pump will rob heat from the space; in an unheated space, the system efficiency will be very low. Even so, such a system will still save energy over year-round electric water heating.

AIR CONDITIONERS

There are two types of air conditioners: central and room units. Central units are located in one part of the building. They cool either by blowing cold air through ducts throughout the building, or by circulating chilled water through pipes to each room where fans blow air across the pipes and into the space. Room units, on the other hand, cool only the space in which they are located. They are either mounted in windows or built into an external wall. Both types of air conditioner contain the following parts: an indoor heat exchanger, which removes heat from the indoors; an outdoor heat exchanger, which delivers the heat to the outdoors; the refrigerant, a special fluid such as Freon that boils near room temperature and takes the heat from the indoors to the outdoors; a compressor, which does the cooling work on the refrigerant; and a thermostat.

If you do not presently have an air conditioner but are thinking about installing one, consider the following points:

Is air conditioning necessary? In many buildings in some parts of the United States, air conditioning is required so infrequently that installation may not pay. This is especially true for buildings that are well shaded, have good cross-ventilation or are made of masonry (concrete, brick or stone). In addition, improving the envelope of a building by adding wall insulation or storm windows, by ventilating the roof space above the insulation or by installing a light-colored roof (if a new one is required) can make the building more comfortable during the summer. Some of these features also have the added benefit of improving comfort conditions during the winter.

Simple changes in habit can also make a building more comfortable during hot weather. The building can be closed up during the day (by shutting windows and drawing drapes and shades) to keep the heat out. At night, windows can be opened and fans turned on to draw in the cooler night air. This is especially effective in masonry buildings because the structure stores the cool temperatures of the night, thereby keeping the building cooler during the day. If a building does not store *coolth* well, it may pay to open the windows during the day and let a breeze blow through or, if need be, run a fan. A breeze can often make hot temperatures more bearable. Substituting fans for air conditioning can save 60 percent or more in energy use and cost. Installation of a dehumidifier with a fan may be advisable because high humidity tends to increase discomfort at high temperatures. Dressing appropriately for hot weather can also reduce the need for air conditioning.

What is the efficiency of the unit? The nameplate of the unit should give this efficiency by listing its EER, or *Energy Efficiency Ratio.* This is a measure of the cooling capacity of the air conditioner for every watt of electricity consumed (in Btu's per hour per watt). The higher the EER, the more efficient the unit is. Units with EERs higher than 8 are now available and generally cost little, if any, more than less-efficient units. Units may also be tagged with a SEER, or *Seasonal Energy Efficiency Ratio,* which gives an average efficiency over the entire cooling season. The SEER tends to be lower than the EER. By federal law, new air conditioners must have tags listing the EER or SEER (see "Appliances," chapter 4).

Is the unit properly sized for the cooling load of the building? This is especially important for central units. Internal building heat sources such as cooking, lighting and people must be considered when sizing a central unit. In addition, air conditioners should not be sized for the hottest days, which occur infrequently. Instead, sizing for fairly hot days will reduce electricity consumption by the unit. While the thermostatically set inside temperature may not be maintained on the hottest day, the resulting discomfort should not be too great. It may pay to receive a detailed cooling load calculation from an energy auditor.

Can the unit be installed in a shaded location? Direct sunshine on the outdoor heat exchanger of an air conditioner decreases unit work efficiency by about 10 percent.

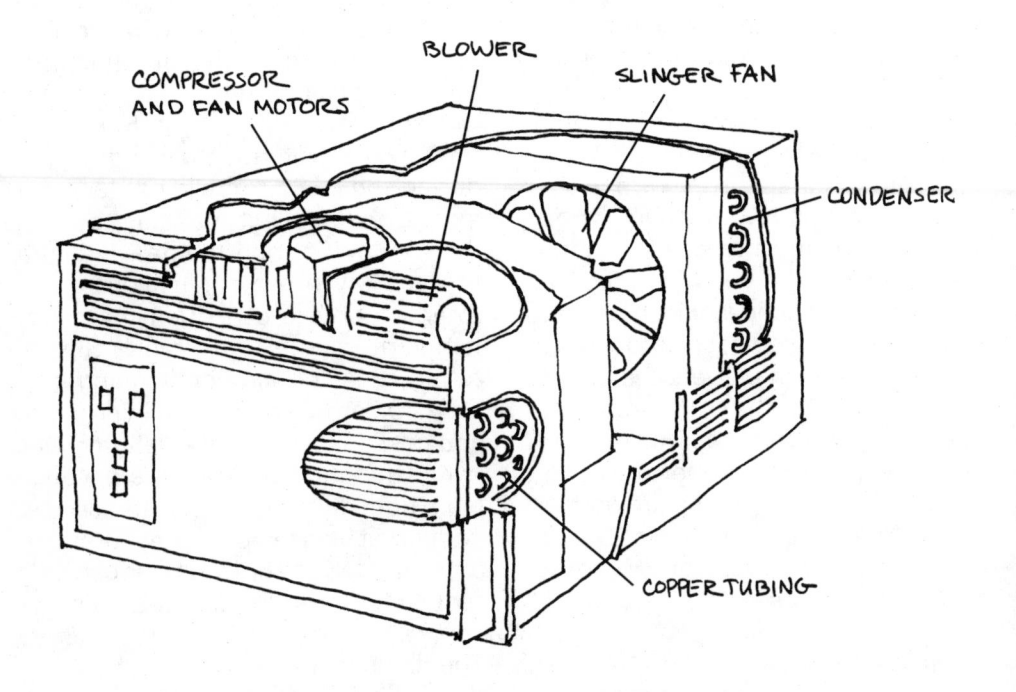

FIGURE 3-13: ROOM AIR CONDITIONER

Maintenance

Regular maintenance of an air conditioner is important to prevent deteriorating efficiency and includes the following:

Making sure air conditioning units, condenser coil and any grilles or registers are not blocked. Otherwise, cooling efficiency will be reduced.

Checking for air leaks either to the outside or to unconditioned spaces in the building. If there are leaks, the air conditioning units will have to expend more energy than necessary to cool the building and their capacity may in fact be exceeded.

Regularly cleaning the air filter on room units and the registers or grilles on central units. On room units, the air filter sits inside the front cover and functions to keep the air clean. It frequently becomes covered with dust and must be examined every two to four weeks and cleaned when necessary. This alone can save 5 to 15 percent of the energy used in cooling. Air registers and grilles serving central units also become covered with dust and must be cleaned to maintain cooling efficiency. Some central units also have air filters.

Balancing the cooling delivered to a building from a central unit. This means that all rooms served by a central unit should be at about the same temperature. Balancing can be achieved by adjusting dampers in air ducts, air register openings, chilled water or fan settings and thermostats.

Cleaning thermostats and checking if the thermometer readings on the thermostats are accurate. Accurate thermostats allow the air conditioner to be used efficiently.

Following manufacturer's instructions for additional maintenance. This will help to increase the lifetime of the unit.

Covering or removing a window unit during the winter. This prevents cold air from passing through the unit into the building.

Modifications

Some modifications can be made to air conditioners that will reduce their energy consumption. Room air conditioners often do not have a thermostat with temperature controls. Instead, the thermostat knob is graded from low to high or 1 to 5 or 10. Since it is difficult to assess room temperature and adjust the thermostat properly without a thermometer, it may make sense to replace the existing thermostat with one graded by temperature or, better yet, to wire the unit to a room thermostat control. If you do not wish to do this, you should leave a thermometer in the room or rooms being cooled so that you can more accurately adjust the thermostat.

You may wish to install a fan-only setting on the air conditioner, if it does not already have one. During moderately warm days, the fan alone should be able to maintain the comfort of the building, thereby saving 60 percent or more of the energy required to operate the entire air conditioning system.

Replacement

If you are considering replacing your air conditioner, you should determine whether the unit is in good shape, how old it is and how efficient it is. The efficiency can be determined from the EER. Older units do not have a tested EER, but this can be calculated by dividing the cooling capacity of the unit (in Btu's per hour) by the power requirement (in watts; one kilowatt equals 1,000 watts). Both of these quantities should be listed on the nameplate. If wattage is not listed, multiply the current rating (in amps) by 110 or 220 volts, whichever is appropriate. If the EER of your unit is 5 or less, and it is not performing well, replacement may be in order.

Table 3-11
Air Conditioners

	Measure	Frequency	Savings Potential (%)	Done By	Cost ($)	Payback (yrs)	Comments
Maintenance	Remove obstructions from front of air conditioners, or their grilles, and registers	Regularly	5-15	Building occupant	...	...	Delivery not effective if vents blocked
	Clean air filter on room units, and grilles and registers of central units	Every 2-3 weeks	5-10	Building occupant	...	...	Delivery not effective otherwise
	Check for air leaks	Yearly	5-20	Building occupant, professional	Varies	Varies	Look for obvious holes or breaks
	Balance the system	Yearly, or as needed	5-10	Building occupant, professional	Varies	Varies	Only for central units; may be part of service contract
	Clean and calibrate thermostats	Yearly	5	Building occupant, professional	Varies	Varies	...
	Follow manufacturer's suggested maintenance	Regularly	5-10	Building occupant, professional	...	...	...
Modifications	Replace thermostat with temperature-graded unit or room thermostat	...	Varies	Electrician	20-100	8	Better control over temperatures achieved; applicable especially to room units
Replacements	Replace old unit with more efficient unit	...	Varies	Building occupant, professional	Varies	Varies	Reduces electricity consumption, and is usually no more expensive than less efficient models

Correction factors: To find fuel savings, multiply by the appropriate factor for your zone. To find payback, divide by the same factor. Zone I: 0.33; Zone II: 0.67; Zone III: 1.00; Zone IV: 1.33.

SERVICE CONTRACTS FOR HEATING AND COOLING SYSTEMS

Regular maintenance of heating and cooling systems (particularly oil and gas) is necessary to maintain operating efficiency. This service can be provided by fuel suppliers, equipment manufacturers or heating engineers.

A service contract on a boiler or furnace should specify the following:

- The system will be tested, tuned and cleaned regularly (at least once a year for oil; every two years for gas). The parts cleaned should include the burner, nozzles, heat exchanger surfaces and flue.

- The technician will use instruments to do the testing.

- The technician will provide before and after efficiency readings, with documentation of what acceptable values should be.

- In addition, the technician will provide before and after readings of net flue temperature, percent carbon dioxide, smoke number (for oil), carbon monoxide (for gas) and draft, with documentation of what acceptable values should be.

You should check to see if the hole needed to make many of these tests is in the flue pipe. If the hole is not there, the contract has not been honored. Keep in mind that fuel suppliers have no strong incentive to maintain heating system efficiencies. They sell fuel, and an inefficient system uses more fuel than an efficient one.

A service contract on a woodstove should specify that the stove and flue pipe will be cleaned at least once a year. Draft readings should also be taken.

A service contract on a central air conditioner should specify that the filters in the central unit will be cleaned regularly, the heat exchanger surfaces cleaned, the thermostatic controls recalibrated, the refrigerant level checked and the system rebalanced, if necessary.

SETBACK THERMOSTATS

Thermostats control the operation of heating and cooling systems by comparing local air temperatures to preset on and off temperatures. By allowing the heating or cooling system to turn on and off periodically, instead of operating all the time, efficiency is greatly increased, and significant energy savings may be realized. Of particular interest are thermostats that can be preset for different temperature settings at different hours of the day or days of the week. These are called *setback thermostats.* Considerable fuel savings can be realized from the use of such thermostats.

In a moderately cold climate, lowering average indoor temperature by one degree from 68°F over an eight-hour period can cut daily fuel consumption by 1 percent. A more typical setback is 10°F for eight hours a day, which can result in fuel savings of about 10 percent. This means that for most setback thermostats, the simple payback will be one year or less.

A good setback thermostat must have a means of precisely determining room temperature. There are three common types of temperature sensors available in thermostats. They are the pneumatic sensor, the solid-state sensor and the bimetallic strip sensor.

Pneumatic sensor. A pneumatic sensor uses a temperature-sensitive air-pressure system to read local air temperature. They are normally used in large buildings because they are inexpensive and capable of turning large systems on and off. Thermostats with pneumatic sensors are simple to install and easy to maintain. They are not generally available for single-family homes, however, because they usually send their on and off signals through a central control system.

Solid state sensors. This type of sensor is a semiconductor chip, a thermistor, a thermocouple (a device made of two different metals in contact that generates a temperature-dependent electric current) or some other solid state electronic device that reads local air temperature. Such a system is generally accurate to $1\frac{1}{2}°F$, with an error of no more than $1°F$, and costs from $40 to $70.

Bimetallic strip sensor. The most common type of temperature sensor is a coil made of two different metals that winds and unwinds as the temperature changes. While these sensors are the least expensive of the three types, they are also the least reliable and the shortest lived, eventually suffering from metal fatigue that causes them to become inaccurate or to break. Bimetallic strip sensors have an accuracy of $2°F$ and an error of 2 to $3°F$. Regular cleaning can help maintain accuracy. Thermostats with this type of sensor generally cost no more than $25.

Conventional thermostats can generally be adjusted over a certain, limited temperature range. An optimum range that allows for maximum energy savings is $50°F$ to 68 or $70°F$ for heating control and 75 to $85°F$ for cooling control. Thermostats normally come with a control subsystem that allows the heating or cooling system to remain on until room temperatures exceed the set temperature by several degrees. This is called the *dead band control.* Another useful control subsystem operates the pumps or fans and fuel burner independently of one another so that the distribution system does not operate before the water or air is hot, and so that the burner is switched off before the thermostatic temperature setting is reached. This subsystem, called the *anticipator,* reduces electricity consumption.

Setback thermostats can provide one, two or more setback controls that allow you to preset the thermostat for different temperature settings at various hours of the day or days of the week. The unit typically has a mechanical or electronic clock, driven by batteries or household current, that switches the temperature setting at a specified time. Two setbacks will allow you to preset different daytime and nighttime temperatures. For example, you might choose to set the house temperature at $68°F$ from 6 AM to 10 PM and $60°F$ from 10 PM to 6 AM. Three setbacks will allow you to add a setting for the period during the day when the house is unoccupied. Thermostats with four setbacks and a weekend and holiday override are also available. These multiple setback thermostats range in cost from $50 to $150.

As a general principle, any indoor thermostat should be placed out of the way of drafts, direct sunlight and heating outlets or radiators. When placed in a room a thermostat should be located at a spot where the temperature is average for the room, and not where the room is hottest or coldest.

For Further Reference

Heating Systems

American Society of Heating, Refrigerating and Air-Conditioning Engineers (ASHRAE). *Handbook of Fundamentals.* New York, 1979.

Berlad, Al et al. *Seasonal Performance and Energy Costs of Oil or Gas-Fired Boilers and Furnaces.* Springfield, Va.: National Technical Information Service, 1977. Order no. NTIS-BAIL 50647.

D'Alessandro, B. "Auxiliary Heating: Is This the Redemption of Gas and Oil?" *Solar Age,* November 1979, pp. 38-42.

U.S. Department of Energy. *Minimum Energy Dwelling Workbook.* Washington, D.C.: U.S. Government Printing Office, 1977. Order no. *SAN 1198-1.*

U.S. Department of Housing and Urban Development. *In the Bank. . . or Up the Chimney?* Washington, D.C.: U.S. Government Printing Office, 1977. (Also published as: *How to Keep Your House Warm in Winter, Cool in Summer.* New York: Cornerstone Library, 1977.)

Wilson, T., ed. *Home Remedies.* Philadelphia: Mid-Atlantic Solar Energy Association, 1981.

Wood

Cooper, J. A. "Environmental Impact of Residential Wood Combustion Emissions and Its Implications." *Journal of the Air Pollution Control Association* 30 (August 1980): 855.

Modera, M. P., and Sonderegger, R. C. *Determination of In-Situ Performance of Fireplaces.* Berkeley, Calif.: Lawrence Berkeley Laboratory, University of California, 1980. Order no. LBL-10701.

New England Governors' Conference. *Heating with Wood/Burning Wood Safety.* Boston, n.d.

Shelton, J., and Shapiro, A. B. *The Woodburner's Encyclopedia.* Waitsfield, Vt.: Vermont Crossroads Press, 1977.

"Special Edition: Wood Heat." *Alternative Sources of Energy* 35.

Talbot, M. "Burning Wood: The New State of the Art." *Alternative Sources of Energy,* September/October 1980, pp. 14-17.

Heat Pumps

Brown, T. "Our Friend the Heat Pump," *Home Energy Digest,* Winter 1980, pp. 114-128.

Leckie, J. et al. *More Other Homes and Garbage.* San Francisco: Sierra Club Books, 1981.

Lunde, M. "Solar Assisted, Liquid-Coupled Heat Pumps for Residential Use." *Solar Age,* June 1980, pp. 39-48.

Sumner, J. A. *An Introduction to Heat Pumps.* Dorchester, U.K.: Prism Press, 1976.

Other Space Heating Technologies

Brookhaven National Laboratory. "Adding a Flue Damper to a Home-Heating System." Upton, N.Y., 1980.

_____. "Adding a Flue Economizer to a Home-Heating System." Upton, N.Y., 1980.

_____. "Converting to a Flame-Retention Burner for More Efficient Heating." Upton, N.Y., 1980.

_____. "High Efficiency Residential Boilers." Upton, N.Y., 1980.

_____. "Upgrading Oil Home-Heating Systems." Upton, N.Y., 1980.

Gay, L. "A Buyer's Guide to Wood and Coal-Fired Furnaces." In *The Energy Consumer's Handbook*. Minneapolis: Home Energy Digest, 1980.

Office of Technology Assessment. *Residential Energy Conservation*. Vol. I. Washington, D.C.: U.S. Government Printing Office, 1979.

Skousen, J. "Multi-Fuel Furnaces." *Home Energy Digest*, Summer 1978, pp. 42-49.

Setback Thermostats

SUN Catalog. Bascom, Ohio.: Solar Usage Now. Reprinted 2-3 times a year.

4 Managing Energy in the Home

Successful home energy conservation requires more than just the technical improvements to the building envelope and heating systems described in the first three chapters of this book. Proper management of energy consumption is also very important. In the typical American home, there exist many opportunities to conserve energy that are neither difficult nor expensive but that can realize significant savings.

For example, when old appliances wear out, they can be replaced with more efficient ones. This requires reading a few labels, making a few easy calculations and comparing similar models of the same kind of appliance. Lighting systems can be modified to reduce electricity consumption. The lighting in a specific area should be suitable for the tasks carried out there. In some cities, there is the opportunity to reduce residential energy consumption and costs through what is called *load management*. This involves regulating your demand for energy through the use of mechanical or electronic devices. Finally, with minor changes in life-style, energy consumption can be reduced. This means using appliances efficiently, turning down thermostats and reducing water consumption. Considering that nearly 20 percent of the energy consumed in this country is used in the residential sector, wise management of this energy can substantially reduce energy consumption in the nation.

APPLIANCES

Home appliances such as refrigerators, water heaters and clothes washers are major consumers of energy. Twenty-five to 40 percent of residential energy consumption in the United States is appliance-related. The operation of some appliances at particular times of the day and year can contribute greatly to a utility's electric *peak power* demand. Examples include operating kitchen ranges in the evening or air conditioners in the summer. Energy conservation and the replacement of appliances with energy-efficient models could have a significant effect on electricity and gas consumption in the home and in the nation as well.

Energy efficiency does not mean bigger, thicker, heavier or more costly appliances; indeed, there is often very little to distinguish an energy-efficient appliance outwardly from its more power-hungry counterparts. Only stan-dardized laboratory tests can measure actual appliance energy efficiency. As an aid to consumers, the federal government now requires that appliances be labeled with tags showing their yearly operating costs. Consumers can use the information on these labels and other information to determine the energy costs of various appliances.

Comparing Energy Costs of Appliances

In order to determine differences among appliances regarding energy costs during their operating lifetimes, you need the following information: (1) a comparison list of prices and annual energy consumption for different appliance models and brands; (2) an estimate of how long you will own and operate various appliances; (3) an estimate of energy costs and cost increases during the operating lifetimes

Table 4-1
Estimated Appliance Operating Lifetimes
(first owner)

Appliance	Years	Appliance	Years
Central air conditioning	15-20	Range/Oven	15-20
Clothes dryer	10-15	Refrigerator	10-15
Clothes washer	10-15	Room air conditioner	7-10
Dishwasher	10-15	Television	5-10
Freezer	10-15	Water heater	10-20
Furnace/Burner	20-30	Others	5-10

Table 4-2
Average Annual Appliance Use in American Homes

Appliance	Hours/Year	Appliance	Hours/Year
Air conditioner	890	Oil burner	768
Clothes dryer	215	Range	100
Clothes washer	210	Refrigerator	350
Dishwasher	300	Television	1,380
Freezer	360	Water heater	1,068

SOURCE: G. Sullivan, *Wind Power for Your Home* (New York: Cornerstone Library, 1978).

of the appliances; and (4) a determination of what alternatives in terms of size and features will save energy and reduce the cost of operating various appliances during their lifetimes.

Comparison Shopping

There are significant differences in energy consumption among different models of appliances made by one manufacturer and different brands of otherwise identical appliances. For example, the annual electricity requirements for new home refrigerators can differ by as much as a factor of five or six among models and a factor of two within specific categories. While some of these differences can be attributed to size, features or style, there can still be sizable variations in virtually identical appliances manufactured by different companies. The difference in purchase price is not likely to be very large, and it is not unheard of for a more energy-efficient appliance to be cheaper than one with higher energy costs. Even if this is not the case, over the operating lifetime of the device, the savings in energy costs can amount to hundreds or thousands of dollars, more than making up the price difference.

When buying an appliance, it pays to comparison shop and to obtain both prices and expected energy consumption and/or energy cost per year. This latter number can often be obtained from the federal energy label on the appliance. The energy cost per year will allow direct comparisons among different appliances. If you only know the annual energy consumption, multiply it by the cost of electricity and gas (assumed here to be 7¢ per kilowatt-hour and 50¢ per therm).

Operating Lifetime

Estimate how long you will own and operate the appliance. Typical appliance lifetimes can be long. Refrigerators, for example, are normally used for 20 years; washing machines for 10 to 20. Try to remember how long you owned the appliance you are replacing. If you cannot make a reliable estimate of appliance lifetime, 10 years is a reasonable guess. (See table 4-1 for estimates of appliance lifetimes.)

Energy Cost

Over the operating lifetime of the appliance, energy costs are sure to increase. While it is very difficult to make any reliable predictions of energy prices over the next 10 or 20 years, it seems safe to say that they will at least double or triple. A 7 percent annual increase results in a doubling of price in 10 years; at a 4 percent annual increase, doubling occurs in 20 years. With a 12 percent annual increase for 10 years or a 6 percent annual increase for 20 years, the cost of energy is tripled.

To calculate the lifetime energy cost, it is sufficient to use present energy costs. While this will underestimate the energy savings you will realize from using a more energy-efficient appliance, it should give you a good indication of which appliance is the best buy.

To calculate the lifetime operating cost, it is necessary to multiply the annual energy consumption (in kilowatt-hours or therms per year) by the energy cost (in cents per kilowatt-hour or therm) and by appliance lifetime. For example, a refrigerator that uses 1,000 kilowatt-hours of electricity per year and has a life expectancy of 20 years will have a lifetime energy cost of $1,400 (1,000 KWH × 20 years × 7¢/KWH). Table 4-2 gives average annual appliance use in American homes.

To compare two appliances properly, you must compare the sum of the purchase price and lifetime energy cost for each, a quantity called the *life-cycle cost*. The appliance with the smaller life-cycle cost is, of course, the better buy. See "Calculating Appliance Life-Cycle Cost" in this chapter for an example of how to calculate this cost.

In some instances, relatively unimportant differences in appliance features can mean significant differences in energy consumption. For example, side-by-side refrigerator/freezer units usually consume more energy than models with the freezer above the refrigerator compartment. Instant-on televisions always draw power if plugged in while other sets do not. Therefore, by settling for slightly less convenience, you may be able to realize significant energy savings.

Federal Energy Guide Labels

Federal law has mandated that certain types of household appliances be labeled with tags that show the yearly operating cost of an appliance. These tags allow consumers to compare the energy costs of different models and brands of appliances, if they know how to read and use them. At the present time, the tags are required on furnaces, refrigerators, refrigerator/freezers, freezers, water heaters, clothes washers, dishwashers and room air conditioners. Central air conditioners and heat pumps will be labeled at a future date. The labels are not yet required on kitchen ranges or ovens, microwave ovens, television sets, clothes dryers, humidifiers and dehumidifiers or home heating equipment other than furnaces. Labels are also not required on small appliances. But you can still calculate the energy costs of unlabeled appliances, provided you know about how many hours the appliance will be operated each year and the energy consumption in kilowatts per hour. A sample calculation is shown in the following box.

There are three types of energy labels: energy cost labels, energy efficiency rating labels and generic labels. *Energy cost labels* are used on refrigerators, refrigerator/freezers, freezers, water heaters, dishwashers and clothes washers (see figure 4-1). Each label has a large number that shows the estimated annual energy cost of the

Calculating Appliance Life-Cycle Cost

How do you calculate the life-cycle cost of an appliance? The life-cycle cost is the sum of the appliance purchase price, the lifetime energy cost and the costs of maintenance and repairs.

Let's say you want to buy a 17-cubic-foot refrigerator/freezer. There are two models that seem to fit your needs:

Refrigerator A, which costs $800 and uses 1,248 kilowatt-hours of electricity per year, and

Refrigerator B, which costs $530 and uses 1,625 kilowatt-hours of electricity per year.

As we shall see, Refrigerator B, despite its lower purchase price, is not the better buy. We'll assume that you will own the appliance for 20 years, and that electricity prices will cost 7¢ per kilowatt-hour during that period. (In reality, of course, the price of electricity will increase.) Annual energy costs for Refrigerator A are found by multiplying annual energy use by the cost of electricity: $87.36. Lifetime energy costs are found by multiplying annual cost by 20 years: $1,747.20. For Refrigerator B, the corresponding numbers are $113.75 and $2,275.

Therefore, all else being equal, the life-cycle cost of Refrigerator A is $800 plus $1,747, or $2,547, and of Refrigerator B, $530 plus $2,275, or $2,805. Refrigerator A is the better buy, even though its purchase price is more than that of Refrigerator B.

appliance based on an electricity cost of about 5¢ per kilowatt-hour. The label also compares the energy cost of the appliance with those of competing brands and models of similar size and with similar features. The label also includes the annual energy cost of the appliance for a range of energy prices. Labels on dishwashers and clothes washers allow you to compare the costs of running the appliance with either a gas or an electric water heater.

The energy cost label can also tell you approximately how many kilowatt-hours of electricity the appliance will consume each year. For example, if the annual energy cost is $90 calculated for an electricity cost of 5¢ per kilowatt-hour, the appliance will consume $90 divided by 5¢ per KWH or 1,800 kilowatt-hours. Knowing this quantity will enable you to calculate and compare life-cycle costs for different appliances and other energy prices.

Energy efficiency rating labels are used on room air conditioners (see figure 4-2). These appliances are labeled with an Energy Efficiency Ratio (EER) or a Seasonal Energy Efficiency Ratio (SEER). The EER is equal to the number of Btu's of cooling supplied per watt of electricity consumed, that is:

$$EER = \frac{\text{cooling capacity of AC (in Btu/hr)}}{\text{wattage}}$$

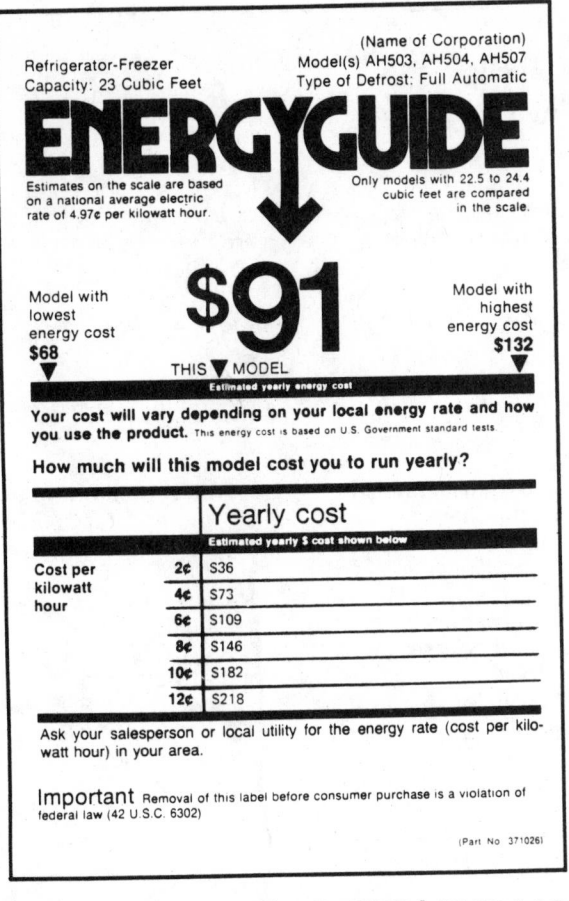

FIGURE 4-1: A SAMPLE ENERGY COST LABEL

The SEER is equal to the number of Btu's of cooling supplied during the year per kilowatt-hours of electricity consumed in the year, that is:

$$SEER = \frac{\text{cooling supplied during the year (in Btu's)}}{\text{kilowatt-hours consumed during the year}}$$

The higher the EER or SEER, the more efficient the appliance. The label includes the range of EERs for competing models and brands of room air conditioners of approximately the same cooling capacity. Also included on the label is a table that tells you the annual energy cost of the appliance, provided you know the cost of electricity and the number of hours you will use the appliance each year. Again, by multiplying the annual energy cost by the number of years you expect to own the air conditioner, you can determine the lifetime energy cost (at current electricity prices). To find out the annual consumption of electricity (in kilowatt-hours) of the appliance, divide the total energy cost (in dollars) by the cost of electricity in the table (in cents per KWH). The SEER averages out the

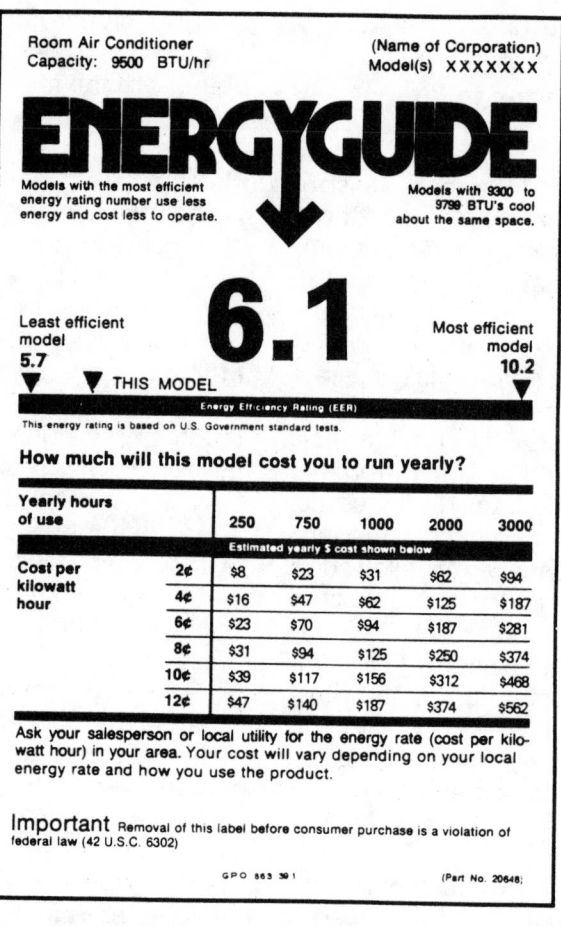

Room Air Conditioner
Capacity: 9500 BTU/hr

(Name of Corporation)
Model(s) XXXXXXX

ENERGYGUIDE

Models with the most efficient
energy rating number use less
energy and cost less to operate.

Models with 9300 to
9799 BTU's cool
about the same space.

Least efficient
model
5.7

6.1

Most efficient
model
10.2

▼ ▼ THIS MODEL   ▼

Energy Efficiency Rating (EER)

This energy rating is based on U.S. Government standard tests.

How much will this model cost you to run yearly?

Yearly hours of use		250	750	1000	2000	3000
		Estimated yearly $ cost shown below				
Cost per kilowatt hour	2¢	$8	$23	$31	$62	$94
	4¢	$16	$47	$62	$125	$187
	6¢	$23	$70	$94	$187	$281
	8¢	$31	$94	$125	$250	$374
	10¢	$39	$117	$156	$312	$468
	12¢	$47	$140	$187	$374	$562

Ask your salesperson or local utility for the energy rate (cost per kilowatt hour) in your area. Your cost will vary depending on your local
energy rate and how you use the product.

Important Removal of this label before consumer purchase is a violation of
federal law (42 U.S.C. 6302)

GPO 863 391

(Part No. 20648)

FIGURE 4-2: A SAMPLE ENERGY EFFICIENCY RATING LABEL

EER over typical operating conditions found during the season, to give a seasonal average EER.

Generic labels are required on furnaces and central air conditioners. These labels contain some simple (and obvious) steps you can take in order to save on heating costs. The label also directs you to pick up a fact sheet for the unit, developed by the manufacturer and containing information on system components and overall efficiency for different combinations of parts. The fact sheet includes annual energy costs for the furnace or air conditioner.

These labels are a step forward in helping consumers make intelligent choices, but they have several deficiencies. They contain no information on appliance lifetimes or data on energy costs of those appliances that fall in between the extremes in a given class. The purchase prices and life-cycle costs of other units are not included, the names of the more and less energy-efficient appliances are not given on the label and consumers have no way of knowing whether the more efficient units are sold anywhere in their area. Consumers are left pretty much on their own to search for the most energy-efficient units.

Energy-Efficient Characteristics of Specific Appliances

Refrigerator/Freezers

In addition to differences in size, refrigerator/freezers come in three different classes: *manual defrost* with a freezer compartment inside the refrigerator which must be manually defrosted at frequent intervals; *partial-automatic defrost* which has separate freezer and refrigerator compartments with the freezer requiring periodic manual defrosting; and *automatic defrost* which never requires defrosting. In most cases, automatic defrosting units use the most energy, followed by partial defrosting and manual defrosting units, although some partial defrosting units consume more energy than some automatic defrosting units, and other partials consume less energy than some manuals. If you opt for a different defrost feature in order to save energy, be sure that your new choice actually does consume less energy.

Another choice of features is the side-by-side refrigerator/freezer versus the top-freezer type. In general, the latter style consumes less energy than the former, although there are exceptions. Smaller units do not always consume less energy than larger ones. Top-loading freezers are more energy-efficient than upright ones because they do not spill out cold air when opened. Another thing to consider is cabinet insulation; the better insulated the cabinet, the more easily the refrigerator will retain coolth, and the less work it will have to do. This does not mean, however, that the best-insulated model uses less energy than other units; it may, for example, use more electricity because of inefficient compressor design. Only careful comparison shopping can determine which models and brands are the most energy-efficient.

Air Conditioners

The efficiencies of different models of air conditioners vary widely. Because room air conditioners must be labeled with an EER, it is fairly easy to distinguish which units are energy-efficient. Energy-efficient units are generally more expensive, but it is possible to eliminate much of the price difference by purchasing an energy-efficient unit on sale.

It is important to purchase an air conditioner of the proper size. If a unit is too large, it will cycle on and off frequently and use extra energy. A room air conditioner with a cooling capacity of 3,000 to 4,000 Btu (EER = 9) is large enough to cool a well-weatherized 200-square-foot room. Other features to look for include a variable cooling setting, automatic thermostat and a fan-only setting. Room air conditioners usually do not have thermostats with specific temperature settings and must be rewired for this (see "Air Conditioners," chapter 3).

Heating Appliances

Ranges and ovens, clothes dryers, water heaters and some dishwashers are included in this category. In addition to individual features, an important energy efficiency and cost question involves the type of fuel the appliance will use—natural gas or electricity.

For example, one million Btu's of heat produced by natural gas burned at 75 percent efficiency (and costing 50¢ per therm) cost about $6.67, while the same amount of heat produced by electricity (at 7¢ per kilowatt-hour) will run $20.51. These costs will vary, of course, depending upon where you live.

Although natural gas prices will rise in the future, it is unlikely that they will ever be as high as electricity prices.

Also, because heating appliances generally consume peak electric power, which is generated by the burning of gas or oil, at least one-third, and possibly more, of the energy content is wasted when heating appliances use electricity instead of natural gas. Therefore, the choice of electricity or natural gas is an important one, because it can reduce both overall gas consumption and your energy costs.

Ranges. Not only are gas ranges less expensive to operate than electric ranges, they also provide heat more quickly and require almost no time to cool off. This cuts down on energy consumption and helps keep your kitchen cooler. Any gas range should have electric ignition, rather than a standing pilot light, since pilots can consume as much as 35 to 45 percent of the total annual energy use of a gas range.

Clothes dryers. Gas clothes dryers should also have electric ignition.

Water heaters. The size of a water heater should be closely matched to the hot water requirements of your home. If the tank is too large, energy will be wasted in keeping the extra water hot; if it is too small, you may have to take lukewarm showers if you're last in line. The tank should be well insulated. If it is not, buy a water heater insulation kit, which costs about $20, or a roll of 6-inch fiberglass, which costs about $13. The payback for this improvement will be very short. Gas water heaters should have electric ignition and automatic (bimetal) flue

dampers. The latter must be added to the unit. The installation of flow restrictors on taps and showers will also significantly reduce hot water consumption.

Dishwashers. Dishwashers are not generally considered heating appliances, but they do, of course, consume hot water. Some dishwashers have a separate heating element to heat water to 140°F. With this option, you can save energy by setting your water heater to 120°F. Also a dishwasher should have a dry cycle manual override that will allow dishes to air-dry instead of being dried by electrically heated air. Washing full loads of dishes whenever possible saves energy, too.

Clothes Washers

A clothes washer should have settings for different-size loads to conserve hot and cold water. Since the machine uses the same amount of electricity to wash a load regardless of size, you should wash full loads of clothing whenever possible. Using the cold wash and rinse setting allows you to conserve a great deal of hot water.

Televisions

In general, there is little or no relationship between purchase price and energy cost for televisions. TVs are typically operated for 1,000 to 2,000 hours each year in the average American household, so energy-efficient models can save significant amounts of energy. Color TVs consume more energy than black and white sets. Sets with vacuum tubes require more energy than solid state sets. Sets with an instant-on feature are actually never off; they are always consuming small

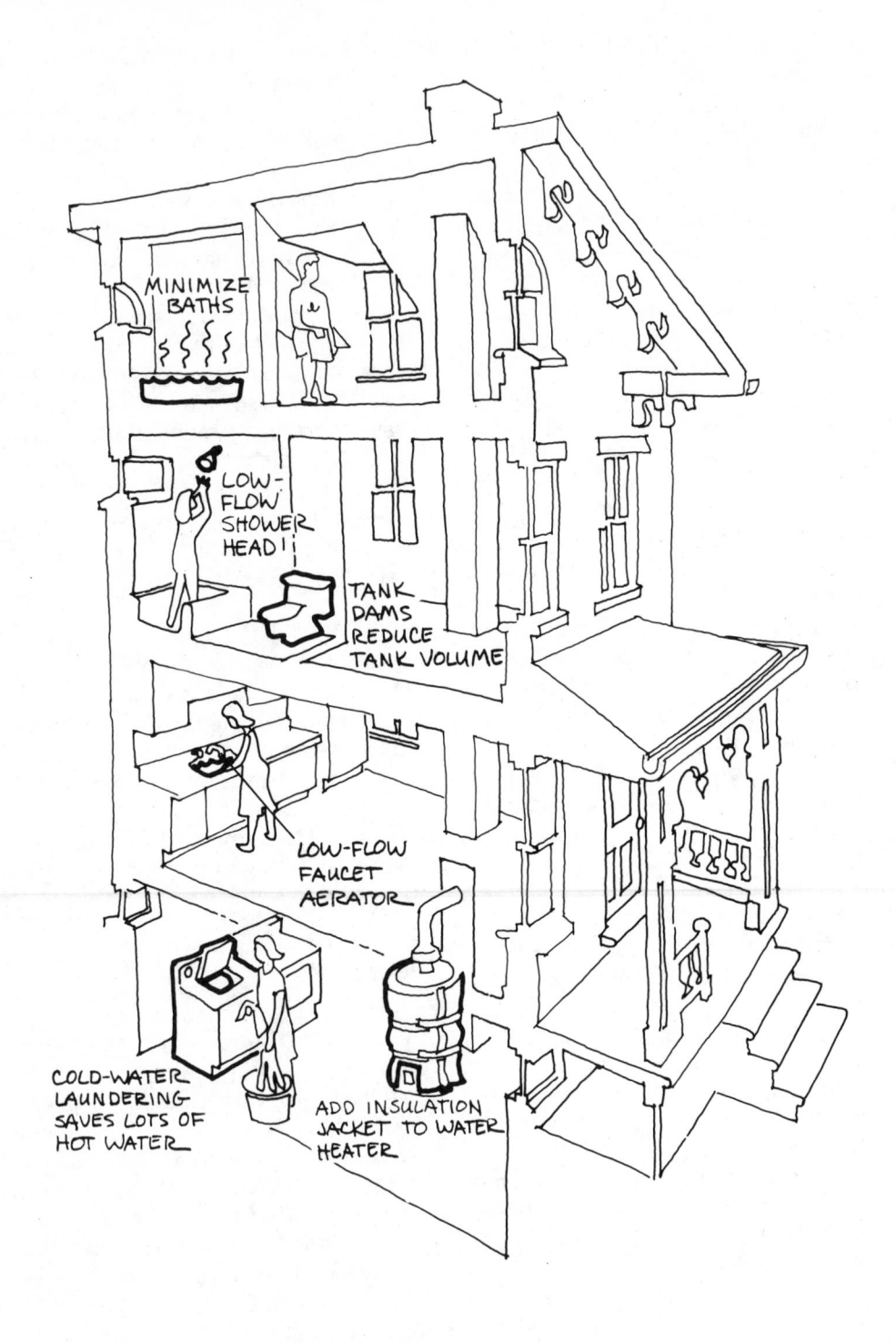

MINIMIZE BATHS

LOW-FLOW SHOWER HEAD!

TANK DAMS REDUCE TANK VOLUME

LOW-FLOW FAUCET AERATOR

COLD-WATER LAUNDERING SAVES LOTS OF HOT WATER

ADD INSULATION JACKET TO WATER HEATER

FIGURE 4-3: WATER CONSERVATION

amounts of power in order to keep the components warm. You can eliminate this hidden consumption by installing a switch on the cord or simply by unplugging the TV when it is not in use.

LIGHTING

Lighting is responsible for about 10 to 15 percent of total residential energy consumption and offers considerable opportunity for energy savings. Through a combination of replacing existing lamps with more efficient ones, decreasing the wattage (or brightness) of some, improving controls, changing fixtures and repositioning lights, lighting energy consumption in your building can be reduced significantly. In addition, lowered lighting levels decrease both internal building heat load and air conditioning load. This can save you energy during warm weather.

Making a Lighting Audit

A lighting audit of a building involves the measurement of lighting intensity and wattage in footcandles (f.c.) and the comparison of present lighting to certain recommended values or limits, followed by appropriate action to modify lighting or alter the lighting system. Some states may require a lighting audit of buildings greater than a certain floor area. The Commonwealth of Massachusetts, to give one example, requires such an audit of all buildings with floor areas of 10,000 square feet or more. The results of the audit must then be filed with the State Building Code Commission.

Even if it is not required in your state, a lighting audit can be very helpful in reducing your lighting costs. An audit is likely to show that some parts of your building are overlit compared to what is needed. Many lighting improvements will not impose any hardships on building occupants. Note that such an audit normally will cover only the common areas in a residential building, leaving other areas unaffected.

System Improvements: Fixtures, Bulbs and Wiring

Replacement

Existing lights can be replaced by bulbs of equal brightness but lower wattage. For the most part, this means replacing incandescent bulbs by fluorescent ones, which can result in energy savings of 60 percent or more. Replacement by fluorescents normally requires installation of a new fixture, but fluorescent bulbs that screw into standard light sockets are now on the market. Existing fluorescents can be replaced by more efficient ones, saving 15 to 50 percent in energy consumption. You may have to replace the fluorescent ballast (transformer) when doing this.

Delamping

Lights can be replaced by bulbs of lower wattage, resulting in energy savings of 10 to 50 percent. This will decrease brightness, but if a lighting audit shows an overlit space, this is the easiest course of action. Use of movable lamp fixtures for specific tasks, such as reading, can allow delamping or turning off of overhead, broadcast lighting fixtures. Particularly good places for delamping are hallways, vestibules and other common areas.

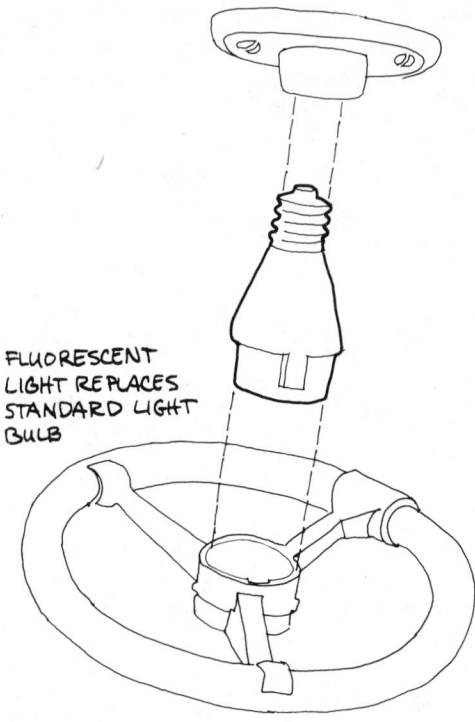

FLUORESCENT
LIGHT REPLACES
STANDARD LIGHT
BULB

FIGURE 4-4: EFFICIENT LIGHTING

In some instances, you may be able to remove bulbs, leaving empty fixtures, without causing serious lighting deficiencies. If fluorescents are removed, you may have to rewire the circuit in order to prevent continuing energy consumption by ballasts. Or you can install "phantom" tubes that draw no energy at all.

Luminaires

A *luminaire* is the glass or metal fixture that reflects or diffuses light from the bulb. Such fixtures generally trap light and also spread it in particular directions. By choosing an appropriate luminaire you can reduce energy consumption by 10 to 40 percent, though they are generally not chosen because of energy considerations, but rather for appearance. Manufacturer's literature can give information on the effectiveness of a particular fixture.

Controls

The simplest way to control lights, of course, is to turn them off when not needed. You can wire lights to time clocks that automatically turn them on or off at specific times. This is particularly useful in unoccupied buildings that must be lit at night for security reasons. Greater energy savings can be realized by placing photocell controls on lighting circuits. Such controls dim or turn off lights when daylight entering through windows reaches a preset level and switch them on at dusk. The payback for such a control is typically three to six years. A manual dimmer switch allows you to control the brightness and hence the energy consumption of a particular lighting fixture.

In large buildings, putting lights in some areas on a centralized load controller, which switches them off when building electricity demand exceeds a certain level, can save on demand charges by reducing peak load. Demand charges, paid by large electricity consumers, represent the utility's cost of keeping the generating capacity available to meet the user's peak demand. (See "Energy Pricing and Rate Structures," chapter 7.) Such control systems are generally suitable only for the common areas of residential buildings and may not be cost-effective there. By appropriate positioning of light switches and by putting switches on individual fixtures, you can more easily control unnecessary lighting, especially in large buildings.

Repositioning

Relocation of light fixtures can often improve the quality of light in a space and decrease the required brightness, for example, by bouncing light off a wall or bringing the light closer to the task. You can use movable fixtures for a similar effect.

Color and Daylighting

Dark-colored walls absorb much of the light that falls on them. White or light-colored walls, on the other hand, reflect and diffuse light throughout a room, creating an even distribution. By painting, wallpapering or adding light-colored hangings to a room, you can decrease wattage requirements by 20 to 40 percent.

Use of daylight can make interior lighting totally unnecessary. Improvements in window placement and addition of windows, skylights and clerestories can improve lighting quality and allow penetration of light deep into living spaces. You can also accomplish this with "bounce-lighting," using white or reflecting venetian blinds in existing windows. In a room or building where the layout of windows cannot be changed, rearranging your furniture can improve both task lighting and general lighting. Daylight should strike a work area such as a desk from the side or above and not from behind for the greatest ease of work.

Proper Maintenance of Lights

Dust and dirt can decrease lighting levels by 20 to 30 percent, and shorten bulb life by increasing the entrapment of heat by the fixture. Therefore, you should clean luminaires and bulbs regularly. Check wiring periodically for shorts, bad connections and loose or frayed wiring. Be sure you do not oversize bulbs. Many fixtures are designed only for bulbs up to a certain wattage, beyond which they may overheat and fail. Correcting these problems will not only save energy, but may also prevent an electrical fire. Fluorescent lamps that begin to flicker or fail entirely should be replaced. The ballast continues to draw energy even when the bulb is working improperly.

The intensity and wattage of existing lighting are not the only important factors here. The quality of light also depends upon its direction and the angle at which it strikes the task or working area, the way light is reflected from the area, the color of the light, the color of the room and the way light scatters from various parts of the room. In general, daylight is the standard against which all artificial light is compared, and every effort should be made to maximize the use of daylight whenever possible.

LOAD MANAGEMENT

Another important opportunity to decrease energy consumption is by controlling your demand for electricity at a particular time. This is called load management and it may allow an electric utility to operate its most efficient and least costly equipment more of the time as well as reduce the need for new generating and transmission equipment. This, in turn, allows the utility to offer lower electric rates at particular times of the day and year. By taking advantage of these lower rates and by using electricity wisely, you may be able to halt or even reverse the otherwise inexorable increases in your electric bill. *Time-of-use rate structures* (TOURS) are just beginning to be used by some utilities though in the near future the practice should become more widespread as utilities seek to smooth out their daily load patterns. Figure 4-5 shows how a time-of-use rate structure can be used to encourage more power consumption during periods when consumption is typically low.

How does load management work? There are two types of demand put on electric utilities. There is always a minimum level of demand day and night,

season after season, known as the *base load*. At certain times during the day and during the year, demand rises above the base load, and electricity usage reaches a maximum. This happens when people return home from work and turn on their appliances or on very hot days, when many air conditioners and fans are operating during the same time period. This type of demand is called a *peak load*.

Different types of power plants provide base and peak load power. Base load plants operate almost continuously because they are either the least expensive to operate or because they are difficult to start up and shut down. Nuclear power plants are almost always used as base load plants. Peaking plants generally burn oil or gas, both expensive fuels, but such plants are fairly easy to start up and shut down. Base load plants therefore are "paying" their way around the clock, while peaking plants must be paid for even when they do not operate. Base load operation is usually (but not always) less expensive than peaking operation and therefore generating costs to the utility are less during off-peak periods than during on-peak periods. It may therefore be in the interest of both the utility and its customers to limit peak demand.

Utilities usually keep a certain amount of generating capacity in reserve in order to allow maintenance of plants. If peak demand increases and eliminates this reserve, it may be necessary for the utility to build new power plants. However, if the utility can shift more demand to base load periods, construction of new plants (coal and nuclear) may not be necessary.

How is it possible to limit your peak demand for electricity? First, of course, you can use less electricity, either by

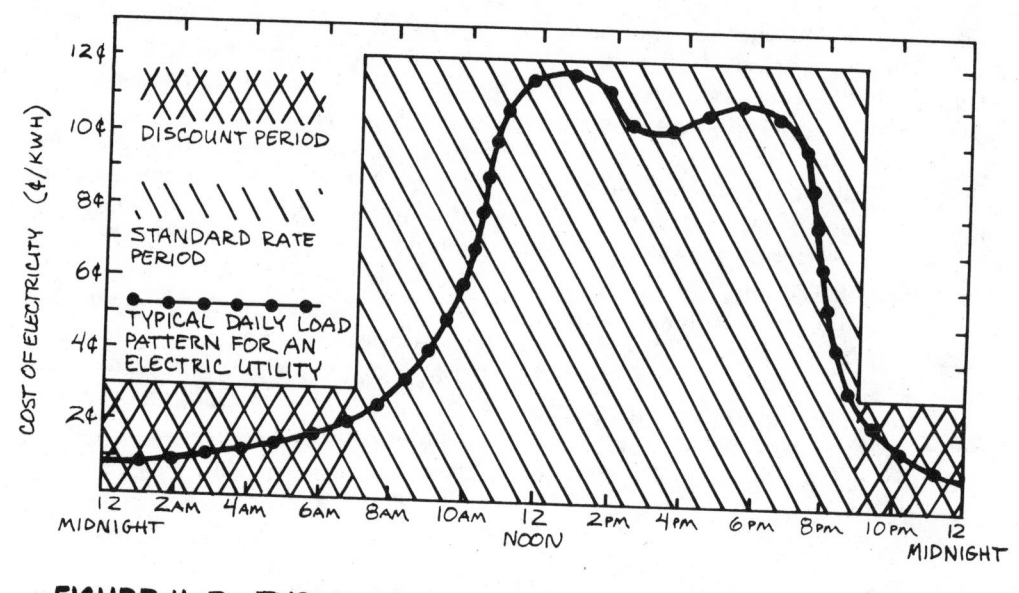

FIGURE 4-5: TYPICAL DAILY TIME-OF-USE RATE STRUCTURE

purchasing appliances that are more energy-efficient, or by shutting them off when they are not needed. Second, you can operate appliances during base load periods. Even if you are not being offered special electricity rates as an incentive to do this, you are helping to limit peak demand and, therefore, the need for new peak generating capacity. Peak demand hours in most large cities generally occur in the late afternoon and early evening, and during summer mid-afternoons due to air conditioning loads.

A device that normally runs continuously or at random times during the day, such as a water heater, a fan, or an air conditioner, can be wired to a clock timer that will switch the appliance off during the peak demand period. Alternatively, for example, an air conditioner can be wired to let the indoor temperatures drift upward during peak hours. In some instances, electric utilities themselves may be interested in cutting on-peak electricity use and may offer to install control devices that will briefly shut off certain appliances during peak periods by remote control.

TOURS rates may vary either across the day or over the seasons, or both. With both daily and seasonal TOURS rates in effect, the most expensive time to operate an appliance would be on a hot summer afternoon.

For consumers with TOURS rates and electric resistance heating or air conditioning already in place, it may pay to install an energy storage device. An energy storage device that has been in use in the United States for many years is the ordinary tank water heater; however, there are other types of energy storage devices just coming on the American market that have long been used in Europe. They enable you to store heat or coolth during off-peak hours, when electricity is cheaper, and take it out of storage for use during peak demand periods. Off-peak air conditioning can use less energy than normal daytime operation, because the compressor and heat exchangers in the unit will be operating against nighttime temperatures, which are generally lower than those during the day. Heat storage devices include oil or magnesia brick storage heaters. These may be substituted for conventional electric resistance baseboard units. Air conditioning storage devices include chilled water tanks, cooled rock beds and ice makers combined with storage tanks.

The opportunities and paybacks for load management tend to be greater for large power users such as apartment buildings. Large consumers may already pay demand charges based on monthly peak demand. (See "Energy Pricing and Rate Structures" in chapter 7.) Because demand charges are high, it can quickly pay to shift or change energy use or to store energy in order to reduce peak demand. In addition, storage can do double duty by also providing economical and socially advantageous backup to solar and other renewable systems you might wish to add later.

Load Management of Domestic Hot Water

How can load management be applied in your home? Let's assume you wish to reduce your domestic hot water costs. If your local electric utility offers time-of-use rates (TOURS), you can reduce costs by heating water during off-peak hours. Say, for example, that on-peak electricity costs 12¢ per kilowatt-hour, while off-peak costs only 3¢ per kilowatt-hour. Your family uses 15 kilowatt-hours per day to heat water. One load management solution you could easily apply would be to install a clock timer on your water heater that would shift two-thirds of the water heating to off-peak hours. (In other words, hot water might not always be available during on-peak hours, and your family would have to shift its water usage habits accordingly.) Such a clock timer costs about $30, but look at its effects! Ten kilowatt-hours of electrical consumption would be shifted to off-peak hours, saving you

$$(12¢\text{-}3¢)/\text{KWH} \times 10 \text{ KWH/day} = 90¢/\text{day!}$$

At this rate, the payback for the timer would be a mere 33 days, and during the first year you would save almost $330.

One interesting consequence of this simple action is that it would shift about 3 kilowatts of power demand from on-peak to off-peak hours. This generating capacity would then be available for other on-peak uses. The cost of this energy conservation measure would be about $10 per kilowatt, compared to the cost of building new electrical generating capacity that runs from $750 to $1,500. Indeed, it would be to your utility's benefit to give you the clock timer free, install it at no cost and pay you a cash bonus, too!

For Further Reference

Farallones Institute. *The Integral Urban House.* San Francisco: Sierra Club Books, 1979.

Goldstein, D. B., and Rosenfeld, A. H. *Energy Conservation in Home Appliances through Comparison Shopping: Facts and Fact Sheets.* Berkeley, Calif.: Lawrence Berkeley Laboratory, University of California, 1978. Order no. LBL-5910.

Leckie, J. et al. *More Other Homes and Garbage.* San Francisco: Sierra Club Books, 1981.

Price, B. L., and Price, J. T. *Homeowner's Guide to Saving Energy.* Blue Ridge Summit, Pa.: TAB Books, 1976.

Rosenfeld, A. H. et al. *Saving Half of California's Energy and Peak Power in Buildings and Appliances via Standards and Other Legislation.* Berkeley, Calif.: Lawrence Berkeley Laboratory, University of California, 1978. Order no. 6865.

5 Renewable Energy Systems

Renewable energy systems derive their "fuel" from the sun. *Solar energy* is the direct use of sunlight to provide space and domestic water heating and electricity. *Wind energy* is the result of uneven heating of the earth's surface by the sun and can provide heat, electricity or mechanical power. *Biomass energy* is derived from trees and crops whose growth is the result of photosynthesis. Biomass can provide heat or liquid fuels from wood, plant waste and alcohol. *Hydroelectric power* is derived from falling water that has been raised to the clouds by solar evaporation. *Ocean thermal energy* is available because of the solar-induced temperature difference between surface and deep waters.

Renewable fuels have the virtue of availability in unlimited supply for as long as the sun continues to shine. Solar energy, wind power and water-related power sources are also clean and nonpolluting, an important consideration for smog-plagued cities. While renewable energy systems can be costly, renewable fuels are essentially free for the taking. They do not require strip-mining or the production of toxic wastes. They do not upset the heat or carbon dioxide balances of the earth. And they are not subject to severe shortages or sudden, unpredictable price increases. At the present time, many renewable energy systems are not cost-effective and cannot compete with conventional fossil fuels or electricity. However, as the prices of conventional energy supplies continue to increase, renewables will become more and more competitive.

Of course, not all of these energy sources are usable in city homes, but those that can be used in urban settings hold great promise for reducing residential consumption of conventional fuels.

PASSIVE SOLAR HEATING AND NATURAL COOLING

Passive Solar Heating

Passive solar heating involves the use of a building's structure in order to capture and store the sun's energy without employing electrical or mechanical devices such as pumps or fans. This type of solar heating has several advantages over *active solar heating* systems, which do require these devices along with a separate thermal storage component. Because there are few moving parts in passive systems, breakdowns rarely, if ever, occur. System maintenance is minimal and operation usually requires a small degree of manual control to manage daily heat gain and minimize nighttime heat loss. The use of existing building characteristics often tends to keep installation and materials costs low relative to the cost of active systems. Except for backup heating, there are no associated fuel costs after the system is installed and, even on the coldest sunny days, a passive heating system can supply a significant fraction of a building's heating load. Finally, a passive system, unlike most active ones, can also collect heat on overcast days, when there is often useful solar gain.

The appropriate choice of a passive system depends upon local climatic conditions, latitude, access to sunlight, the building's construction and its location relative to other buildings, trees and so on. The four building prototypes described in the first chapter all have characteristics that make them suitable for some passive solar modifications. This section discusses the basic concepts of passive solar heating and the different types of generic systems that can be used. Passive systems use natural heat transfer processes to move captured energy to the point of need. These processes are *conduction*, the movement of heat through materials, *natural convection*, the transfer of heat by air currents and *radiation*, the direct transfer of heat from hot to cold surfaces.

In order to supply heat when the sun isn't shining, a passive solar system requires *thermal mass* to store extra heat collected when the sun is shining. The most common thermal mass materials are masonry (stone, brick, concrete) and water in containers. The amount of energy a given volume of thermal mass can store for a given temperature rise is called its *heat capacity*. This is the amount of heat (in Btu's) required to raise the temperature of the material one degree Fahrenheit. For example, it takes 62 Btu to raise one cubic foot of water $1°F$, while it takes 22 Btu to raise a pound of concrete 1°F. This means that water can store about three times the heat that concrete can, per unit of volume. Heat can also be stored when a material changes from a solid to a liquid state and released when the substance solidifies. This process is called a *phase change*. The heat stored when melting takes place— the *heat of fusion*—is much greater than the heat capacity of the material. One of the more common phase-change materials used in solar heating systems is called Glauber's salt, which belongs to a class of *eutectic salts*. The heat of fusion for a cubic foot of Glauber's salt is 9,140 Btu, which illustrates the large heat storage capability of these materials.

Passive solar heating systems generally have three components: (1) the *collector*, a structure usually covered by glass or plastic that allows the sun's energy to enter the system but inhibits it from leaving; (2) the *space*, which is the area to be heated; and (3) the *thermal mass*, which stores heat for later use. These three components are usually assembled in direct gain, indirect gain and remote thermal storage systems.

In a *direct-gain system* sunlight passes through the collector (a window or skylight) directly into the living space and heats the space and the storage thermal mass, which may simply be masonry walls or floors. At night, the space is heated by radiation and convection from the mass.

In an *indirect-gain system* sunlight passes through the collector (usually a glass wall over existing mass) but not into the living space and heats an intervening airspace and the storage mass, which is often a concrete or water-filled wall. The intervening airspace may be thin as in a Trombe wall or thick as in a solar greenhouse. During the day, the sun heats the hot airspace, and the hot air rises and passes through vents into the living space. At the same time, the sun also heats the storage mass, which later on radiates its heat into the living space. At night, the storage mass continues to radiate heat into the living space.

In a *remote thermal storage system* the collector (which can be located away from the building) and the storage mass are physically isolated from the space to be heated. During the day, the air heated in the collector passes into the living space and can also flow to the storage mass. At night, the mass radiates and possibly convects its heat to the living space. If a mechanical part such as a fan or pump is added to force air or water circulation, the system becomes a *hybrid*. A hybrid system is one that uses passive components to collect heat and active components, such as fans, pumps and ducts, to distribute the heat (through forced convection) to the living space. In general, the addition of these active components adds only minimally to system energy requirements.

Solar Collectors

There are four basic types of solar collectors: (1) windows vertically mounted in walls; (2) skylights and clerestories which are windows installed on a roof; (3) passive wall or window box collectors; and (4) attached solar greenhouses.

Windows. Windows allow light to enter a space, but they can also be prodigious losers of heat as a result of air leakage through cracks, conduction through the glass and radiation to the glass from warm objects. The greatest

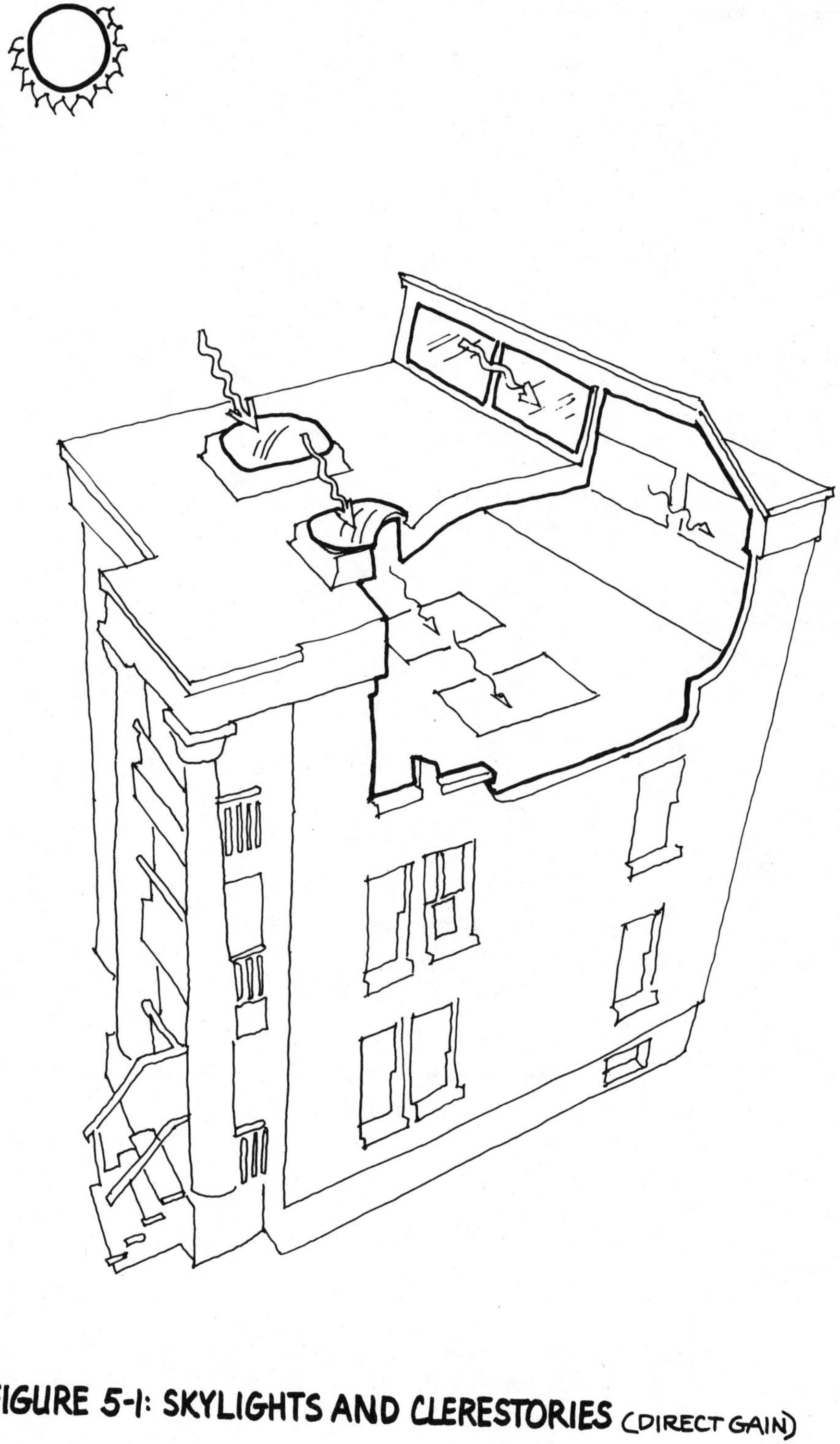

FIGURE 5-1: SKYLIGHTS AND CLERESTORIES (DIRECT GAIN)

heat loss through windows occurs at night through those that face north because they never receive direct sunlight and only lose energy. Passive solar heating seeks to take advantage of sunlight entering through windows, while minimizing heat loss through the windows. Existing windows must be upgraded to prevent air leaks and also covered at night with some form of *movable insulation*. If new windows are installed they should be placed mainly in south-facing walls. You might also consider eliminating some unneeded windows on the north side to further reduce the building's heat load. All windows should be weather-stripped and have two layers of glazing, either double-pane glass or single-pane with a storm window. Triple glazing absorbs and reflects more incoming solar energy than it saves in reduced heat loss. Thus, double glazing should be used on south-facing windows (perhaps with movable insulation) and triple glazing elsewhere (for information on upgrading windows, see the sections on windows and doors, chapter 2).

Skylights and clerestories. Solar energy can also be collected by south-facing skylights or clerestories which are especially useful when the southern walls of a building are shaded, but the roof is not. The best use of these windows requires that they open directly onto the living space to allow deep sunlight penetration. These windows are also useful because they can supply light to dark spaces in the building interior.

A variation on solar windows is the *solar attic*, in which roof shingling is removed and replaced with glass or plastic glazing. Although the light is not admitted directly into the interior of the house, the attic can collect the entering heat, which then can be transferred to the rest of the house, usually with a fan and ductwork.

External wall and window box collectors. These air-heating collectors use metal *absorber plates* placed in boxes behind glass or plastic glazing. The heat they collect is vented into the interior of the building.

Solar greenhouses. An attached solar greenhouse is located on the south side of a building next to the living space to be heated. Such a greenhouse can efficiently collect solar energy and also can provide space for growing food.

Using Windows for Solar Heating

Glazing: How much is needed? Before installing a passive solar heating system it is important to estimate the solar heating potential of existing south-facing windows. One square foot of south-facing glazing can collect as much as 1,300 Btu of solar energy during a sunny, 30°F day. On such a day, a 1,500-square-foot, well-insulated house may require between 200,000 and 300,000 Btu. Not very much glazing is required to supply a significant portion of heat demand, particularly if some precautions, such as installation of movable insulation, are taken.

However, a house with too much glazing and no heat storage will be too hot during the day and too cold at night. In general, a system with only solar glazing and no added thermal mass should have 0.1 to 0.2 square feet of glazing for each square foot of heated floor area (see table 5-1). Direct and indirect gain systems have varying requirements for glazing and thermal mass.

Orientation. Windows facing away from true south can still be useful. An unshaded window oriented 25 degrees from true south will still receive more than 90 percent of the energy received by a south-facing window, while a window facing southwest or southeast will collect about 75 to 80 percent. Beyond 45 degrees from true south, windows rapidly lose their effectiveness as collectors (see table 5-2).

Alteration or addition of windows. If vertical windows must be added, they should be recessed slightly into the adjacent wall in order to reduce infiltration from crosswinds and to provide summer shade, although additional provision for shading and summer venting is usually required. To recess a window, install replacement windows with narrower frames. A new window sash should be wooden, because such sashes generally provide lower heat loss and tighter fit. If the sash is metal, it should contain a nonconducting thermal break to reduce heat loss. South-facing windows should be double-glazed while all others should be triple-glazed, since they are not needed as heat collectors. If skylights and clerestories are added, alterations to the roof structure and the installation of special bracing may be necessary. These alterations can become quite costly, but they may be worth doing as part of a major building rehabilitation.

Table 5-1
Glazing Area Required for Passive Solar Windows

Average Daily Temperature (January)	Ft² of South Glass per Ft² of Heated Floor Area	
	5 Btu/Ft²/DD* (a well-sealed and insulated building)	10 Btu/Ft²/DD* (average building)
20°F (Zone IV)	0.1	0.2
30°F (Zone III)	0.075	0.15
40°F (Zone II)	0.05	0.1

*Building heat loss rates.

Table 5-2
Effect of Window Orientation on Solar Energy Collection

Window Orientation	Degrees from True South	Percentage of Maximum Possible Insolation Falling on Window
True South	0	100
South-Southeast; South-Southwest	22.5	90
Southeast or Southwest	45	75
East-Southeast; West-Southwest	67.5	58
East or West	90	35

Table 5-3
Glazing Materials

Type	Manufacturers	Thickness (ins)	Weight (lbs/ft²)	Strength
Glass (regular, low-iron)	Widely available	0.090-0.125	1.20-1.60	Good to poor
Fiberglass-reinforced polyester	Kalwall,* Vistron Corp., others	0.040	0.25-0.29	Very good
Polyethylene (thin plastic film)	Du Pont, 3M	0.001-0.007	0.02-0.053	Fair to good
Rigid plastics (acrylic)	Swedlow, Du Pont	0.125	0.73-0.77	Very good
Insulating panels				
Acrylic (double wall extrusion)	Rhom & Haas, CYRO	2 layers	0.25-1.00	Good
Glass	ASG†, others	2 layers	4.5	Good

SOURCES: Farallones Institute, *The Integral Urban House* (San Francisco: Sierra Club Books, 1979); P. Temple and J. Kohler, "Glazing Choices." *Solar Age*, April 1979.

*Kalwall Corp. is now Solar Components Corp.

†ASG is now AFG.

Glazing materials. Large expanses of glass may provide an inviting target for vandalism. Solar collectors need not be glass or even transparent. Most other glazing materials will work relatively well, are more resistant to breakage and often cost less. In certain situations, translucent glazing materials are better than glass because they diffuse the incoming light which, in turn, allows for more efficient absorption by heat-storage materials (see table 5-3 for types and costs of glazing materials).

Movable insulation. Even with double or triple glazing and weather stripping, a window will lose large quantities of heat through radiation and convection. In order to minimize such heat losses, windows should be covered at night with insulating shades or shutters. Movable insulation can eliminate the radiation of heat from warm objects and bodies to the cold window surface and can cut heat losses by more than 50 percent (see table 5-4). This type of insulation can triple the R-value of a window and may be made of rigid foam inserts or shutters covered with wood or fabric with an insulating core. A variety of indoor and outdoor blinds and doors operated by ropes and/or pulleys are also available. There is an elegant but costly insulating system called Beadwall, manufactured by Zomeworks Corporation in Albuquerque, New Mexico, which uses a vacuum system to fill a double-

Shortwave/Longwave Transmittance		Thermal Expansion	Cost ($/ft²)	Remarks
0.84 (regular)- 0.90 (low-iron)	0.03	Low	0.99- 2.35	Long lifetime unless broken
0.86- 0.88	0.06- 0.12	Moderate to high	0.60- 1.00	Melts at 300°F
0.89- 0.96	0.43- 0.80	High	0.05- 0.58	Limited lifetime Melts at 150-300°F
0.92- 0.93	0.06	High	0.81- 1.14	Melts at 200-230°F
0.88- 0.93	Low	High	1.25- 2.50	Melts at 230-300°F
0.90	0.03	Low	2.99	Long lifetime unless broken

glazed window with tiny polystyrene balls. Descriptions and typical costs of different types of movable insulation are given in table 5-4.

All movable insulation should be installed with a good seal around the edge. In the absence of such a seal, warm air will be drawn past the top of the insulating device, down the cold window surface and back into the room, resulting in heat loss and, occasionally, noticeably cold drafts. Shades and shutters may be sealed to a window edge with Velcro or magnetic strips such as Nightwall clips by Zomeworks. Also, shades may be installed with restraining tracks. The test of insulation and edge seals is the temperature of the inside glass after a night of use. Good insulation and edge seals will cause the temperature to stabilize close to that of the outside. In addition, a poor (or absent) edge seal may reveal itself by condensation and/or frost on the windowpane. Prior to installing any type of movable insulation, be certain that prime and storm windows fit tightly.

Solar Heat Storage

To provide around-the-clock heating, a solar system requires a means of storing surplus daytime heat for nighttime use. The thermal mass component of the system, which may be masonry, rock, water or phase-change materials, provides such storage. It is necessary to avoid overheating when the sun provides more than 25 to 30 percent of the building heat load on an average day. Thermal mass lessens temperature variations in the living space by absorbing the surplus heat and releasing it when there is no direct solar input.

Table 5-4   Types, Characteristics and Cost Ranges of Movable Insulation

Type	R-Value*	Reduction in Heat Loss (%)	Cost Range	Comments
Curtains & Drapes				
Heavy drapes (to floor)	2	63	. . .	Some good insulating value with valance
Sheer curtains	0	0	. . .	Useless
Thermal curtains	2-4	68	Moderate	Good with valance and restraints
Shades				
Folding quilted shades	2-4	68	Low	Roman do-it-yourself shades
Metallized plastic layers with tracks	10-13	87	Moderate	Must have good edge seals
PVC thermal shades	5	78	Moderate	. . .
Vinyl shades	0.5	46	Low	No good unless sealed at edges to create air space
Shutters				
Isocyanurate (Thermax board with aluminum foil facing ½-1'' thick)	3-8	74	Low	Do-it-yourself; potential fire hazard; also available commercially at higher cost
Plastic shell with urethane core	10-12	87	Moderate	. . .
Polystyrene (¾'' thick)	3	69	Low	Do-it-yourself; potential fire hazard
Wood shutter with polyurethane core	8-9	85	High	. . .
Beadwall (polystyrene pellets)	6-12	83	High	Elegant, complicated and very expensive
Blinds & Doors				
Garage door	Low	48	Moderate	Ordinary garage door, no insulation
Immovable slated blinds	0.5	48	Moderate to high	. . .
Movable slatted blinds	1-1.5	58	Moderate to high	Good for reflecting sunlight, poor insulator
Rolladen (PVC or aluminum with wood core)	1.5	60	Moderate to high	European product; used more for shading and privacy than insulation

SOURCES: William K. Langdon, *Movable Insulation* (Emmaus, Pa: Rodale Press, 1980); manufacturers' literature; issues of *Solar Age;* Lawrence Berkeley Laboratory, *Windows for Energy Efficient Buildings,* no. 2 (Berkeley, Calif., 1980).

*R-value includes double-glazed window and dead air space between glazing and insulation.

The most common storage materials, masonry and water, are not expensive, but installation may be costly if structural alterations to the building are required. The relatively high density of these materials may result in a typical storage component weighing as much as five to eight tons.

Floor areas adjacent to load-bearing walls may have a weight load limit of as much as 1,000 pounds per square foot; elsewhere, the load limit may be as low as 40 to 50 pounds per square foot. Before placing an extra-heavy load on any floor, be sure to consult an expert.

Masonry (brick or concrete). Masonry can be installed either as a floor or a wall. The relatively low heat capacity and large mass of masonry materials mean that a storage system will be quite heavy. Some thermal storage configurations can involve floor loadings in excess of 1,000 pounds per square foot. Use of masonry is advisable only if installed (or already in place) as part of an exterior wall, next to a foundation or a load-bearing wall or a ground-level concrete slab.

Water. Water absorbs heat in a uniform, constant manner and, when placed in a container, temperatures at the surface change more slowly and over a narrower range than is the case with masonry storage. (A narrow range of surface temperatures is desirable because a surface that is very warm will lose more heat back through the glazing.) In addition, a cubic foot of water weighs about half as much as the same volume of bricks but because of its greater heat capacity can store three times as much heat. If water storage is a possibility, it may be a better choice than masonry. Water is inexpensive, but containers may not be. Containers also limit versatility, and

the possibility of leaks or flooding is an ever-present one. Freezing can cause some containers to crack.

Rocks. Rocks have the same general properties as masonry but slightly greater heat capacity. However, because of the large volume of air spaces in a rock storage system, its actual storage capacity is essentially identical to masonry. The two, however, are used for different purposes: masonry for direct heating by sunlight, rock for convective heating by air. Rock storage is generally used either in remote thermal storage systems or when the storage component cannot be near the collector. Rock storage is the common choice for air-heating collector systems because a rock-filled bin is very porous, offers low resistance to airflow and has a lot of surface over which to transfer heat.

Phase-change materials. Phase-change materials (PCM), such as eutectic salts and paraffin, are chemical compounds that melt between 80 and 120°F. The heat of fusion of these materials is great in comparison to the heat capacity of water or masonry; therefore small volumes of PCMs can provide heat storage equivalent to much larger volumes of water or masonry (see table 5-5). Phase-change materials are presently marketed in a variety of formulations under a number of different brand names. They are generally packaged in containers or incorporated into ceiling tiles. Pound for pound, phase-change materials are a good deal more expensive than the more common storage media, although their greater heat storage potential lessens the cost differential. Phase-change materials are not useful, however, in systems whose maximum temperatures do not often exceed the material's melting point. Also, because they have been commercially available for only a short time, the long-term performance of phase-change materials is not guaranteed. In fact, there have been problems with some types of materials failing to solidify after many cycles of heating and cooling, but this has generally happened only in large containers. Another problem is that phase-change materials in large containers may not melt even after a full day's exposure to the sun. Such large containers are more suitable for forced hot air flow, and it may be best to choose storage containers that have a high surface area-to-volume ratio.

Direct-Gain Systems

In a direct-gain system, sunlight passes through the glazing, heats objects and people in the living space and falls upon the storage component. With large areas of glazing these systems can easily overheat rooms if they don't contain sufficient thermal mass to maintain a relatively constant temperature within the living space. The thermal mass is generally made part of the floor or northern wall and is positioned to receive direct or reflected sunlight

Table 5-5
Thermal Storage Materials

Material	Weight (lbs/ft³)	Specific Heat (Btu/lb/°F)	Heat Capacity* (Btu/ft³/°F)	Approximate Storage Costs† ($/100,000 Btu)	Weight (lbs)	Volume (ft³)
Brick	140	0.20	28 (24)	2,000-6,000	35,000	250
Concrete	140	0.23	32	450-900	31,800	230
Phase-change material (calcium chloride hexahydrate)	90-104	Heat of fusion: 45-114 Btu/lb	...	1,000-4,500	1,160	11.4
Rock	170	0.21	36 (31)	450-900	34,000	200
Water	62.4	1.00	62.4	240-400 (including container)	5,000	80 (600 gal)

SOURCES: M. Riordan, "Thermal Storage," *Solar Age*, April 1978; P. F. Kando, "Eutectic Salts," *Solar Age*, April 1978.

NOTE: This table provides information to determine the quantity of heat-storage material required for a passive solar heating system.

*First number assumes no voids; second assumes 15%.
†Materials cost only.

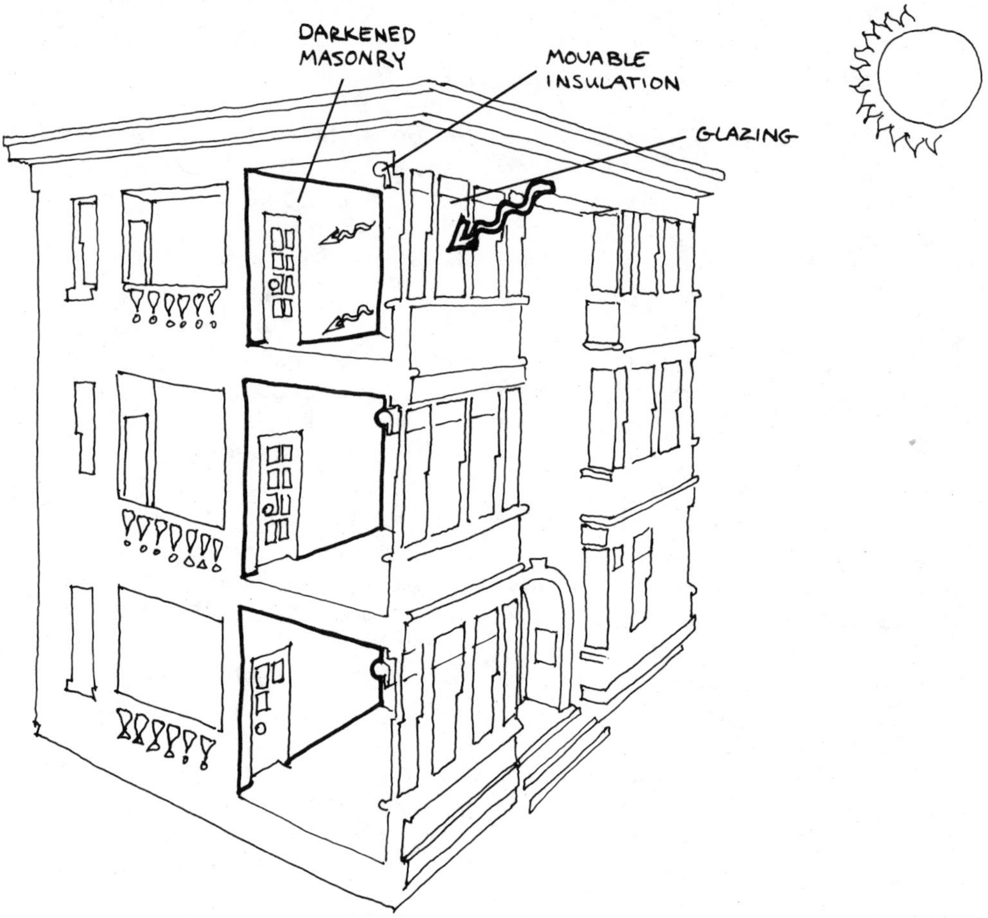

DARKENED
MASONRY

MOVABLE
INSULATION

GLAZING

FIGURE 5-2: MASONRY HEAT STORAGE
(PASSIVE SYSTEM)

for most of the day. As long as the space is warmer than the mass, or the mass is in direct sunlight, heat will tend to flow into the mass. When the space temperature falls below that of the mass, heat radiates from the mass into the living space. In a way, the mass acts like a thermal shock absorber, dampening temperature variations within the living space. Depending upon climate, for a well-insulated house with storage, 0.10 to 0.35 square feet of glazing are required for each square foot of floor area, if the system delivers heat around the clock. In contrast, a direct-gain system supplying only daytime heat requires about 0.1 to 0.2 square feet of glazing for each square foot of living space floor area (see table 5-6).

System efficiency and occupant comfort depend upon how well the glazing, living space and storage are interfaced. A large glazed area collects large quantities of heat, but if the heat storage capacity of the directly exposed mass is too small, the space will overheat rapidly on a sunny day. On the other hand, if the newly installed heat storage capacity is too large, the cost of the system may be higher than is necessary, and the living space temperature may not get as high as desired.

With masonry thermal mass, about 4 cubic feet of exposed masonry, placed along the back or side wall of the room or as an exposed floor, should be used for each square foot of glazing that admits direct sunlight (i.e., 4 square feet of an 8- to 12-inch-thick wall or 8 square feet of a 4- to 6-inch-thick wall). This quantity of masonry will limit indoor temperature fluctuations to about 15°F. Larger amounts will result in smaller fluctuations. The wall should be a dark color, with flat black being the best for heat absorption. If you prefer to have some color, dark blue or red is also acceptable. If a thermal mass is also an exterior wall, it should, of course, be insulated on the outside in order to minimize heat loss in that direction.

For a water wall, temperature fluctuations depend upon water volume and surface area in relation to window area, space heat loss and container surface color. The wall should receive direct sunlight four or more hours daily and should contain at least 1.1 cubic feet of liquid (8 gallons) per square foot of glazing. Opaque containers should be painted a dark color—black, if possible—in order to maximize absorption. If transparent or translucent containers are used, a dark dye should be added to the water.

A major difficulty with retrofitting a large direct-gain system to an existing building can be the need to provide adequate heat storage. Unless the system is placed next to a structural or exterior wall, most floors not on grade will be incapable of supporting a full-size storage system without installation of additional bracing. Phase-change materials incorporated into structural materials (e.g., ceiling tiles and wall panels) are now becoming available, opening the way for more extensive direct-gain retrofit possibilities using lighter weight thermal mass. In row houses and apartments, interior masonry fire walls can be used as an in-place storage component, although it is necessary to strip away interior finishing in order to expose the masonry to sunlight. These walls are sometimes directly exposed to outdoor air in hidden ways, for example, through the roof. Be sure to check out and, if necessary, eliminate this source of heat loss.

If the system is sized to provide a significant fraction of space heating requirements, it will probably be necessary to install a distribution system that can transfer heat to other rooms in the building. This can be accomplished by installing vents or ducts through walls to adjacent rooms and using fans to force airflow. It may be advisable to install both supply and return vents and ducts so that air can be recirculated more readily.

Table 5-6
Glazing Area and Storage Mass Required for Direct-Gain System

Average Daily Temperature (January)	Glazing Area Required per Ft² of Heated Floor Area		Storage Mass Required per Ft² of Glazing	
	5 Btu/Ft²/DD* (ft²)	10 Btu/Ft²/DD* (ft²)	Masonry (lbs or ft³)	Water (lbs or ft³)
20°F (Zone IV)	0.18	0.36	560 lbs 4 ft³	67.1 lbs 1.1 ft³
30°F (Zone III)	0.14	0.28	Same	Same
40°F (Zone II)	0.10	0.20	Same	Same

SOURCE: Edward Mazria, *The Passive Solar Energy Book* (Emmaus, Pa.: Rodale Press, 1979).

*Building heat loss rates.

Indirect-Gain Systems

One way of avoiding the problems of glare, overheating and fading of furnishings associated with a direct-gain system is by placing the heat storage component just behind the glazing and in front of the living space. It may be possible to use a masonry exterior wall as storage by installing a double layer of glazing over it. The indirect-gain system resulting from this arrangement (which has no insulation on the storage wall despite its being an exterior one) should provide adequate radiative heating to the space when it is most required. The thickness of the wall is critically important. A wall that is too thick will cause lower-than-desired living space temperatures and will heat it at the wrong time; one that is too thin will cause overheating.

Nonvented thermal storage wall

A thermal storage wall consists of masonry storage placed 3 to 4 inches behind a glazing. There is no venting through the wall into the living space; all heating of the living space is by radiation from the wall. Such a wall also blocks off incoming sunlight and the view, unless the wall contains windows. During the day, sunlight heats the wall directly. The heat takes some time to pass through the wall (called *thermal lag*) and if it is of proper thickness, heat will radiate from the interior side at night. A disadvantage of this arrangement is that the room is heated only slightly during the day. Daytime heating is provided primarily by leftover heat collected during the previous day.

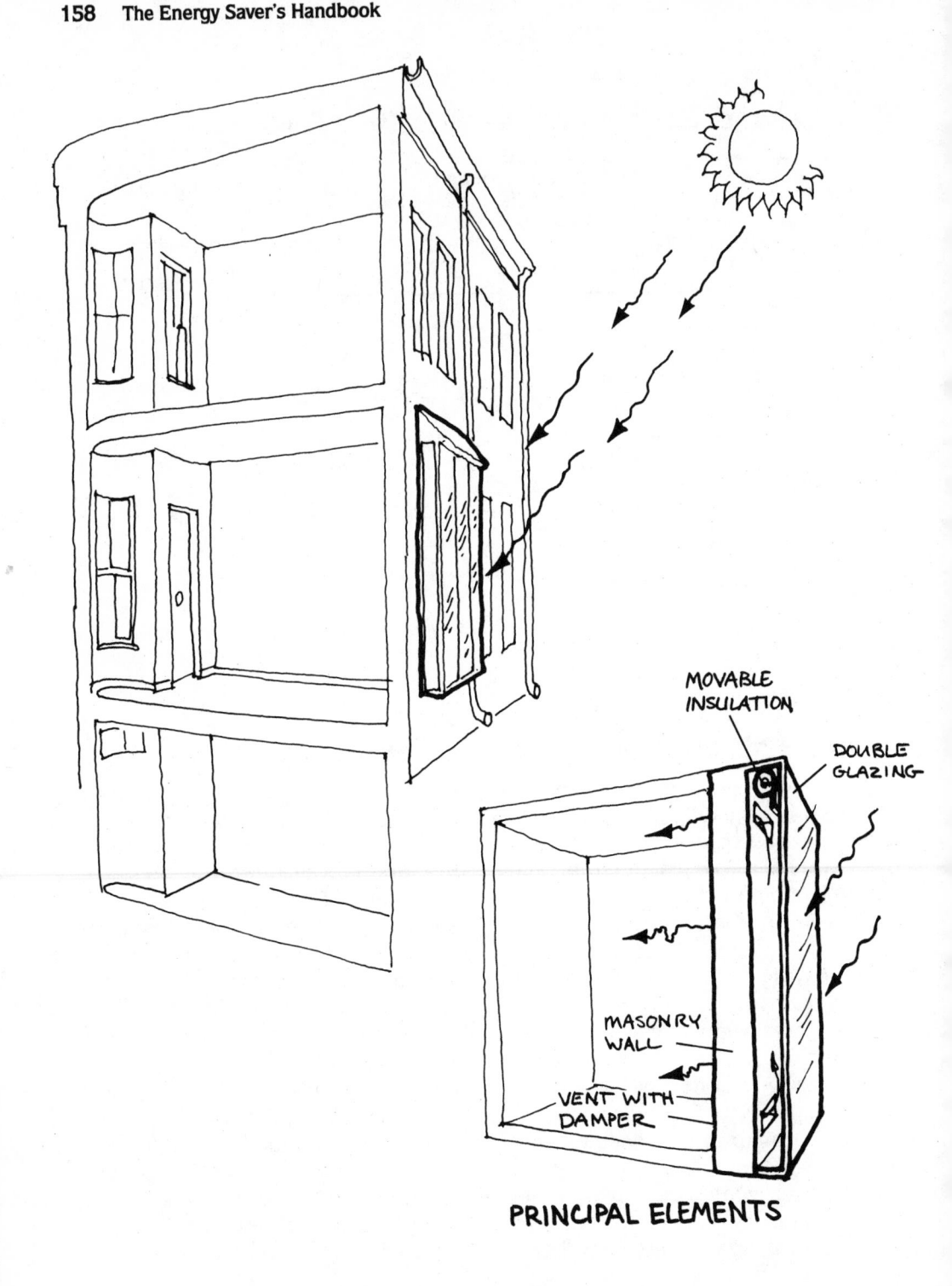

MOVABLE
INSULATION

DOUBLE
GLAZING

MASONRY
WALL

VENT WITH
DAMPER

PRINCIPAL ELEMENTS

FIGURE 5-3: TROMBE WALL (PASSIVE SYSTEM)

The Trombe Wall

A Trombe wall is almost identical to the nonvented thermal storage wall except that it contains vents near the floor and ceiling that allow the living space to be heated both day and night. During the day, the vents are left open. Heated air rises through the air space between the glass and the wall and passes into the living space through the upper vent. Cooler air from the living space is returned to the airspace through the lower vent. Simultaneously, the wall is absorbing solar energy and heat is moving by conduction toward the interior face of the wall in a manner identical to the unvented wall. At night, the vents are closed, and radiative heating from the wall warms the living space. To further ensure that heated air moves only in the direction of the living space, anti-backdraft dampers may be placed across the vents.

The appropriate size and thickness of a Trombe wall depends upon latitude, climate and heat loss, with the latter two factors being the most important. In a moderately cold climate (6,000 degree-days) about 0.4 to 0.8 square feet of double-glazed, south-facing wall are required for each square foot of floor area (see table 5-7). This amount will keep average space temperatures within the range of 65 to 75°F and will supply about 60 to 70 percent of seasonal space heating requirements, although the actual range of temperature fluctuation will be greater. With the use of movable insulation and light-colored or reflecting surfaces to direct sunlight onto the wall, the required area can be reduced by 15 to 30 percent. Slightly oversizing the wall will provide enough stored heat to carry the space through a cloudy day; however, if you oversize the wall

Table 5-7
Glazing Area and Storage Mass Required for Indirect-Gain System

Average Daily Temperature (January)	Glazing Area Required per Ft² of Heated Floor Area*		Storage Mass Required per Ft² of Glazing	
	Masonry (ft²)	Water (ft²)	Masonry (lbs or ft³)	Water (lbs or ft³)
20°F (Zone IV)	0.60-1.20	0.48-0.96	145 lbs / 1 ft³	33.3 lbs / 0.5 ft³
30°F (Zone III)	0.39-0.78	0.32-0.64	175 lbs / 1.25 ft³	40 lbs / 1.30 ft³
40°F (Zone II)	0.23-0.46	0.20-0.40	210 lbs / 1.5 ft³	48 lbs / 1.55 ft³

SOURCE: Edward Mazria, *The Passive Solar Energy Book* (Emmaus, Pa.: Rodale Press, 1979).

*Smaller of the two numbers represents a building heat-loss rate of 5 Btu/ft²/DD; larger number, 10 Btu/ft²/DD.

and the glazing by too much, over-heating during the day and excessive heat losses at night will occur.

It is also important to size the thickness of a Trombe wall properly if a new wall is being built. An 8- to 12-inch-thick concrete wall will have a proper thermal lag of 6 to 10 hours and will maintain the space temperature within a fairly comfortable range. In addition, at least two square feet of venting (1 ft² at the top, 1 ft² at the bottom) are required for each 100 square feet of wall area. The exterior of the wall should be a dark color chosen for high absorptivity (check paint specifications which can be supplied by the manufacturer to ensure this), while the interior can be any color. Finally, if desired or necessary, space temperatures can be more carefully controlled by adding movable insulation to the interior side of the wall. Figure 5-3 shows a Trombe wall.

Water Wall

Water in containers can be used instead of masonry in an indirect gain system. For identical wall size and storage capacity, water is slightly more efficient than masonry because its surface temperature does not go as high for the same amount of heat absorbed, and therefore less heat is lost back through glazing. (In general, the higher the temperature of a storage substance, the more heat it loses through glazing.) On the other hand, nighttime surface temperatures in a water wall are higher than in a masonry wall, thus leading to greater heat loss. Because of convective circulation in the storage tank, water thermal mass can more quickly transfer heat to the living space, thereby reducing thermal lag. Some types of containers

can be arranged to admit daylight into the space.

In a moderately cold climate (6,000 degree-days) each square foot of heated floor area requires about 0.3 to 0.65 square feet of south-facing water wall surface area (see table 5-7). Thickness is less critical, although an 8- to 12-inch wall will limit space temperature variations to about 13 to 18°F. The exterior of the wall should be a dark color; the interior side can be any color. If transparent or translucent containers are used, a dark dye should be added to the water.

One advantage of the indirect-gain system is its adaptability to masonry or brick buildings. A sealed, glazed window wall may be erected over part or all of a south-facing exposure, thereby allowing the brick or masonry wall to be used as the heat storage component. Windows may be left intact to provide daytime heating and lighting. Also, if the window wall extends over two stories, windows may be used as vents and the entire system can function as a Trombe wall. The following caveats should be noted: Any wood or heavy fabric interior wall finish should be removed from the walls being used as storage so as not to block radiation into the room. Some masonry buildings are constructed with three-layer walls. Older buildings have an outer layer of hard-faced brick, an inner layer of soft-faced brick and a middle layer of rubble. This is especially common in row houses. Newer buildings are generally constructed of brick over cinder block with an air gap in the middle. In addition, cinder blocks often have open centers. In both cases, thermal conduction through the wall will be small or nonexistent. Be sure to check wall construction before you proceed!

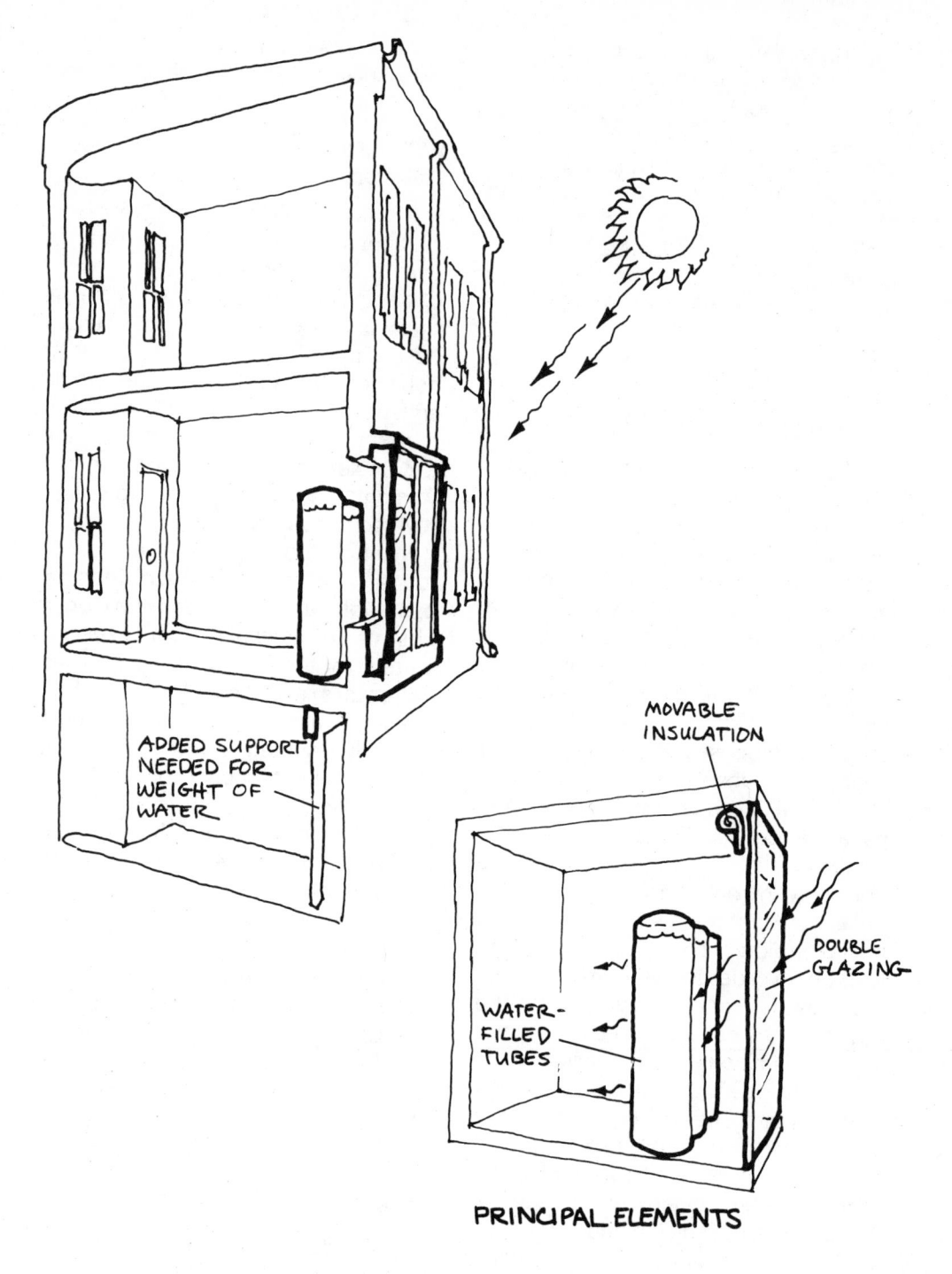

MOVABLE
INSULATION

DOUBLE
GLAZING

WATER-
FILLED
TUBES

ADDED SUPPORT
NEEDED FOR
WEIGHT OF
WATER

PRINCIPAL ELEMENTS

FIGURE 5-4: WATER WALL (PASSIVE SYSTEM)

An indirect-gain system can be installed on a frame building; however, a large masonry or water wall may require structural reinforcement. In that case, replacement of an exterior wall or construction of an adequate foundation may be necessary. Frame walls are usually considered better candidates for air-heating collector systems, which are discussed next.

Air-Heating Collector Systems

An air-heating collector system does not use windows or glazed walls. Instead, collectors that are completely separated from the living space are either hung on a south wall or placed next to the building. These systems depend upon either natural or forced convection to move air through the collector and into the living space. In an air heating system, a flat plate collector composed of glazing mounted over a dark absorber heats the air. If there is a storage component, it is usually a rock-filled bin placed near the living space (for example, under the floor). With rock storage it is usually necessary that the distribution system be fan or blower-powered (forced convection) in order to overcome the rock bin's resistance to airflow. During the day, air in the collector is heated by the sun and is blown into the living space or the storage bin if the space doesn't require heat. The cooler air from the living space or the storage bin is pulled back into the collector. At night, the space can be heated by blowing air through the storage bin and extracting its heat. These different air handling modes are thermostatically controlled by dampers that are placed in the duct-

work connecting the collectors, the living space and the storage component.

Wall and Window Box Collectors

Modular wall collectors consist of a sheet of glass or plastic mounted over a black metal absorber. As the sun shines on the collector, air behind the black absorber is heated, rises and passes through vents at the top of the collector and into the building. Colder air from the living space flows through the vents at the bottom into the collector to complete the natural convection loop. At night, the vents are closed to prevent reverse convective airflow.

Window box collectors are similar to wall collectors in construction and operation but are mounted in windows. They can be portable and are especially suited to tenant ownership. These collectors are limited in available collector area, and the number of properly oriented windows determine the number of collectors that can be installed.

Both styles can be built on site and are easily retrofitted to frame and masonry structures as substitutes for additional direct-gain window area. The only building modification required for wall collectors is the installation of vents. When wall collectors supply more than 25 to 30 percent of a building's heating load, thermal storage may be required in order to save energy and avoid overheating.

The Attached Solar Greenhouse-Sunspace

If a suitably sized, sunny location is available along the southern side of

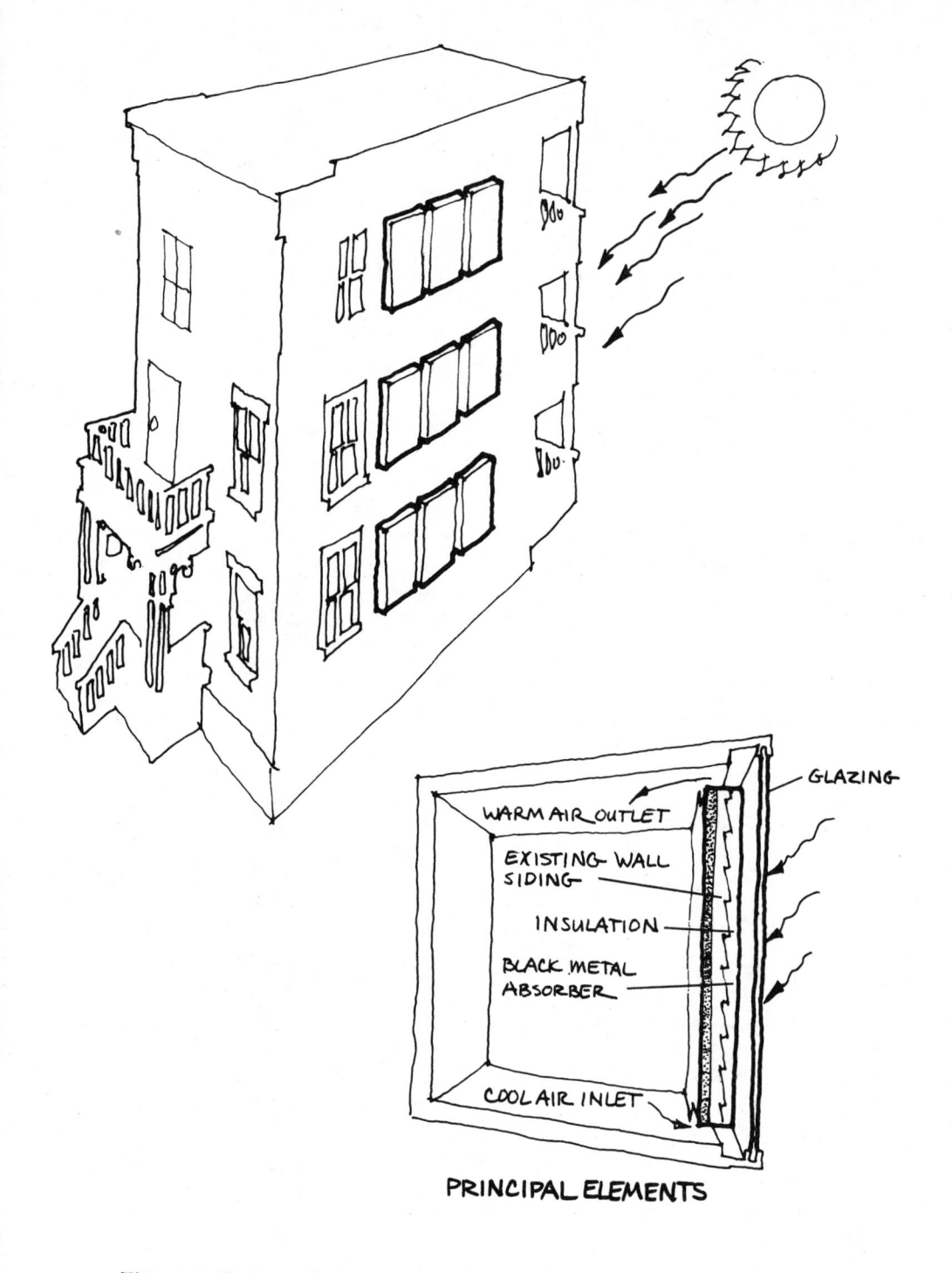

GLAZING

WARM AIR OUTLET

EXISTING WALL SIDING

INSULATION

BLACK METAL ABSORBER

COOL AIR INLET

PRINCIPAL ELEMENTS

FIGURE 5-5: AIR-HEATING COLLECTOR (PASSIVE SYSTEM)

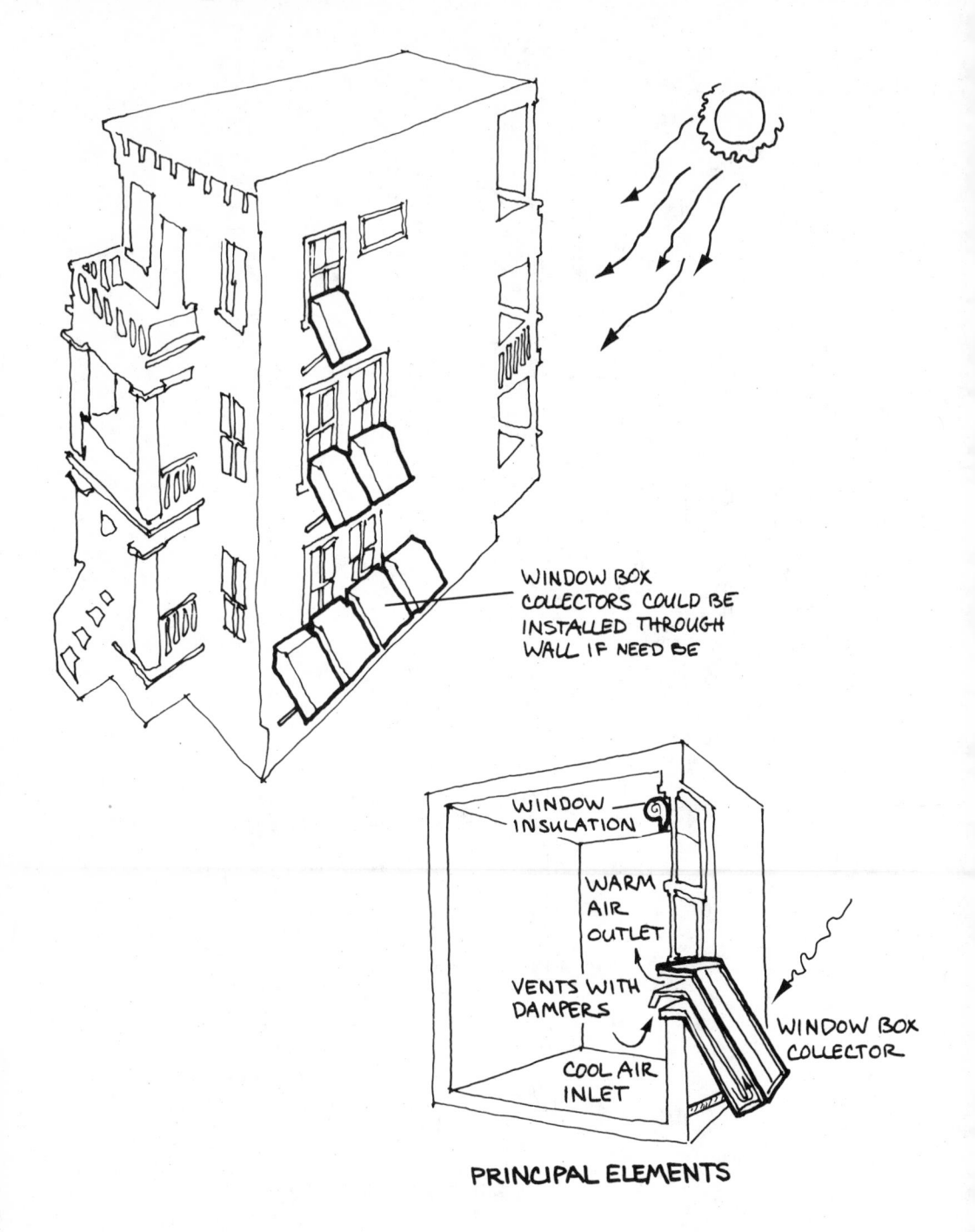

WINDOW BOX
COLLECTORS COULD BE
INSTALLED THROUGH
WALL IF NEED BE

WINDOW
INSULATION

WARM
AIR
OUTLET

VENTS WITH
DAMPERS

COOL AIR
INLET

WINDOW BOX
COLLECTOR

PRINCIPAL ELEMENTS

FIGURE 5-6: WINDOW BOX HEATER (PASSIVE SYSTEM)

your building, you may be interested in building an attached solar greenhouse or sunspace. The heat collected by a greenhouse can supply a significant part of a building's heating load, and the enclosed space can also be used for growing food and as a living space.

On a clear, cold day, a greenhouse will collect much more heat than it loses. Traditional all-glass greenhouses lose enormous quantities of heat through walls that do not face south, so they must be heated in winter. If a greenhouse is placed along a south-facing wall of a building, heat losses through the north wall are greatly reduced, and the greenhouse can become a heat source for the living space.

An attached solar greenhouse or solarium can be as simple as a glassed-in porch or a lean-to greenhouse on the south side of a building, or it can be a very complex, expensive structure. It is also possible to glaze a south-facing porch. Whichever it is, the structure must be thoroughly sealed and double-glazed and should be fitted with movable insulation for use at night. If double glazing is too expensive or difficult to install, a functional second layer may be made by layering sheets of plastic film (such as Monsanto 602) over the frame, leaving an air space between the two layers of glazing. Ordinary polyethylene plastic can be used but it will have to be replaced every year or two. The greenhouse should also have roof or gable vents in order to prevent overheating in the summer, and if it is

Table 5-8
Glazing Area Required for Solar Greenhouse

Average Daily Temperature (January)	Glazing Area Required per Ft² of Heated Floor Area*	
	Masonry (ft²)	Water (ft²)
20°F (Zone IV)	0.95-1.55	0.83-1.31
30°F (Zone III)	0.67-1.06	0.60-0.92
40°F (Zone II)	0.55-0.66	0.42-0.62

SOURCE: Edward Mazria, *The Passive Solar Energy Book* (Emmaus, Pa.: Rodale Press, 1979).

NOTE: This table assumes solar greenhouse is used to provide heat to adjacent living space. Glazing area should provide enough heat to keep both living space and greenhouse at an average temperature of about 65°F. This table also assumes movable insulation is used at night on greenhouse and that the structure is double glazed.

*Smaller number represents a building heat-loss rate of 5 Btu/ft²/DD; larger number, 10 Btu/ft²/DD. Storage mass requirements are the same for the indirect-gain system.

to be used to heat the building, it should have vents, windows or doors in the common wall, opening into the living space.

Ideally, a solar greenhouse should be sized to meet its heating requirements and those of the adjacent space(s). If no heat storage is provided, the greenhouse should have about 0.25 to 0.35 square feet of south-facing, double-glazed surface for each square foot of floor space in the adjacent space(s). If the system includes heat storage, it should be sized as described in table 5-8.

It is usually desirable to provide for heat storage in a greenhouse. Then, if movable insulation is used at night, the greenhouse will not drop below freezing, and the plants will be protected. If the greenhouse is attached to a masonry building, the common wall can be used to provide heat storage for both the greenhouse and the space(s). Heat storage can also be provided by containers or drums of water, a rock bin installed beneath plant beds or a concrete, brick or gravel floor. In placing the thermal mass take care to ensure that it is located below the level of plant beds and in direct sunlight, if possible, so that heat will also be provided to the area.

More sophisticated designs can increase the flexibility of an attached solar greenhouse. If, for example, the greenhouse is two stories tall, a convection loop can be set up from the greenhouse into the upper story and back out of the lower one. At the cost of installing some ducting and blowers, a greenhouse can even be built on the building roof, supplying heat to the building and perhaps to a storage bin located in the basement. Figures 5-7, 5-8 and 5-9 show some possible solar greenhouse configurations.

Fitting Passive Solar Heating Systems to Building Prototypes

Each arrangement of glazing, space and mass has advantages and disadvantages that affect its suitability for retrofit to particular buildings. In new construction, you can design the building to fit the system. In retrofit, however, the passive system must be designed to fit the building. Some of the systems described here are applicable to all four building types, while others are not. Table 5-9 offers a brief overview of which systems are most appropriate to the four building prototypes.

[*Continued on page 174*]

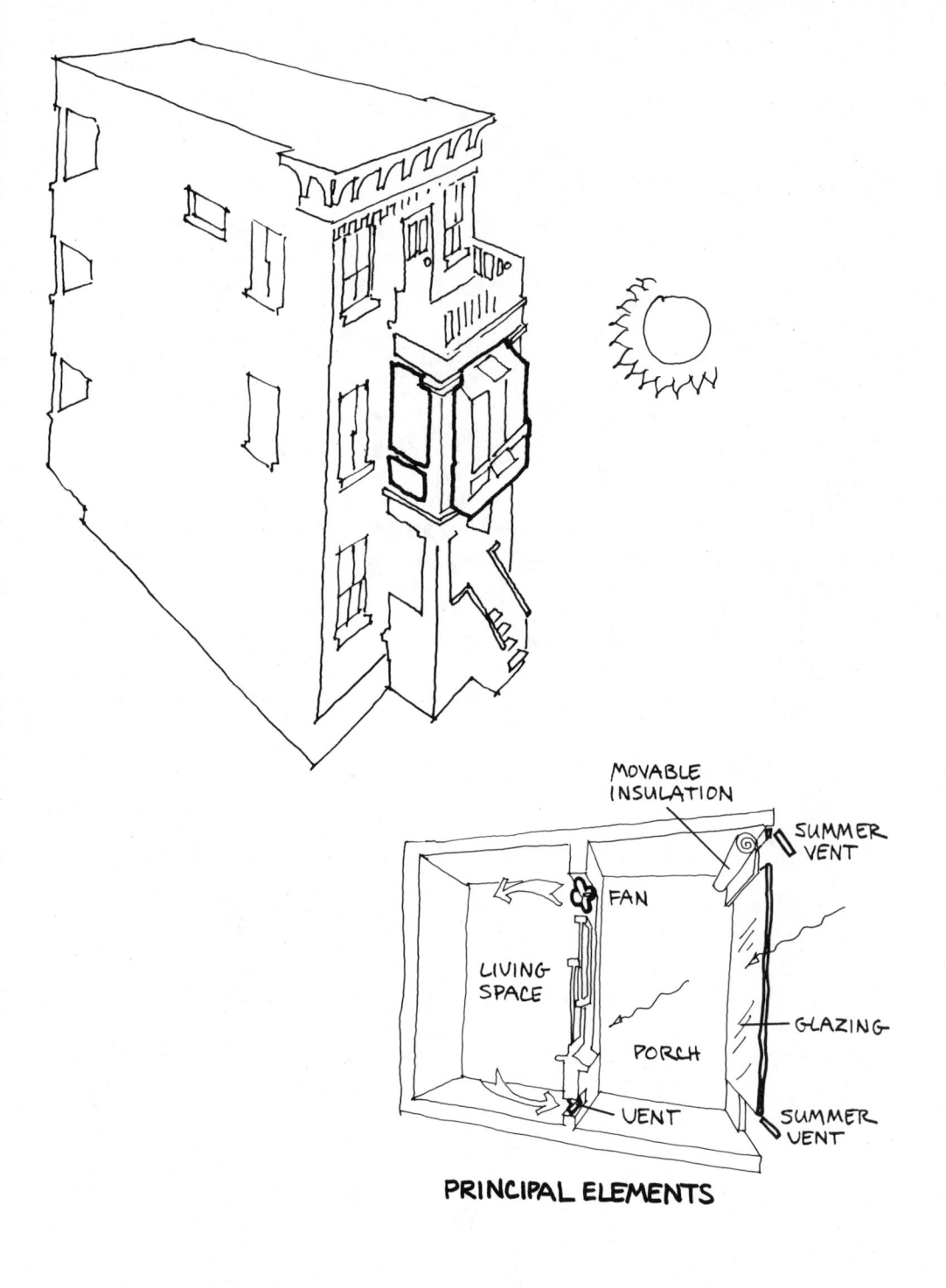

MOVABLE
INSULATION

SUMMER
VENT

FAN

LIVING
SPACE

GLAZING

PORCH

SUMMER
VENT

VENT

PRINCIPAL ELEMENTS

FIGURE 5-7: SOLAR PORCH OR BALCONY

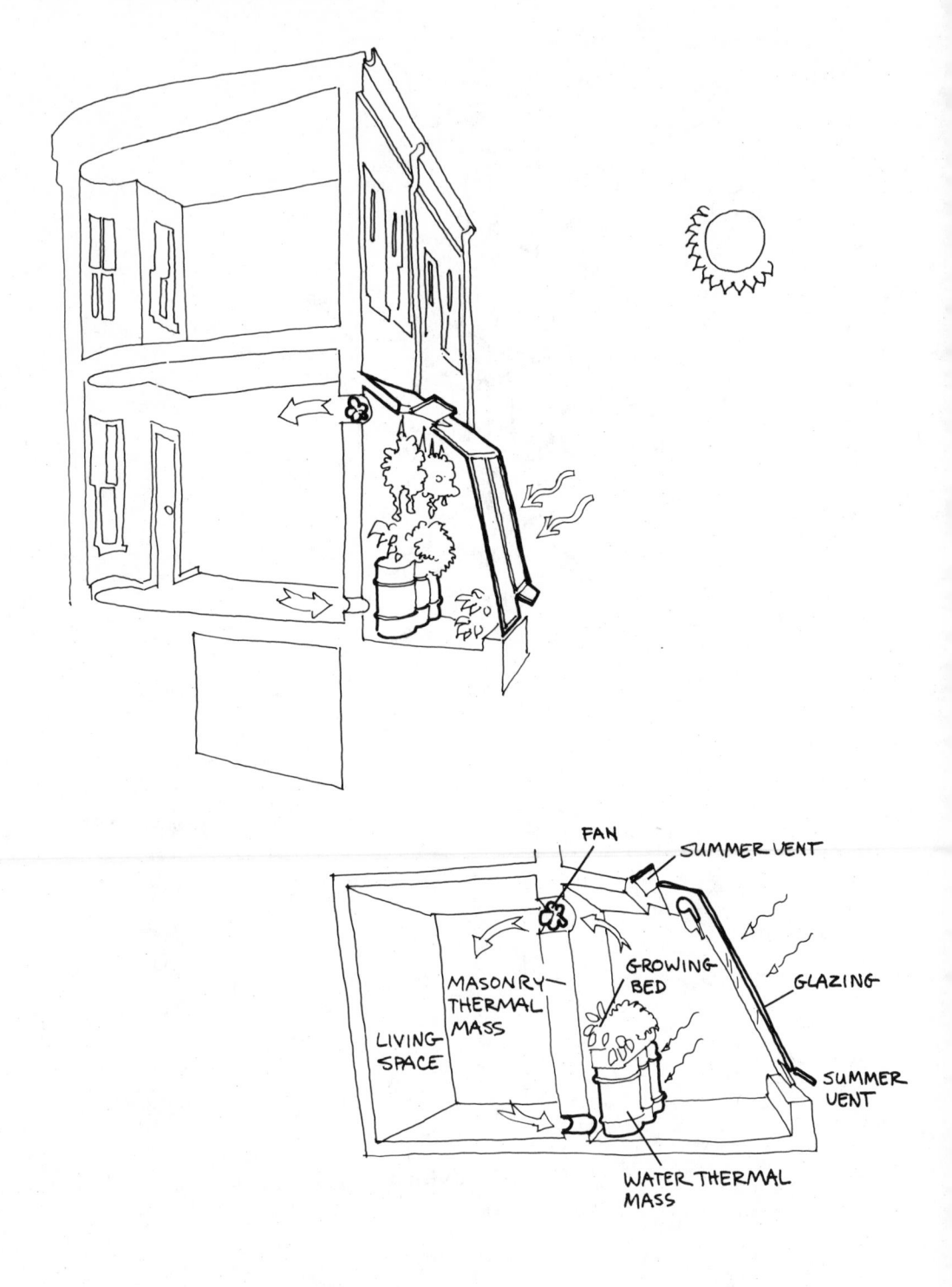

FAN

SUMMER VENT

MASONRY THERMAL MASS

GROWING BED

GLAZING

LIVING SPACE

SUMMER VENT

WATER THERMAL MASS

FIGURE 5-8: ATTACHED SOLAR GREENHOUSE

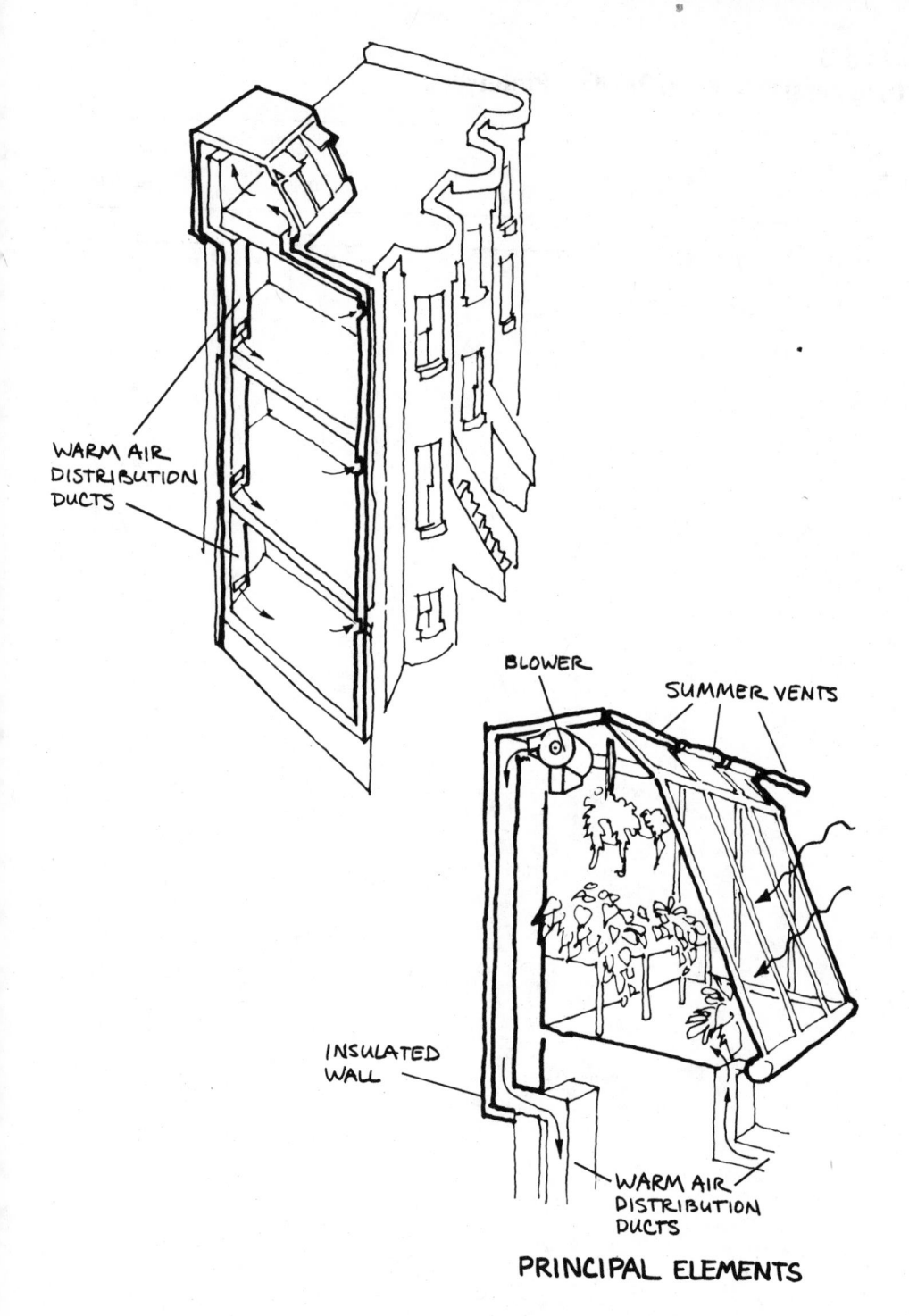

WARM AIR
DISTRIBUTION
DUCTS

BLOWER

SUMMER VENTS

INSULATED
WALL

WARM AIR
DISTRIBUTION
DUCTS

PRINCIPAL ELEMENTS

FIGURE 5-9: ROOFTOP SOLAR GREENHOUSE

Table 5-9
Fitting Passive Solar to Building Prototypes

System	Schematic of Solar System	Building Type			
		Row House	Triplex	Single-Family	Apartment
Roof					
Clerestory		Y	Y	M	Y
Dormer		D	D	Y	M
Skylight		Y	Y	Y	Y
Solar attic		M	M	Y	D

NOTE: Y=yes; M=maybe; D=doubtful; G=ground level; SP=possible structural problems.

Table 5-9—*Continued*

System	Schematic of Solar System	Building Type			
		Row House	Triplex	Single-Family	Apartment
Greenhouse/Solarium					
Ground		Y	Y	Y	Y
Porch		Y	Y	Y	Y
Rooftop		Y	M, SP	D	Y, SP
Direct Gain					
Masonry storage		Y	Y, G, SP	Y, G, SP	Y, SP
Rock bed storage		Y, SP	Y, G, SP	Y, G, SP	Y, SP

[Continued on next page]

Table 5-9—*Continued*

System	Schematic of Solar System	Building Type			
		Row House	Triplex	Single-Family	Apartment
Direct Gain—*Continued*					
Solar window	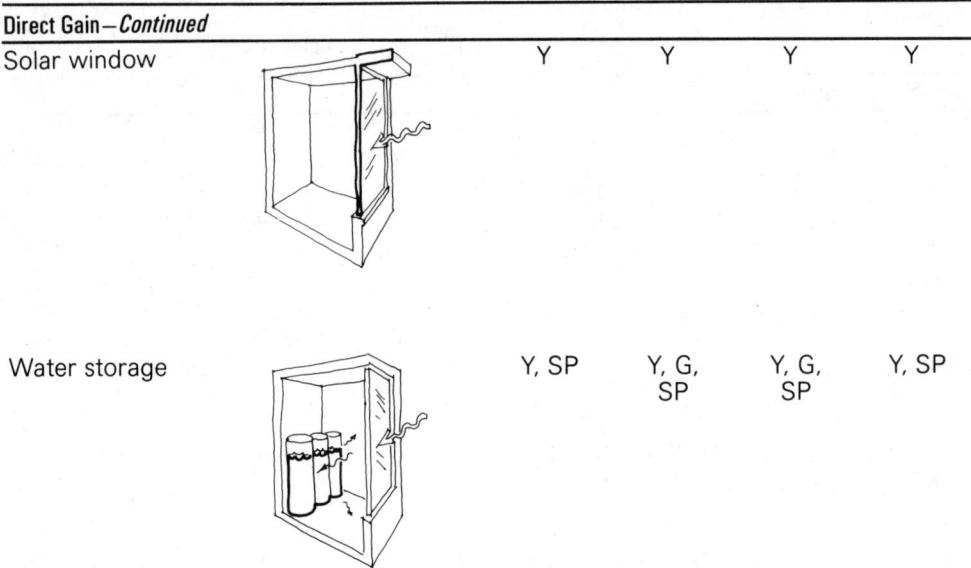	Y	Y	Y	Y
Water storage		Y, SP	Y, G, SP	Y, G, SP	Y, SP
Indirect Gain					
Glazed exterior		Y	D	D	M
Thermal wall		M, SP	M, SP	M, SP	M, SP

NOTE: Y=yes; M=maybe; D=doubtful; G=ground level; SP=possible structural problems.

Table 5-9 — *Continued*

System	Schematic of Solar System	Building Type			
		Row House	Triplex	Single-Family	Apartment
Indirect Gain — *Continued*					
Trombe wall		Y, SP	M, SP G	M, SP G	Y, SP
Water wall		Y, SP	M, G	M, G	Y, SP
Air-Heating Collector					
Large collector, with remote storage		M, SP	M, SP	M, SP	M, SP
Modular wall collector		Y	Y	Y	Y
Window box collector		Y	Y	Y	Y

Passive Solar Economics

Passive systems have a wide range of paybacks. Table 5-10 shows estimates of the range of paybacks for most of the systems discussed in this section. The length of payback is to a large degree dependent upon materials and labor costs and, of course, the cost of the fuel that solar energy is replacing.

Table 5-10
Passive System Cost Estimates and Simple Paybacks

System	Estimated Cost per Ft^2 ($)	Range of Simple Paybacks (yrs)*		
		Heating Oil @ $1.25/gal	Natural Gas @ $0.50/therm	Elec. @ $0.07/KWH
Greenhouse	10-20	6.2-12.3	11.1-22.2	4.9-9.8
Solar attic	5-10	3.1-6.2	5.5-11.1	2.4-4.9
Solar porch	0.50-5	0.3-3.1	0.6-5.5	0.2-2.4
Solar skylight	10-15	6.2-9.3	11.1-16.6	4.9-7.3
Solar wall	3-5	1.9-3.1	3.3-5.5	1.5-2.4
Wall collector	4-10	2.4-6.2	4.4-11.1	2.0-4.9
Window box	4-5	2.4-3.1	4.4-5.5	2.0-2.4
Direct gain Concrete thermal mass	7-15	4.3-9.3	7.5-16.6	3.4-7.3
Water thermal mass	5-10	3.9-6.2	5.5-11.1	2.4-4.9
Indirect gain Concrete or water thermal mass	12-15	7.4-9.3	13.3-16.6	5.8-7.3

*Simple payback assumes the seasonal heating value of 1 square foot of south glazing to be roughly 100,000 Btu. A dollar value can be calculated to show how much money 1 square foot of glazing saves per heating season, depending on the cost of the fuel it replaces, and the average efficiency at which that fuel is used:

$$\text{For } \$1.25/\text{gal oil: } \frac{100,000 \text{ Btu} \times \$1.25/\text{gal.}}{0.55 \times 141,000 \text{ Btu/gal}} = \$1.61 \text{ (at 55\% efficiency)}$$

$$\text{For } \$0.50/\text{therm gas: } \frac{100,000 \text{ Btu} \times \$0.50/\text{therm}}{0.55 \times 100,000 \text{ Btu/therm}} = \$0.98 \text{ (at 55\% efficiency)}$$

$$\text{For } \$0.07/\text{KWH electricity: } \frac{100,000 \text{ Btu} \times \$0.07/\text{KWH}}{1.00 \times 3,413 \text{ Btu/KWH}} = \$2.05 \text{ (at 100\% efficiency)}$$

Living with Passive Solar Heating

Unlike conventional heating systems, passive solar systems alone don't afford the owner quite as close control over temperature fluctuations within the living space. But these fluctuations can be smoothed by using your back-up heat whenever there isn't enough solar energy available to maintain comfort. Proper design is of utmost importance in getting the most from your solar system. In that vein it's important to consider the whole building in your energy improvement plans. Weak points in the building envelope will naturally reduce your solar benefits. It may also be necessary to adjust your living habits a little in order to minimize your need for back-up heating. Temperature is not the only parameter determining comfort. Others are humidity, drafts, activity and dress. Comfort can be maintained at temperatures lower than you are accustomed to if attention is paid to the following matters:

Eliminate drafts. Air circulation during the day and at night, especially across uncovered glass, can make a room seem colder than it really is and can also rob the room of heat. Elimination of drafts by caulking, weatherstripping and use of window valances can make the space more comfortable.

Maintain moderately high humidity. The body cools itself by perspiration and a dry atmosphere will cause this perspiration to evaporate rapidly, leaving a feeling of coolness. Humid air prevents rapid evaporation and also retains more heat than dry air. By keeping the relative humidity of your home at a moderately high level, a lower air temperature will be acceptably comfortable. Humidity can be increased by using a humidifier, by having many plants around the house to give off moisture or by placing pans of water on radiators.

Keep cold surfaces covered. Because warm bodies radiate heat to cold windows, covering windows with movable insulation will reduce heat loss from the room and also increase the comfort level of the room.

Dress appropriately. If you are willing to dress a little more warmly than you are accustomed to doing, you can remain comfortable at lower-than-normal temperatures with little or no discomfort.

Natural Cooling

Some passive heating systems can be used to help with summertime cooling. But whatever the system, it is important that it be designed so as not to add heat to the building outside the heating season. If you feel that summertime cooling is important or essential, you may wish to choose a solar heating system that can also provide some cooling.

The simplest way of keeping heat out is by using shade. Any solar collector that provides direct heating to a space should have some provision for keeping out summer sun, such as an awning, an overhang at the top of the collector or a movable shade that covers the exterior of the collector. It is very important that shading be outside the collector. Otherwise, sunlight will pass through the glazing and be trapped inside. Very effective shading can also be provided by strategically placed deciduous trees. In the winter, the bare branches will cause only minimal blockage of the sun, but in the summer, direct sunlight will be almost completely eliminated. Even ordinary window shades will provide some relief; they will be most effective if they have a reflective or light-colored outer surface.

Insulation can help greatly in reducing the need for cooling. Attic insulation, in particular, isolates the warm attic from the rest of the house and can keep the upper floor(s) of a building significantly cooler. Wall insulation reduces heat conduction through the walls. Weather-stripping, caulking and adding storm windows will slow down the loss of cool air to the outside if you have air conditioning. If supplied with appropriate venting, a number of passive heating systems can help to cool a home. In fact, even ordinary windows can be used to provide some degree of nighttime cooling. This is done by opening lower sashes on the side of the house facing the prevailing winds and upper sashes on the downwind side. Hot air will then be forced out of the house. Clerestories and skylights can be equipped with vents to allow hot air to be exhausted to the outside. Again, opening windows facing the prevailing breeze will assist in circulation. With appropriately placed inlet and outlet vents, a Trombe wall can also provide some daytime cooling. The inlet vent should be placed near floor level on the north side of the building and the outlet at the top of the glazed window wall. The bottom vent through the wall should also be kept open. As air is heated in the air space, it will rise through the outlet vent. Air from the building will then be drawn into the air space, and cooler air from the outside will flow in through the inlet vent to replace the exiting air.

A hybrid, rock-storage system can also be used to provide daytime cooling. Cooler night air can be blown through the bin to cool the rocks. By day this stored coolth is released to the house through the duct system. At night, heat collected in the bin is once again expelled by cool night air.

ACTIVE SOLAR HEATING

An active solar heating system is differentiated from a passive one by the presence of mechanical devices such as pumps or fans that move solar heat from the point of collection either to storage or to the living space. Having this separate circulation system means that both the storage and the living space can be remotely located from the collector system. (In this regard, an active solar system can most be likened to an isolated-gain passive solar system.) This greatly increases the variety of available design options, and retrofit to existing buildings may in some cases be easier than retrofitting some passive systems. Active systems, however, are generally more expensive than their passive counterparts. The major points to keep in mind when deciding between an active and a passive system are the cost-benefit factor and the degrees of difficulty in installation and maintenance. Active systems generally have higher maintenance costs.

The key component of an active solar system is the collector—a glazed structure usually mounted on the roof of a building. *Collector fluid* (air, water, an antifreeze mixture or a refrigerant) flows through the collector. As the fluid passes through the collector, it absorbs heat. The fluid is then pumped through pipes, in the case of water, antifreeze or refrigerant, or through ducts, in the case of air. Warm air can be circulated directly into the building's air distribution system or to storage, generally a rock bin or a phase-change salt chamber. Water is normally pumped

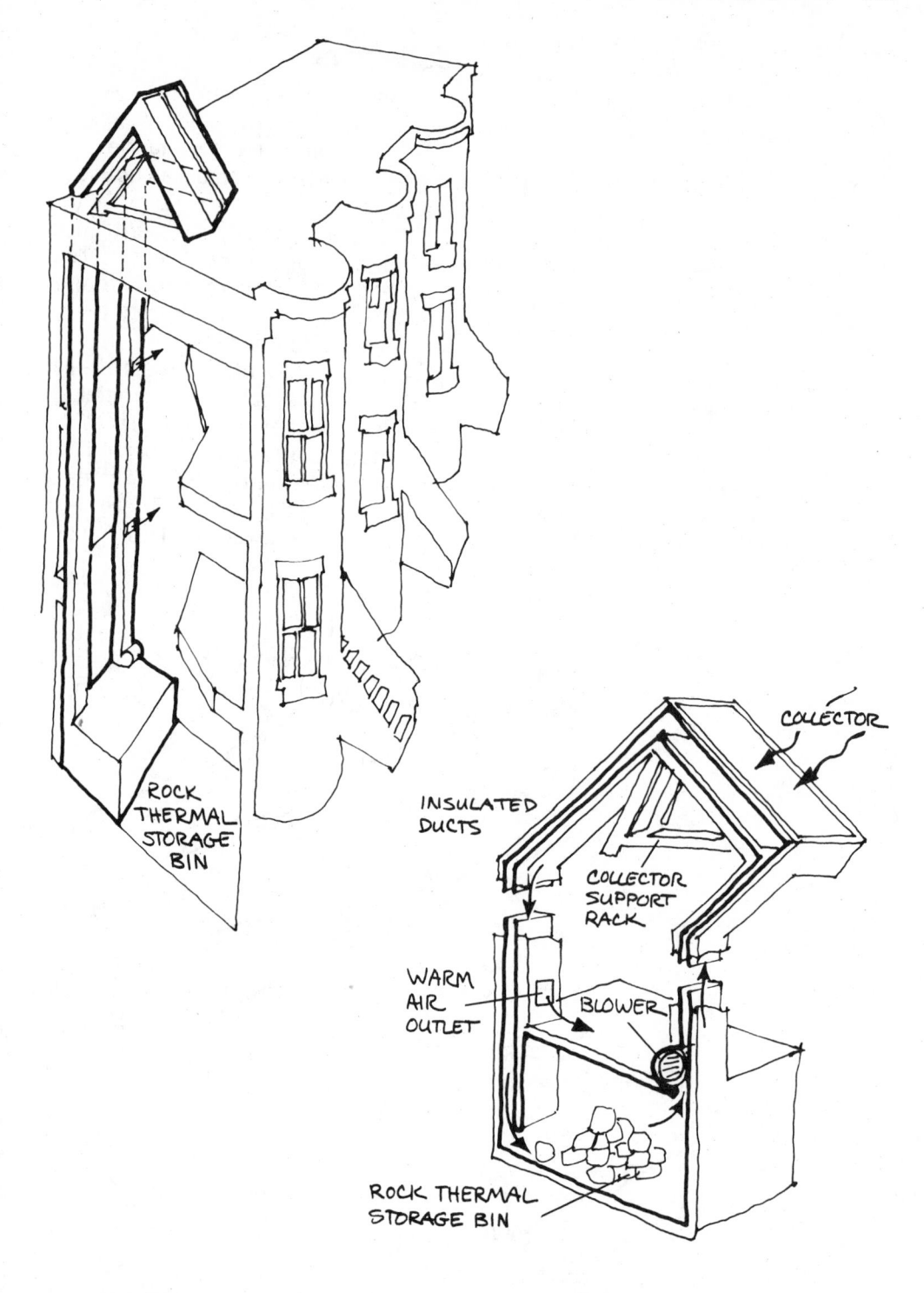

ROCK THERMAL STORAGE BIN

COLLECTOR

INSULATED DUCTS

COLLECTOR SUPPORT RACK

WARM AIR OUTLET

BLOWER

ROCK THERMAL STORAGE BIN

FIGURE 5-10: AIR-HEATING COLLECTOR (ACTIVE SYSTEM)

directly into a water-filled storage tank. The hot water distribution system draws its water from this same tank. Refrigerants and antifreeze mixtures do not mix directly with the water in the storage tank but are instead piped through a heat exchanger immersed in the tank. A heat exchanger consists of loops of pipe that allow heat to be transferred from the collector fluid to the storage tank, while preventing mixing of the two liquids. This is necessary in order to prevent contamination of the storage tank water, which may also be used for domestic hot water purposes. Use of a heat exchanger also allows the necessary quantity of antifreeze mixture or refrigerant (generally quite expensive) to be kept to a minimum.

Collectors

Active solar systems come with a number of different collector designs. The most common type, and the most likely to be used on an urban house, is the glass-covered flat plate collector. Other types include concentrating trough and tube-over-reflector collectors, and parabolic collectors. The latter is quite expensive and complex to use and is not discussed here.

A *flat plate collector* consists of a flat black absorber plate in a metal or wooden box that transfers heat to the collector fluid. Insulation behind the absorber plate and around the sides of the box reduces heat loss. There is the glazing, which can be flat glass, translu-

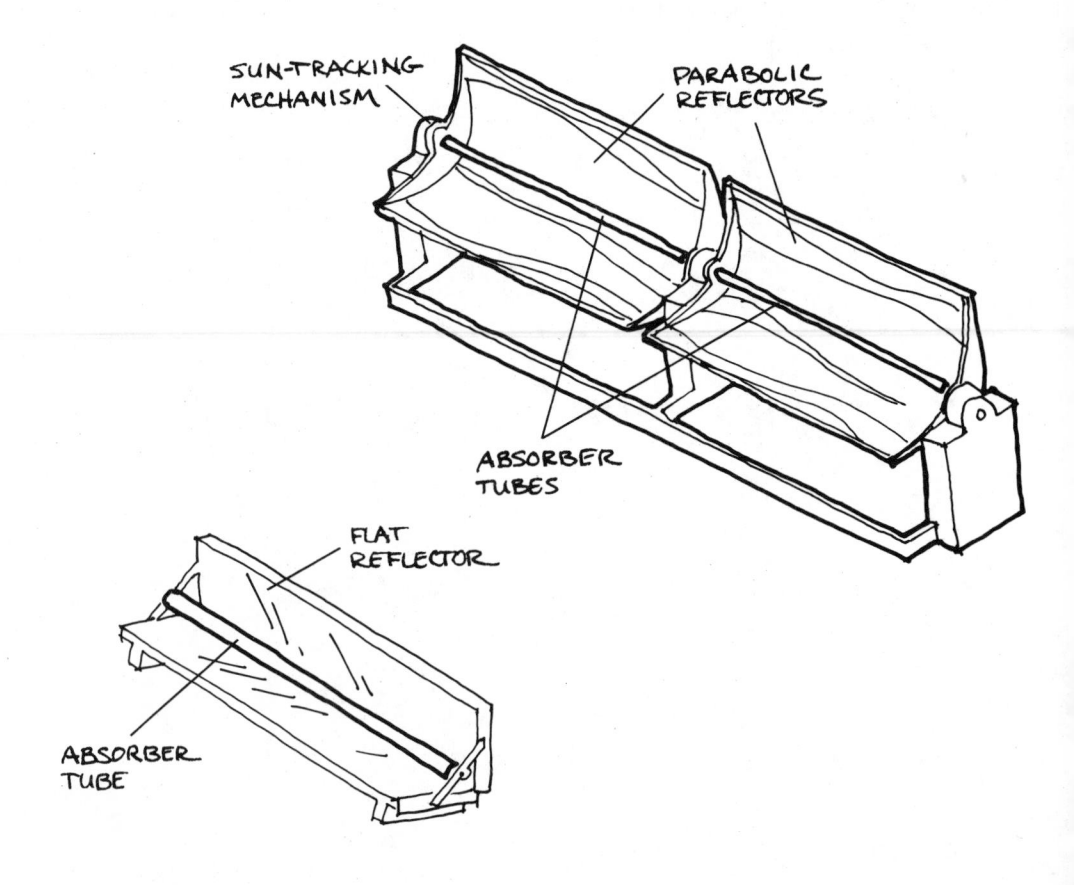

SUN-TRACKING MECHANISM

PARABOLIC REFLECTORS

ABSORBER TUBES

FLAT REFLECTOR

ABSORBER TUBE

cent fiberglass or clear plastic, on the sun side of the collector, 1 to 2 inches above the absorber. If the collector fluid is a liquid, there are runs of black pipe attached to the metal absorber. If the fluid is air, the absorber is often textured, roughened or corrugated in order to increase the area over which the air can pick up heat. The absorber can also be a blackened screen material through which the air can flow or an extruded EPDM rubber sheet with parallel water tubes running through it, known as an *absorber mat*.

It may be desirable to concentrate sunlight onto an absorber in order to reach higher temperatures and to decrease collector area. Most *concentrating collectors*, however, are very expensive and are therefore probably not useful except for large-scale commercial and industrial applications. Also, unlike flat plate collectors which can absorb diffuse radiation, concentrators work only with direct sunlight. Of the many possible designs, two appear to have outstanding advantages in terms of relative efficiency and simplicity:

The *tube-over-reflector collector* consists of a network of tubes running in a north-south direction and lying horizontally over a scallop-shaped, concentrating reflector surface. Each tube absorber is enclosed in a glass pipe. This system can pick up sunlight even when the sun is very low, and is therefore useful throughout the day and the year.

The *trough collector* consists of two mirrored sheets of glass set at right angles to make a trough. These sheets reflect sunlight onto a blackened pipe set about two pipe diameters from the sides of the trough. The absorber can be enclosed in glass pipe, or the entire trough can be covered with a flat glass plate.

Liquid Systems

Liquid solar systems can be divided into several subtypes that can be used for both space and domestic water heating systems. In a *closed-loop system* the collector-to-storage loop is pressurized by a pump and isolated from the heat distribution part of the system. This type of system is always full of liquid, and in climates where freezing weather occurs, it must use an antifreeze mixture. Otherwise, the liquid in the outdoor parts of the system would freeze and burst the pipes. The collector fluid in this case is passed through a double-walled heat exchanger in the hot water storage tank in order to prevent mixing between the potable water and the antifreeze.

In a *drain-down system*, the collector loop is pressurized by the municipal water supply line pressure, but this system drains out as freezing temperatures approach. This is done on a signal from a thermostat that causes a motorized valve to simultaneously close off the line pressure and open a drain line, allowing all the water outside the building envelope to drain away either into a storage tank or a drain. In this system no heat exchanger is needed as potable water is circulated through the collector and distribution system.

In a *drain-back system*, the collector loop is filled by the pump, as it is in the closed-loop system. Therefore, when the pump shuts off, the fluid automatically drains back into an unpressurized storage tank. Sometimes a pressurized heat exchanger collects heat from this tank and delivers it to the living space.

A *refrigerant fluid* involves liquid and vapor circulation in an active (pumped) or passive (thermosiphon, natural convection) system. The collector fluid is a refrigerant (typically Freon) that is

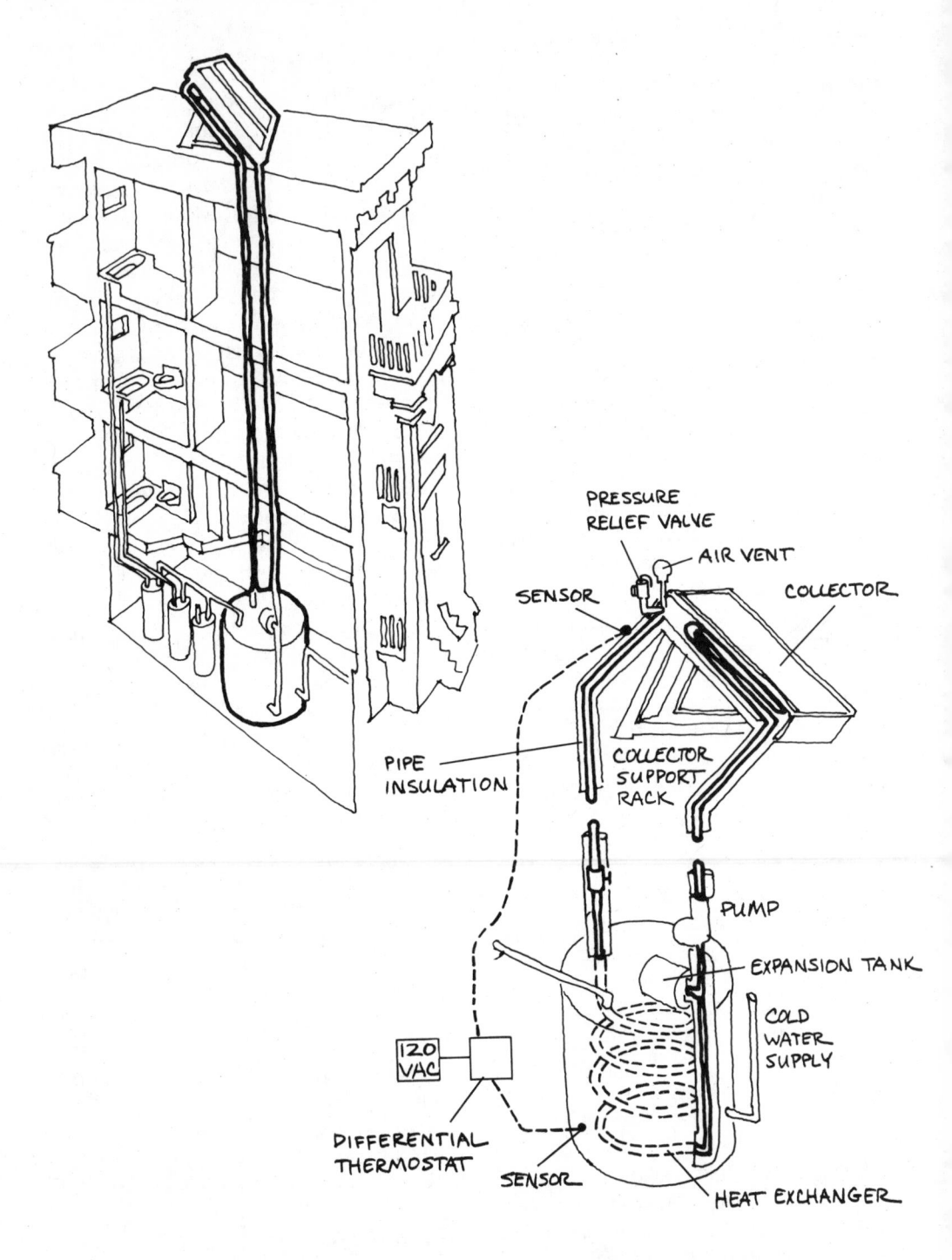

FIGURE 5-12: WATER-HEATING COLLECTORS
(ACTIVE SYSTEM FOR DOMESTIC WATER-HEATING)

easily vaporized at low temperatures. In order to lower the boiling point of the collector fluid, it is kept at a low pressure. Thus, even when exposed to moderate temperatures, such as might occur on cloudy days, the fluid will vaporize. Once vaporized, the fluid is carried through a heat exchanger immersed in a pressurized storage tank, where it condenses and transfers its heat to the water.

Selecting an Active System

In choosing an active solar system, there are a number of considerations including the following:

- Deciding for what use you want solar heat (space heating, domestic hot water or both);
- Determining whether your site is suitable and what measures must be taken to ensure proper system operation;
- Choosing the appropriate type and size of the system to suit your needs and to be compatible with climate conditions;
- Deciding on the appropriate type of back-up system and how it should be integrated with the solar system;
- Choosing whether to build and install the system yourself or to hire out all or part of the job;
- Sizing the system and storage properly;
- Ensuring the reliability of your installer, if you use one, and getting the best possible warranty on the system.

Site Suitability

Assessing your site is critical to selecting the appropriate system and installing it properly. In addition to identifying potential obstacles to solar access such as trees and other buildings, you must also look at building orientation, roof pitch angle (if any) and the need for collector racks and lightning protection.

Because active solar systems are essentially modular in nature, they can be installed in any suitable south-facing location. Orientation is clearly important in this regard, but so is angle of installation. If collectors are to be mounted on a flat roof, they can be rack-mounted and oriented to true south. If the system must be installed on a pitched roof, the roof plane should not face more than about 30 degrees from true south. Beyond this angle, collector efficiency decreases significantly during the coldest months. In such a case, it might be possible to tilt collectors up from the roof to bring them closer to a true south orientation. If collectors are to be rack-mounted in rows, be sure that one row won't shade another.

Optimum *collector tilt angle* depends upon your latitude. For space heating only, collectors should be mounted at an angle from the horizontal equal to your latitude plus 5 to 15 degrees (see figure 5-13). For systems that also supply domestic hot water, the mounting angle should equal your latitude.

If it doesn't seem feasible to mount collectors at this angle, you can either accept decreased collector efficiency or you can increase the collector area. If you roof-mount your collectors, it may be necessary to install additional bracing in the roof in order to provide structural support. Also, if your roof does not have lightning protection, be sure to include lightning rods near your system.

Deciding on System Characteristics

Collector type and fluid. The simplest system to build and operate is an air-heating collector system. Air systems never freeze or develop water leaks that can damage building structure.

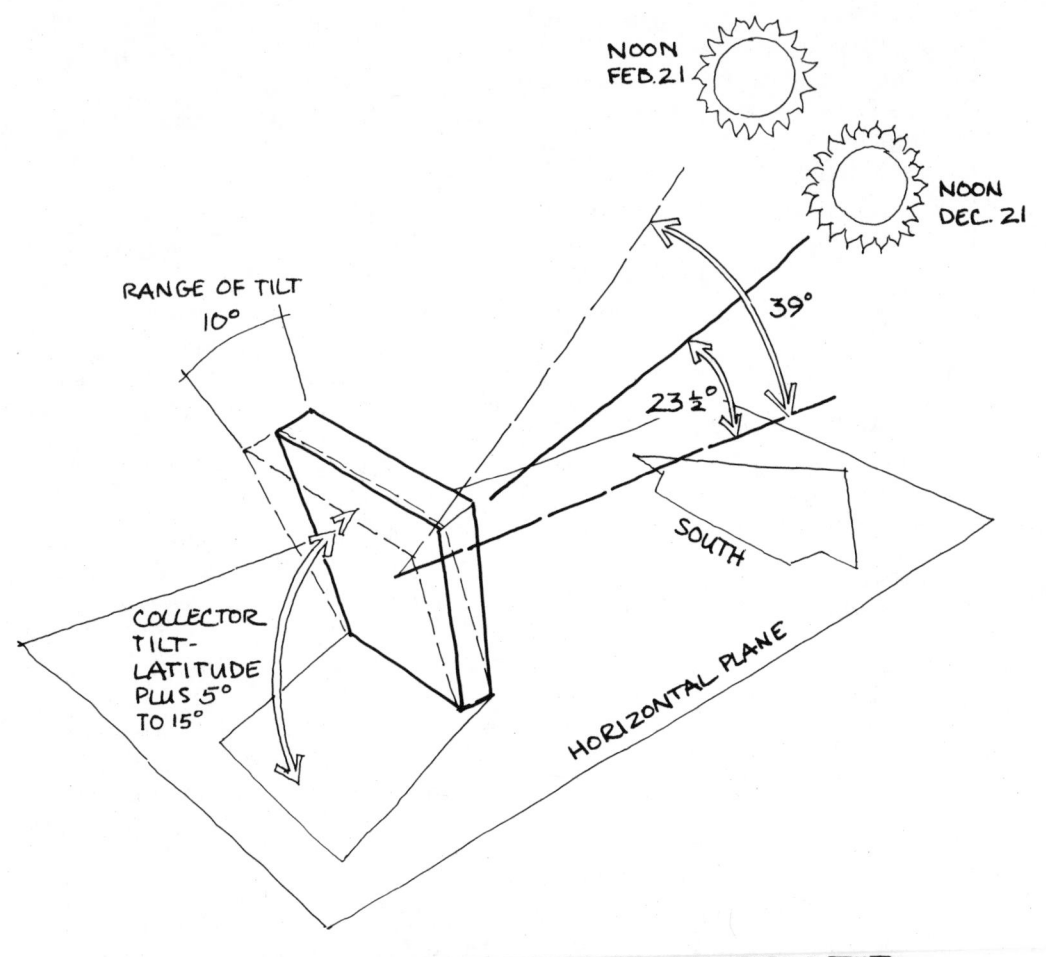

FIGURE 5-13: OBTAINING PROPER COLLECTOR TILT
SUN ANGLES ARE FOR 42° NORTH LATITUDE (BOSTON, CHICAGO)

They must, however, be virtually airtight to ensure the most efficient operation. Finding and eliminating air leaks can be difficult after the fact, so it's important that construction proceeds very carefully with much attention to detail. Water-heating drain-back systems never freeze if properly designed and require no heat exchangers or antifreeze for space heating. Some of the problems that can occur with liquid systems include electrolytic corrosion, which happens when two different metals come in contact in the collector and distribution loops; freezing, which can occur if piping is insufficiently slanted to allow free drainage to the storage tank; and rusting of cast-iron pipes or pumps due to dissolved oxygen in the water.

Glazing for active solar systems is basically restricted to three choices: glass, fiber reinforced plastic (FRP) and thin plastic film (see table 5-3 for details about glazing materials and characteristics). Glass is frequently the glazing of choice because it has very low thermal expansion when heated and lasts virtually forever unless it is broken. Tempered glass is more resistant to breakage and should be considered where vandalism is a chronic problem. FRP glazing has moderate to high thermal expansion, which can be a problem if the glazing isn't adequately supported. Corrugated FRP is stronger longitudinally and less prone to sagging.

Thin plastic film (Teflon) is generally used as an inner glazing when the collector is double glazed.

Collectors can be single or double glazed, depending upon climate and collector operating temperatures. Double glazing is generally used in climates that have more than 6,000 heating degree-days. In milder climates the reduced transmission caused by the second glazing layer exceeds the increased heat gain it causes.

Choice of absorbers is fairly limited. In the case of a water system, you can choose either a metal (copper or aluminum) absorber plate or a special extruded rubber absorber mat. The mat comes in long rolls and can be cut to the desired size. In some applications—for example, for heating swimming pools—the mat can be used without glazing or framing. More commonly, however, an absorber mat is substituted for the metal absorber plate. It is less expensive in comparison to standard absorbers and is well suited to site-built systems.

The collector box can be built with wood or metal. Wood framing is less expensive than metal but may not last as long and may ignite under high stagnation temperatures if it's not protected.

The minimum operating temperature of an active solar system should be equal to the minimum operating temperature of the backup distribution system (assuming both solar and backup use the same distribution loops). Typical minimum temperatures are 95°F for air-heating systems and 140°F for water-heating systems. Most water drain-back systems, however, operate at a temperature between 85 and 120°F. Generally speaking, the lower the operating temperature of the system, the higher the collector efficiency. This is so because heat loss from the collector decreases as the difference between the temperature of the collector and ambient temperature decreases.

Back-Up System

For a space-heating solar retrofit, it may be appropriate to choose a solar system based upon the type of heating and heat distribution systems already in place. Using the same distribution system obviously has its advantages in terms of convenience and cost, but there is no real reason why you cannot have two separate distribution systems. Be aware, however, that adding a second distribution system can be expensive.

If you are junking your present heating system and must decide on a new backup to a solar system, the choices are greater. Besides conventional oil or gas systems, you might wish to install one of the new systems described in chapter 3. In particular, a heat pump can be compatible with a solar heating system. Or, if you are assured of a reliable fuel supply, a wood furnace might provide an appropriate backup and could be compatible with either an air- or liquid-heating system. You would then have a heating system completely based upon renewable energy sources.

System Cost

Active solar systems vary widely in cost. Site-built systems can run as low as $15 per square foot installed, while factory-built systems can cost $30 per square foot and more. Federal tax credits (see chapter 8) allow you to deduct 40 percent of the system cost, or $4,000, whichever is less. Taking these credits, if you are eligible, can reduce costs to $9 per square foot for a site-built system and $18 per square foot for a factory-built one. (These figures do not take into account financing costs.) Since a typical single-family house requires about 600 square feet of collector for space-heating purposes, the total cost of an active solar system will range from $9,000 to $18,000 or if tax credits are taken, $5,000 to $14,000. Your state may also offer renewable energy tax credits. Contact your state energy office for information about such credits.

Another important consideration in evaluating system cost is associated costs such as maintenance, repairs, replacement glazing (if necessary), new pumps and the cost for the power to operate the system. One active system may cost less than another, but if the first has a life expectancy of only 20 years, while the second is expected to last 30, the life-cycle costs of the second may be less than that of the first. This is a rather complicated calculation that requires knowing such costs beforehand; in practice, it may be impossible to foresee all associated costs. As a general rule of thumb, the typical associated costs for an active system will be 2 to 4 percent of the system cost per year.

Back-up fuel cost is also important, because it determines the simple payback on the system and, if it rises unexpectedly, can greatly reduce the payback period. If your system is backed up by electric resistance heating, the most expensive household energy, the payback will be much shorter than if it is backed up by natural gas, which is still the cheapest fuel. But if you have a choice, don't choose a more expensive back-up fuel just to reduce the payback period. Choosing the cheapest fuel lowers your month-to-month operating costs, which is more desirable than shortening the payback.

Site-Built or Factory-Built?

A site-built solar system is one for which the separate components (glazing, absorber materials, framing, piping and so on) are brought to the site and assembled into a system. A factory-built system consists of collectors already assembled and shipped to the site to be installed. Both types of systems have advantages and disadvantages. In making a comparison, you should realize that a properly assembled site-built system will give virtually the same performance as a factory-built one, and it will last as long if materials of factory-built quality are used. The cost of a site-built system can be less than that of a factory-built one, even though labor costs may be greater than those for a factory-built one. Factory-built packages are often more expensive because the consumer has to pay for all the basic materials plus factory labor, profit and other overhead.

Another difference between the two types of systems is that a factory-built system usually has a warranty from the manufacturer that provides some limited protection against material or system failure. A site-built system may only have a more limited warranty from the contractor on installation, which may or may not cover all the components. A factory-built system only requires installation at the site, while with a site-built system more first-hand involvement in design, construction and installation is possible. Site-built systems generally use conventional materials and construction techniques, which can be very important if you plan to maintain the system yourself. In order to maintain the validity of the warranty, a factory-built system will usually have to be repaired by an authorized repair person.

System and Storage Sizing

The appropriate sizing of an active solar system depends upon two factors: solar collection and distribution efficiency and the heating requirements of your building. Storage sizing in turn depends upon solar system size and the number of days of heat storage you wish to have in reserve.

The operating efficiency of a solar collector depends in part on the system operating temperature. The lower the operating temperature, the higher the collector efficiency, because heat loss through the collector glazing decreases as the difference between the temperature of the collector fluid and the outside air decreases. This quantity is called the *delta T*. This change can be seen in the graph in figure 5-14. The slanted lines indicate different amounts of sunlight falling on the collector (300 Btu per square foot per hour is a sunny day; 50 Btu per square foot per hour is cloudy). The bottom of the graph shows delta T. As the delta T increases (toward the right on the graph), the collector efficiency, indicated on the left side of the graph, decreases. As might be expected, efficiency is greater on sunny days than on cloudy days. On very cold, sunny days, efficiency declines

as the delta T increases. The example in the graph involves the following factors to arrive at an efficiency rating: collector fluid temperature is 90°F; outside temperature is 46°F; delta T is 90°F − 46°F = 44°F. On a partly cloudy day the available solar energy is 100 Btu per square foot per hour; therefore, collector efficiency is 40 percent.

At any given instant, system efficiency is always less than that indicated in a collector efficiency graph because of heat losses in other parts of the system, such as from piping or storage. Daily and seasonal efficiencies will always be lower than those measured at noon (when collector efficiencies are usually determined) because the amount of sunlight falling on the collector is usually less before and after noon.

Collector area sizing. Using rules of thumb for solar efficiency, it is possible to estimate the required system size for the four building types. Given common climate conditions and present solar system capital costs, it is most economical to size the system to carry from 60 to 80 percent of your building's seasonal heating requirements. (You should not try for 100 percent because the very cold periods do not occur frequently enough to make 100 percent economical.) In other words, sizing the system in this way will provide about 70 percent of your building's heating requirements on an average winter day. The collector areas required to meet this condition are given in table 5-11.

It is important to realize that these collector areas are only approximate. Before actually sizing a system, a careful analysis using calculator or computer and actual system data should be performed in order to obtain more precise results.

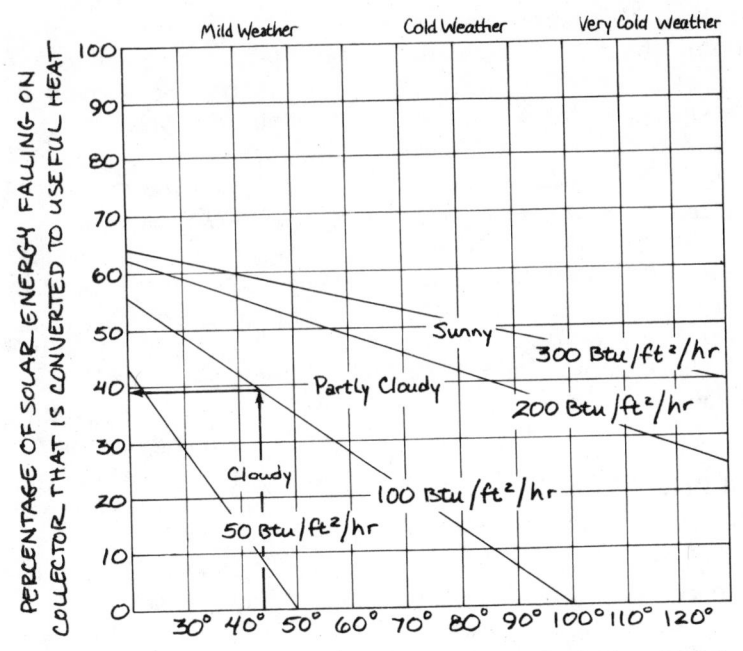

FIGURE 5-14: TYPICAL COLLECTOR EFFICIENCY CURVE

Table 5-11
Active Solar System Collector Areas Required for Four Building Prototypes

Climate Zone	Average Daily Heating Season Temperature (°F)	Collector Area Required (ft²)*			
		Row House	Triplex	Single-Family	Apartment
I	48	300	400	305	1,190
II	38	480	630	485	1,885
III	31	600	800	610	2,375
IV	27	670	890	685	2,650

*Collector area needed to supply 70% of seasonal heating load with system collection and distribution efficiency of 50%, sunny-day insolation rate of 2,000 Btu/ft², 0.6 cloudiness factor and weatherized building heat loads of: row house, 15,120 Btu/DD; triplex, 20,060 Btu/DD; single-family, 15,460 Btu/DD; apartment, 59,850 Btu/DD.

Heat storage sizing. Heat storage for an active solar system should be sized to provide sufficient capacity to carry the system through at least one sunless day. In addition, because the system collects energy and supplies heat directly to the living space for only about one-third of each day, sufficient storage must also be supplied to meet 60 to 80 percent of the load over the remaining two-thirds of the day. In other words, storage system capacity should be equal to about one and two-thirds of your solar system's heating output for an average winter day. Of course, the storage can be larger, but you may be restricted by lack of space, by storage cost or by collector area. Based on the heating requirements and collector areas given in table 5-11, the storage requirements for the four building types are shown in table 5-12.

Installing the System

One of the most common reasons for active solar system failure is improper installation. It therefore pays to ensure that your installer is reliable and that system installation details are done correctly. The best way to find a reliable installer is to contact other individuals who own active solar systems. Ask them whether the installer followed the manufacturer's or the system designer's instructions. Was the system installed according to all design specifications? Has the system operated well? Are there any problems with leakage, rust, corrosion or faulty parts?

In any event, it will pay for you to be aware of certain installation details. Solar collectors can be either roof-mounted or rack-mounted. With roof-mounting, structural support against strong winds is not required; with rack-mounting, it is. The need for rack-mounting depends largely upon the roof angle. If it is too shallow, rack-mounting and cross-bracing will be necessary. Rack-mounting requires that parts of the roofing be removed down to the sheathing for proper attachment. When the collectors are fastened directly to the rafters, without rods, the array must be made watertight and flashed around the edges. Exposed pipes should be insulated with materials resistant to ultraviolet radiation and

Table 5-12
Heat Storage Requirements to Supply 1⅔ Days of Heat at 70% of Load

Climate Zone	Row House		Triplex		Single-Family		Apartment	
	Rock (ft³)	Water (gal)	Rock (ft³)	Water (gal)	Rock (ft³)	Water (gal)	Rock (ft³)	Water (gal)
I	295	905	390	1,240	300	955	1,165	3,695
II	470	1,440	620	1,965	480	1,515	1,850	5,865
III	590	1,810	780	2,475	600	1,910	2,330	7,390
IV	660	2,025	875	2,770	675	2,135	2,605	8,255

NOTE: This table is based on an average temperature increase in thermal storage of 40°F.

weathering. All other distribution pipes should be insulated as well, in order to decrease heat loss.

For the drain-back and drain-down systems, the necessary controls include a temperature sensor at the bottom of the storage tank, a sensor in the collector array at the hottest spot (at the top of an absorber plate), a differential thermostat, which uses the information supplied by the sensors to turn the system on and off and a two-stage space-heating thermostat. Thermometers for continual or occasional temperature monitoring should be provided in the collector supply and return pipe lines, the storage tank and, where desired, in other parts of the system.

The two-stage thermostat sets the indoor temperature at which the solar and back-up heating systems are turned on. Usually, the solar system is turned on first, for example, at a thermostat setting of 68°F. The temperature setting for the back-up system is usually lower, for example, set to 65°F. When the indoor temperature falls just below 68°F, the solar system turns on and heat is drawn either from the collectors, if the system is collecting, or from the storage, if the system is not collecting. If it is able to maintain an indoor temperature of at least 65°F, the solar system will continue to heat alone. Should the indoor temperature fall below 65°F, the back-up system will switch on.

Because the system may cease operating on a hot day either due to storage temperature limits or system failure, it is essential that collectors be designed to survive high stagnation temperatures. In some types of systems, the storage tank liner may degrade above certain temperatures, which sets an upper limit on the amount of heat

that can safely be collected. In such a system, it is a good idea to have a means of "dumping" excess heat. A moderately insulated, single-glazed, flat black collector should peak at 230 to 280°F under stagnation conditions. At these temperatures, paints, insulation, caulks, weather stripping and gaskets may outgas, or release gases trapped inside the materials, which can degrade the materials and also fog the glazing. Rigid or batt fiberglass or foil-faced isocyanurate rigid foam will survive high temperatures. Silicone caulk and EPDM or neoprene rubber gasketing are the best choices for seals. Wood, aluminum or fiberglass can be used for the box, provided the material is separated from the hottest parts of the collector by insulation. Insulation should also be placed between the collector and roof rafters.

Finally, documentation of the system is essential. Both site-built and factory-built systems should be documented with plans, specifications and a clearly written operating manual. Make sure the installer gives you a complete explanation of the system, including information on troubleshooting minor problems.

Warranties

When buying or building an active solar system, it is important that you obtain a warranty for the entire system. For factory-built systems, it should be possible to get at least a five-year limited warranty on the collectors, although many collector warranties are for shorter periods. Other components of a system, such as pumps, usually have warranties that expire before the collector warranty. You should insist that the installer give you a limited one-year warranty on the entire system.

Site-built systems generally do not carry a factory warranty, but some of the system components may be guaranteed. For example, some fiberglass reinforced glazings have a 20-year guarantee against changes in ability to transmit light and against loss of strength. The majority of problems with these systems arise not from failure of parts but from improper installation, so it is important that the installer give at least a one-year guarantee covering the work performed in installing the system.

Active Solar System Economics

The value of a particular active solar system is determined by the amount of conventional energy that the system displaces. This can be determined either by comparing fuel consumption prior to and after installation of the system or by theoretical calculation. The former method bears more relation to the real world because it is based on actual system operation. Assuming, however, that the installation of your solar system has been accompanied by extensive weatherization of your building, it may be difficult to separate improvements in fuel consumption due to each.

As an alternative, table 5-13 shows theoretical results for a low-temperature, single-glazed, flat-plate system with a seasonal collection and distribution efficiency of 50 percent. Depending upon climatic conditions and length of the heating season, a square foot of collector can displace the equivalent of about 1 to 1½ gallons of heating oil, 1⅓ to 2⅓ therms of natural gas or 21 to 37 kilowatt-hours of electricity.

For a 500-square-foot site-built system costing $12 per square foot (installed, after taking federal tax credits), the paybacks are roughly 6 to 10 years if the fuel being displaced is heating oil, 11 to 19 years for gas and 5 to 8 years for electricity. More expensive factory-built systems will, of course, have longer paybacks (see table 5-14).

Table 5-13
Fuel Savings and Value of Active Solar System Collector per Heating Season

Climate Zone	Seasonal Value of 1 Ft² of Collector* (Btu)	Equivalent Fuel Savings†			Value of Fuel Savings ($)		
		Oil (gal)	Gas (therm)	Elec. (KWH)	Oil (@ $1.25/ gal)	Gas (@ $0.50/ therm)	Elec. (@ $0.07/ KWH)
I	72,000	0.93	1.31	21.1	1.16	0.65	1.48
II	90,000	1.16	1.64	26.4	1.45	0.82	1.85
III	108,000	1.39	1.96	31.6	1.74	0.98	2.21
IV	126,000	1.62	2.29	36.9	2.02	1.14	2.58

*Assuming 55% conversion and distribution efficiency for oil and gas; 100% for electricity.
†Assuming 50% collection and distribution efficiency, sunny-day insolation rate of 2,000 Btu/ft², 0.6 cloudiness factor.

Table 5-14
Simple Paybacks for Active Solar Heating Systems
(in years)

Climate Zone	Site-Built System* Displacing:			Factory-Built System† Displacing:		
	Oil	Gas	Elec.	Oil	Gas	Elec.
I	10.3	18.5	8.1	19.0	33.8	14.9
II	8.3	14.6	6.5	15.2	26.8	11.9
III	6.7	12.2	5.4	12.6	22.4	10.0
IV	5.9	10.5	4.6	10.9	19.3	8.5

*System costs $20/ft². A 500-ft² system costs $10,000. After tax credits are taken, system cost is reduced to $6,000 or $12/ft².
†System costs $30/ft². A 500-ft² system costs $15,000. After tax credits are taken, system cost is reduced to $11,000 or $22/ft².

Living with Active Solar Heating

Depending upon the sophistication of its control system, an active solar system is able to operate with a minimum of monitoring or other involvement by its owner, although this does not mean that you can ignore the system entirely. Because of the intermittent nature of sunlight, it is necessary to have a back-up system. When the sun is not shining and the storage temperature is too low, the backup will operate to keep you warm. You should remember that if the solar system fails, the backup will still operate; your house will not become cold, and the system failure may not be noticed.

Thus it is important that you periodically check your active solar system to ensure that it is operating properly, that none of the subsystems (e.g., pumps or fans) are malfunctioning and that no leaks have developed. The temperatures of the various parts of the distribution and collector loops should also be checked at regular intervals. With liquid systems, fluid levels must be checked to ensure that they are high enough. A pump can sometimes fill with air and stop working; if this is not remedied in time, the pump may burn out. Some of these problems may be avoided by proper design; others might happen regardless of how the system is set up. The important thing to remember is that you cannot completely ignore your solar system. If you want your system to operate at maximum efficiency, ensuring that it is operating without obvious flaws is the best way to keep maintenance, the cost of repairs and overall system cost to an absolute minimum.

SOLAR DOMESTIC HOT WATER (DHW)

Active Solar DHW

Although active solar space-heating systems are capable of heating domestic hot water, you may find it more attractive to invest in an active solar domestic hot water system, because these systems are much smaller and easier to place on a roof, are less expensive and have shorter paybacks. Active solar hot water systems are available as complete packages that include collectors, storage, pumps and controls. Such systems are similar to active solar space-heating systems.

Collectors and Storage

Almost all commercially available solar domestic hot water systems use flat plate collectors that are capable of heating water to a temperature sufficiently high for most purposes. Systems can be built with concentrating collectors if temperatures higher than about 140°F are required, but in most cases the cost of these collectors is too high to make such a system economical. The same type of hot water storage used with active space-heating systems is used with domestic hot water systems.

Active solar domestic hot water systems are available with air, liquid or liquid vapor collector fluids, though liquid is the most common. There are certain advantages and disadvantages to each type.

Air systems. Production of hot water with an air-heating collector is usually done with large space-heating systems. It requires the installation of an air-to-water heat exchanger in an air duct or inside a rock storage bed. It may also be possible to place the entire hot water tank inside the rock bed. In any case, losses during the heat transfer process are generally great, and air systems therefore do not perform as well as liquid ones for heating domestic water. Also, air systems are generally capable only of preheating water from 70 to 95°F, providing only 50 to 60 percent of the required energy. Normal domestic hot water temperatures must be between 110 and 130°F. Many air-heating collector manufacturers do, however, offer packaged hot water systems as part of their space-heating packages.

Liquid systems. Liquid systems can use either ordinary water or a water/antifreeze mixture. The closed-loop water/antifreeze package is more common in colder parts of the country where freeze protection is needed. The closed-loop water/antifreeze system is not necessarily the cheapest, simplest or most efficient. A properly designed drain-back system that automatically empties out when not collecting is simpler to operate because it is essentially valveless and therefore almost fail-safe. A water-only system does not have the toxicity concerns associated with water/antifreeze systems. Since water/antifreeze systems must have heat exchangers with double-walled pipes in order to guard against water contamination, they tend to gather heat less efficiently than drain-back systems.

Mixed-fluid systems. Packaged hot water systems using a refrigerant as a collector fluid are available, but they are less common and not always cost-effective. They are safe and freeze-resistant and can collect down to fairly low temperatures.

Selecting an Active DHW System

The points to consider when selecting an active domestic hot water system are nearly identical to those previously mentioned for selecting an active space heating system. They include how the system will be used, the site suitability, operating characteristics of the system, a back-up system, cost, sizing the system, installation and warranties.

Domestic hot water is used on a year-round basis and therefore pays back on a year-round basis. It may be worthwhile to consider an active solar space-heating system that can also provide domestic hot water. This is much more costly than installing a system just for solar domestic hot water heating and does not pay back as quickly, but the overall energy savings are greater. Also, hot water systems are generally not site-built, as it is easier to purchase the system as a package. They are, therefore, fairly expensive on a cost-per-square-foot basis.

Site suitability. If the system is to be mounted on a rooftop, the building's orientation is, of course, critical. DHW systems can be ground-mounted in a suitable location without major problems since they do not require much space. Optimum collector angle depends upon your latitude. Because domestic hot water is required all year, the optimum angle from horizontal is equal to your latitude. If it is impossible to roof-mount collectors at the desired angle, you can rack-mount the collectors to increase the tilt. Also, be sure to include lightning protection on your system.

System characteristics. The simplest hot water system is the water drain-back system. Mixed-fluid systems are also freeze-proof, but can be complex and somewhat more expensive than the drain-back system.

Hot water systems can be subject to freezing, leakage and corrosion if improperly installed. In drain-back systems, pipes should be sufficiently slanted to provide free drainage to the tank, all piping should be of the same metal (preferably copper), and pumps should be made of brass or stainless steel. Pumps with cast-iron impellers can rust out if used in an unpressurized system. Since they are usually factory-built, active hot water systems generally come with glass cover plates, although fiberglass reinforced glazing may be available with some systems. If higher temperatures are desired or if it will be used in a cold climate, the collector may be double glazed.

Solar DHW systems should have a minimum service temperature of 110 to 125°F, to be compatible with temperatures provided by collectors. If higher temperatures are required, the back-up gas or electric water heater can do the boosting.

Back-up system. An active solar domestic hot water system can be backed up either by electricity or natural gas. In a building with an oil-fueled heating system, electricity is the more common backup because it is easier to install. Remember, however, that natural gas is less expensive than electricity, if it's available in your building. It may also be possible to install a domestic hot water heat pump (see "Heat Pumps," chapter 3).

System cost. Active solar hot water systems are generally quite expensive in terms of cost-per-square-foot of collector area, but certainly much less expensive in total cost than active solar space-heating systems. For example, a package system with 60 square feet of collector (adequate for a family of four or five) costs about $2,000 to $3,500 installed, or about $33 to $58 per square foot. Because installation costs do not increase very much as collector area increases, a system with 120 square feet of collector costs about $25 to $40 per square foot. With federal tax credits, these costs are reduced to $1,200 to $2,100 for 60 square feet of collector (or $20 to $35 per square foot) and $1,800 to $3,000 for 120 square feet (or $15 to $25 per square foot).

Sizing the system. The appropriate sizing of an active solar domestic hot water system depends upon solar collector efficiency and the hot water requirements of the system. Storage sizing also depends upon hot water use. Generally about one gallon of storage capacity is needed for every square foot of collector area.

System efficiency. The collecting efficiency of the solar collectors in a hot water system is identical to that in a space-heating system. Efficiency will vary depending upon water temperature. In general, the operating efficiency of a liquid DHW system is less than that of a liquid space-heating system because of higher operating temperatures.

Collector area sizing. As a rule of thumb, a typical liquid active collector will provide about 1.1 gallon of hot water (with a temperature rise of 75°F) per square foot per day. It is most economical to provide about 60 to 70 percent of the individual daily hot water requirements, although more than 90 percent can be provided if desired. About 14 square feet of collector area are required per person. Most packaged systems are sized this way. Table 5-15 shows approximate collector areas required to provide 70 percent of the daily individual hot water requirements for the four building types.

Installation. One of the more common reasons for active solar domestic hot water system failure is improper installation. Problems with plumbing are quite common and may cause continuing malfunctions (see Installing the

Table 5-15
Sizing Solar Domestic Hot Water Systems

Building Type	Number of Occupants	Required Collector Area (ft²)*	System Cost ($)†	Payback (yrs)‡		
				Oil ($1.25/ gal)	Gas ($0.50/ therm)	Elec. ($0.07/KWH)
Row house	10	140	3,420	6.9	14.5	5.4
Triplex	12	168	3,924	6.6	13.9	5.2
Single-family	4	56	1,908	9.7	20.3	7.6
Apartment	50	700	20,000	8.1	17.0	6.4

NOTE: The sizes given here should be considered rough approximations. A precise sizing requires actual system data.

*System assumed to provide 70% of daily individual hot water requirements, or 15 gallons per day with a temperature rise of 75°F. Also assumes sunny-day insolation rate of 2,000 Btu/ft², system collection and distribution efficiency of 50% and cloudiness factor of 0.6. There are 14 ft² of collector per building occupant.

†Assumes cost of $30/ft² plus $1,500 installation cost for row house, triplex and single-family; $3,000 installation for apartment. Federal tax credit of 40% taken on each system.

‡Payback calculation assumes oil-heated hot water at 55% conversion and distribution efficiency, gas-heated hot water at 65% and electric-heated at 100%.

System, under Active Solar Heating, in this chapter). The installer should pressurize the system—whether normally pressurized or not—in order to test for leaks. In closed systems, the quality and acidity of the antifreeze should be checked on a bi-monthly or quarterly basis in order to avoid corrosion problems. The system should be adequately documented, and it is helpful if temperature monitors and flow meters can be incorporated in critical parts of the system.

Warranties. Warranties available for solar hot water systems are generally the same as those for active space-heating systems. Because hot water systems are generally factory-built, warranties on parts and operation are usually available. The installer should give a limited one-year warranty on the entire system.

Solar Hot Water Economics

The value of a solar hot water system is determined by the amount of conventional energy that the system displaces. A typical hot water system collects the equivalent of about 64 kilowatt-hours of electricity (100 percent efficiency), 3.7 therms of natural gas (65 percent efficiency) or 2.8 gallons of oil (55 percent efficiency) per square foot per year. The value of the collected energy is about $4.49 per square foot per year for electricity (at $0.07/KWH), $1.85 per square foot per year for natural gas (at $0.50/therm) and $3.50 per square foot per year for heating oil (at $1.25/gallon). Table 5-15 shows the paybacks for systems installed on the four building types. Paybacks are most advantageous where the solar system displaces electric domestic hot water; however, with federal and state solar tax credits, other paybacks are reduced accordingly.

Passive Domestic Hot Water Systems

Passive DHW systems tend to be less complicated than active systems, but they must be constructed and installed with care. There are two types of passive systems frequently used: the storage or "breadbox" water heater and the thermosiphoning water heater with solar panels.

Storage water heater. This type of preheat system consists of a blackened collector-storage tank in which water is heated by the sunlight that falls on the tank. Cold water from the municipal water line is fed into the tank at the bottom, and warmed water is then removed from the top and fed into a standard hot water tank where it is boosted to the necessary temperature.

Unless the system is insulated, however, it loses heat at night and does not provide much heat on cloudy days. To avoid this, the storage tank (usually an old hot water tank) can be placed in an insulated box with a glazed top and side. This is often referred to as a *breadbox water heater*. During the day, the top and side doors of the box are left open, allowing sunlight to fall on the tank. At night, the box is closed, keeping in the heat. The interior side of the doors may be covered with reflective material in order to increase the amount of sunlight falling on the storage tank. Pipes running to and from the breadbox should be insulated to prevent freezing, if the tank is to be used in the winter.

The weight of the breadbox water heater is considerable; a 50-gallon tank will weigh about 450 pounds, so it is

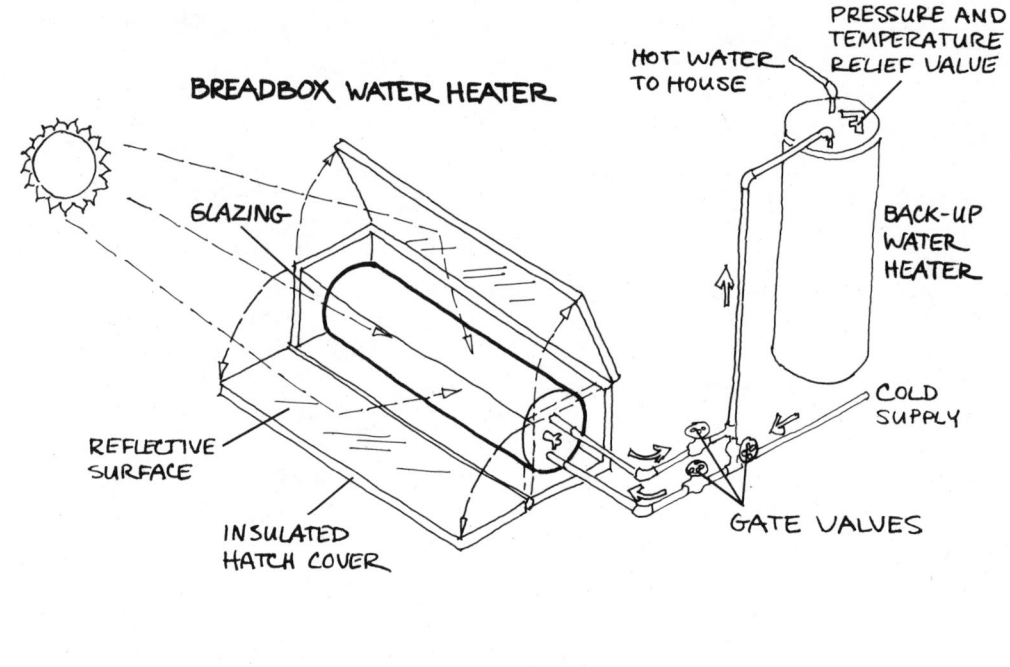

BREADBOX WATER HEATER

GLAZING

REFLECTIVE SURFACE

INSULATED HATCH COVER

HOT WATER TO HOUSE

PRESSURE AND TEMPERATURE RELIEF VALVE

BACK-UP WATER HEATER

COLD SUPPLY

GATE VALVES

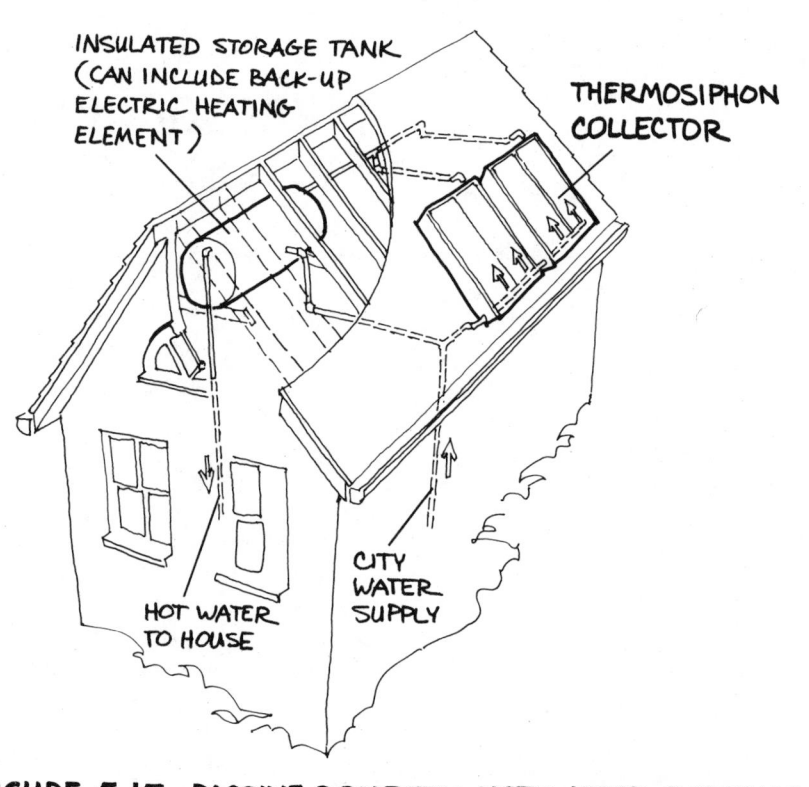

INSULATED STORAGE TANK (CAN INCLUDE BACK-UP ELECTRIC HEATING ELEMENT)

THERMOSIPHON COLLECTOR

HOT WATER TO HOUSE

CITY WATER SUPPLY

FIGURE 5-15: PASSIVE DOMESTIC HOT WATER SYSTEMS

best to place the system on the ground. If it must be placed above ground level, be sure to provide adequate support. A breadbox system built with off-the-shelf components will cost about $500. By using recycled materials the cost of the system can be greatly reduced.

Water heated by a properly constructed breadbox heater should only require significant boosting by a standard water heater during the middle of winter. During the remainder of the year, water from the breadbox should be sufficiently hot for direct use.

Thermosiphoning systems. The thermosiphoning solar hot water system is widely used in Israel and Japan, where collectors are a common sight on rooftops. The system used in those countries is all water and is therefore not suitable for year-round use in cold climates. An all-water system could be used in colder regions about six or seven months of the year. Systems are available that use refrigerant (mixed liquid/vapor) in place of water which makes them freeze-proof.

The all-water system consists of an insulated storage tank mounted at least one foot higher than several single-glazed, flat plate collectors. Cold water from the water main is piped into the bottom of the storage tank. Water is then fed from the bottom of the storage tank into the lower end of the collector panels. As it is heated in the panels, the water rises and enters the top of the storage tank. Hot water is withdrawn from the top of the storage tank. No pumps are required. The water in the tank will remain hot all night if the weather is mild and the demand for hot water is not too great. It is also possible to mount the panels on the slanted roof and the storage tank inside the attic, thus providing additional protection against freezing. If the tank

is installed in an attic, it may be necessary to reinforce the joists to carry the additional load.

The refrigerant system operates similarly, but heat is transferred through a double-walled heat exchanger inside the hot water storage tank. The refrigerant never freezes and this system can be used throughout the year with no worries.

Commercial all-water systems cost roughly $1,000 to $1,500 before tax credits are taken and $600 to $900 after tax credits are taken. This may not be particularly advantageous if the system can only be used for part of the year. The refrigerant system is considerably more expensive ($3,500 for a family of four before tax credits, $2,100 after tax credits) but can be used all year.

PHOTOVOLTAIC SYSTEMS

Photovoltaic or solar cells generate electricity directly from sunlight because of the properties of the semiconductor material (the same material used in transistors) of which it is made. A typical solar cell is composed of two layers of specially treated silicon. When sunlight strikes one layer of silicon, negatively charged electrons are knocked loose from their orbits around atoms and move through the semiconductor material or an external electric circuit, if one is provided. Since electricity is simply the flow of electrons, an electric current results. Usually only 8 to 12 percent of the energy falling on a solar cell is actually converted into electricity.

The maximum possible efficiency of a silicon solar cell bathed in ordinary

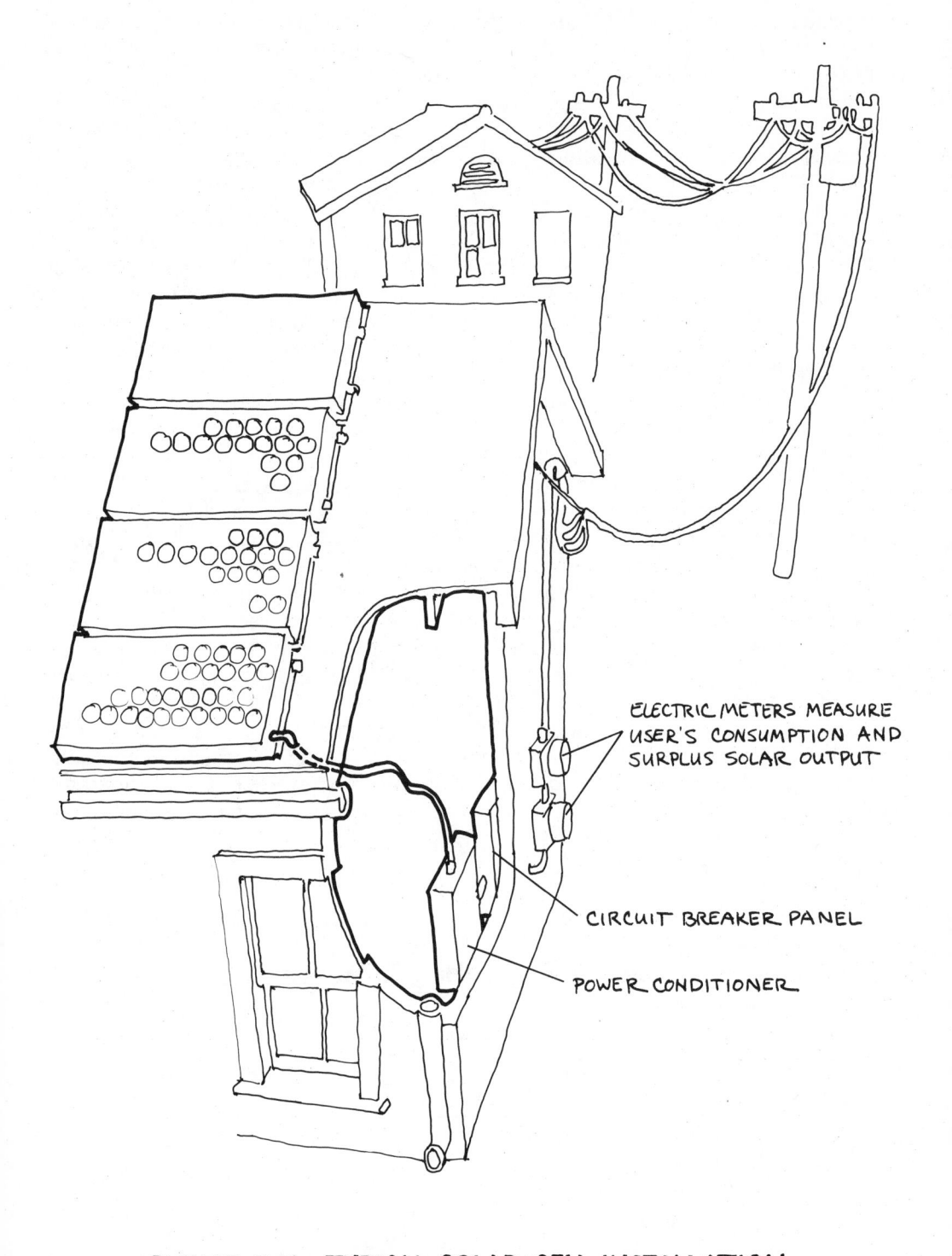

ELECTRIC METERS MEASURE
USER'S CONSUMPTION AND
SURPLUS SOLAR OUTPUT

CIRCUIT BREAKER PANEL

POWER CONDITIONER

FIGURE 5-16: TYPICAL SOLAR CELL INSTALLATION

sunlight is about 27 percent. Until fairly recently, this was considered the maximum efficiency but it appears possible that with the use of new materials, special arrangements of cells, filters and simple concentrators, solar cell efficiencies of 40 percent or more may be achieved in the future. This is, in fact, somewhat higher than the efficiencies routinely achieved by fossil-fueled or nuclear power plants.

Installation of Solar Cells

Solar cells, usually manufactured as discs about three inches in diameter, are simplest to use in flat arrays. A typical array consists of many cells (10, 50, 100, 1,000 or more, depending upon power needs) mounted on a metal or plastic backing, wired together and covered with a protective glass or clear rubber coating. These sheets are rack- or roof-mounted at the desired angle, connected to a power conditioner and wired into the building's electric distribution system. Since solar cells generate electricity on a year-round basis, the optimum angle for mounting them is equal to your latitude.

A typical silicon solar cell generates a maximum of about one watt of power under a noonday sun. Thus about 500 square feet of solar cell array are required to produce 5 kilowatts (5,000 watts) of power at noon on a sunny day. Such an array generates about 600 kilowatt-hours (KWH) of electricity per month, an amount roughly equal to the monthly consumption of a typical single-family home. From one-half to the entire roof area of each of the four building types would have to be covered with solar cells in order to generate the required amounts of electricity. Solar cells clearly require more space than active solar space-heating collectors. It is more economical, therefore, to supply sufficient electricity to meet only the average hourly needs of your building, which usually equals about one-fourth of your peak demand. If more power is required, it is drawn from the utility. The solar cell areas required to supply this amount of power are shown in table 5-16.

Conservation measures can reduce your heat and electricity needs and, therefore, the required collector areas. Because conservation measures are

Table 5-16
Solar Cell Area Required for Four Building Prototypes
(supplying 25% of peak demand)

Building Type	25% of Peak Demand (kw)	Required Solar Cell Area (ft²)
Row house	1.9	190
Triplex	1.6	160
Single-family	1.3	130
Apartment	9.3	930

NOTE: Peak demands taken from table 5-24. One-fourth of peak demand supplies average hourly demand. It is more economical to draw peak power from the utility.

much less expensive than solar collectors or cells, conservation should always come first. Solar cells and active collectors can also be combined, in effect becoming a *cogeneration* system, with the cells acting as an absorber surface for an air- or liquid-heating system. This is presently being done with some commercially available equipment in which concentrators are used to focus sunlight on solar cells. This reduces collector area, and the heat produced by the cells is drawn off for space-heating purposes. A few flat-plate combination solar cell/active heating systems are available on the commercial market, although all are of fairly recent design. It would be easy to design a site-built system incorporating these components, but the costs would probably be high.

Using Solar Cells

Solar cells are relatively simple devices, but because they generate direct current (DC) rather than alternating current (AC), their electricity is not directly compatible with many home appliances. You could convert some appliances to DC and serve them with a separate electrical system, but the rewiring or installation of new wiring for this purpose is likely to be very expensive. Some appliances, particularly those with motors and electronic circuitry, are not convertible to DC. The alternative is to convert solar cell DC to AC using a solid-state electronic *inverter*. With other equipment it would also be possible to sell surplus electricity to your utility and only use utility power when solar cell output was insufficient. This way you can use your utility as "storage" and backup, selling and buying power as the need arises.

As an alternative, you can use the solar cell electricity to operate electric resistance coils immersed in a water storage tank or to power a heat pump in order to provide space heating or cooling. This stored energy can then be used at a later time.

Solar Cell Economics

At the present time, solar cells cost about $5 to $7 per watt, or about $25,000 to $35,000 for the 500-square-foot array described earlier. This is clearly uneconomical, unless you live in a remote area, miles from the nearest electric transmission lines. The Department of Energy has set goals for reducing solar cell array costs to $2 per watt in 1982, 50¢ per watt by 1986, and 10¢ to 30¢ per peak watt by 1990. This would reduce the cost of 500 square feet of cells to the range of $500 to $1,500. Such enormous cost reductions have been routinely achieved by the electronics industry, and there is general confidence that these prices can be reached by the projected dates.

At 50¢ per peak watt, without battery storage, solar cells can supply electricity at about 10¢ per kilowatt-hour. This is about twice as much as average 1981 electricity prices, but we can expect electricity costs in many parts of the United States to rise to this level within the next several years.

In the future, costs may be reduced through the use of new materials and manufacturing techniques under development. At the present time, solar cells are individually sliced from large cylinders of silicon, an expensive and wasteful process. New processes, in which the cells are cut from large flat sheets, like cookies, are likely to greatly reduce costs and waste. New semiconductor materials and different ways of using these materials may, at some point, greatly reduce solar cell costs.

Should some technological breakthrough occur, solar cell electricity may become economically practical much more quickly than presently anticipated.

WIND POWER

Wind-generated electricity is a technology that is well developed and available now. Deciding whether wind power is suitable for you may be difficult because of several factors: Adequate wind must be available; the cost of electricity from a wind system must compare favorably with the cost of conventionally generated electricity; your annual power requirements must be compatible with the output of a commercially available machine; you must have access to adequate financing and various legal and institutional barriers must be hurdled. This collection of factors underlines the need for a very careful analysis of your specific application.

A wind machine captures energy from the wind and converts it to mechanical or electrical energy. First, the wind turns a rotor, which may have two, three, four or more blades. Then, the rotor turns a drive shaft connected either to an electrical generator or to a reciprocating rod which, if connected to a pump, can lift water or perform other mechanical functions. Because conventionally generated electrical energy is generally more expensive than conventionally produced mechanical energy, the use of wind machines to produce electric power offers the greatest potential for cost-effective operation.

There are two basic types of wind machines available: horizontal axis and vertical axis (see figures 5-17 and 5-18). Horizontal axis machines are generally similar to the traditional, fanlike windmills with horizontal drive shafts. In vertical axis machines, the drive shaft is vertical. Horizontal axis machines are usually self-starting, while most vertical axis machines must be given a push to get them going. Vertical axis machines will operate regardless of wind direction, while horizontal axis machines must be rotated into the direction of the wind. The vertical axis allows the generator to be placed at ground level, while the horizontal axis design usually requires the generator to be mounted on a tower. At the present time, most commercially available wind machines are horizontal axis, and the following discussion will focus on this type of wind machine.

Harnessing the Energy in the Wind

The ability of a wind machine to remove energy from the wind depends upon rotor area, blade design and conversion efficiency. The area swept by the rotor determines how much energy can be produced. The greater the area, the greater is the amount of energy that can be extracted from the wind. Because the area of a circle increases as the square of the radius, the amount of energy that can be extracted from the wind increases as the square of the blade length. For example, if the blade length is 5 feet, the blade area is 5 feet by 5 feet by 3.14 or about 75 square feet.

The efficiency of a wind machine (that is, the fraction of available wind energy converted into electricity) is dependent upon the quality of the blade, the transmission design (the link

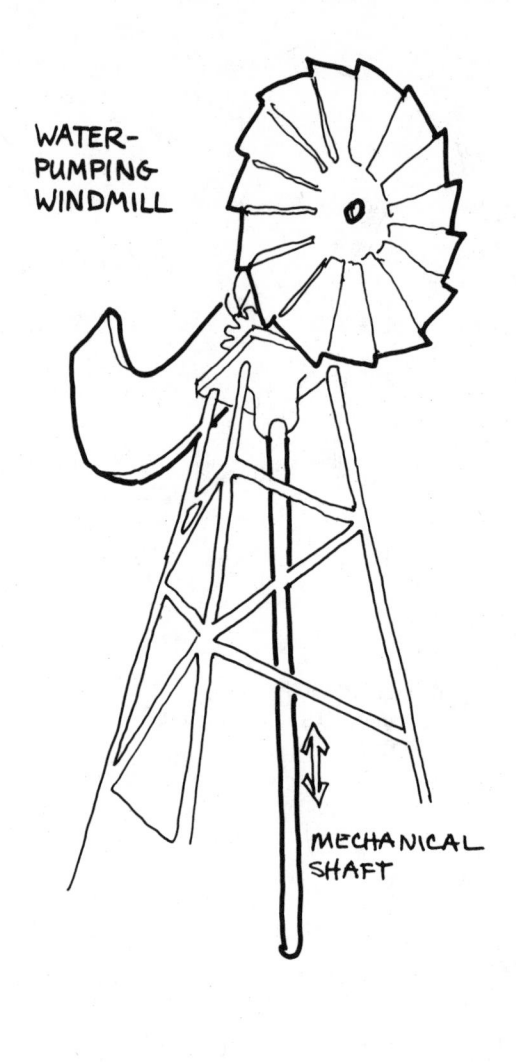

WATER-
PUMPING
WINDMILL

MECHANICAL
SHAFT

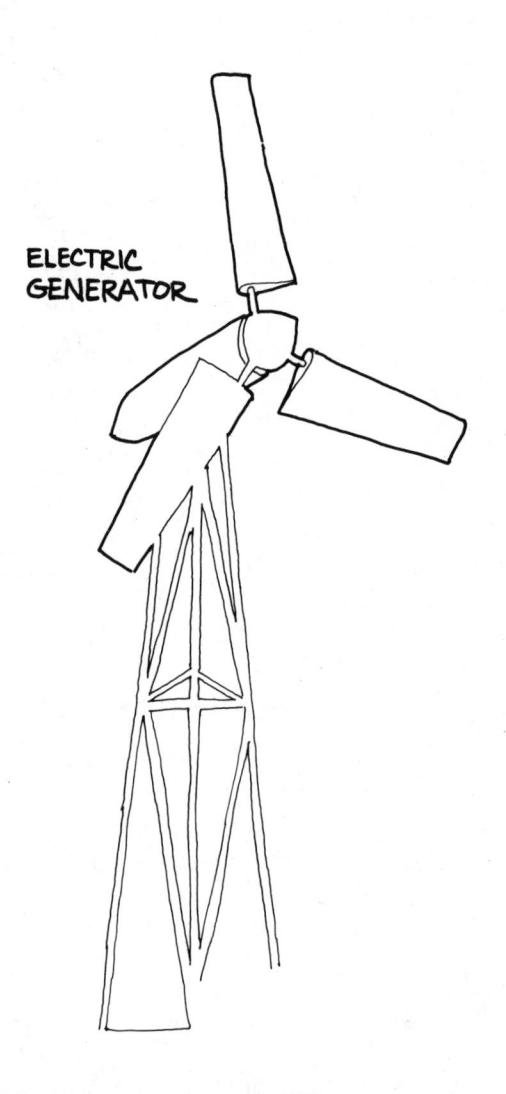

ELECTRIC
GENERATOR

FIGURE 5-17: HORIZONTAL AXIS WIND MACHINES

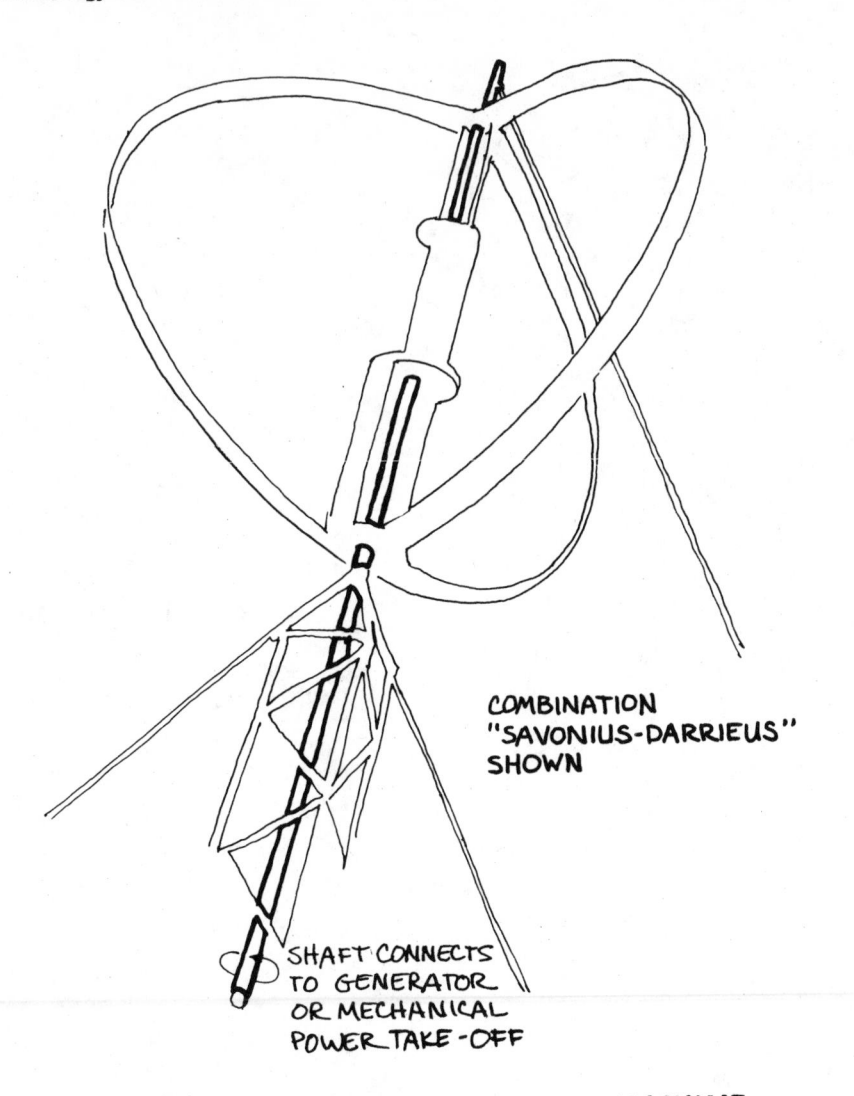

COMBINATION
"SAVONIUS-DARRIEUS"
SHOWN

SHAFT CONNECTS
TO GENERATOR
OR MECHANICAL
POWER TAKE-OFF

FIGURE 5-18: VERTICAL AXIS WIND MACHINE

between the rotor and the generator) and the design of the generator. Laws of physics dictate that a wind machine can extract at most about two-thirds of the energy available in the wind. Wind machines generally do not attain efficiencies this high because some energy is lost due to friction in converting wind energy to electricity. Furthermore, most horizontal axis machines will not begin turning until the wind reaches a predetermined *cut-in velocity*, which is usually around 7 to 10 mph. Wind machines do not reach their maximum power and efficiency until the wind is blowing at the *rated velocity*, generally around 22 or 23 mph. Above the rated velocity, some machines maintain a fairly constant efficiency, while others can actually lose efficiency. At very high wind speeds, a wind machine usually must stop operating, either automatically or mechanically, in order to guard against damage. The velocity at which this occurs is called the *cut-out velocity*. A graph of typical wind machine behavior is shown in figure 5-19.

Electricity produced by the generator on a wind machine is sent through a cable into the electricity distribution system of a building or perhaps directly to an electric utility. Many available wind generators produce only direct current (DC), rather than the standard 110- to 120-volt, 60-cycle alternating current (AC) which is commonly used in homes in the United States. Some wind machines, however, do generate AC power. While electric lights and some appliances can operate on DC without modification, many appliances, such as televisions, stereos, refrigerators and electric clocks, require AC power to operate. Therefore, it is necessary to convert DC to AC, which is done with an inverter, the same type that can be used in a photovoltaic system. Another useful conversion device is the *synchronous inverter*, which converts DC to AC and mixes in utility-

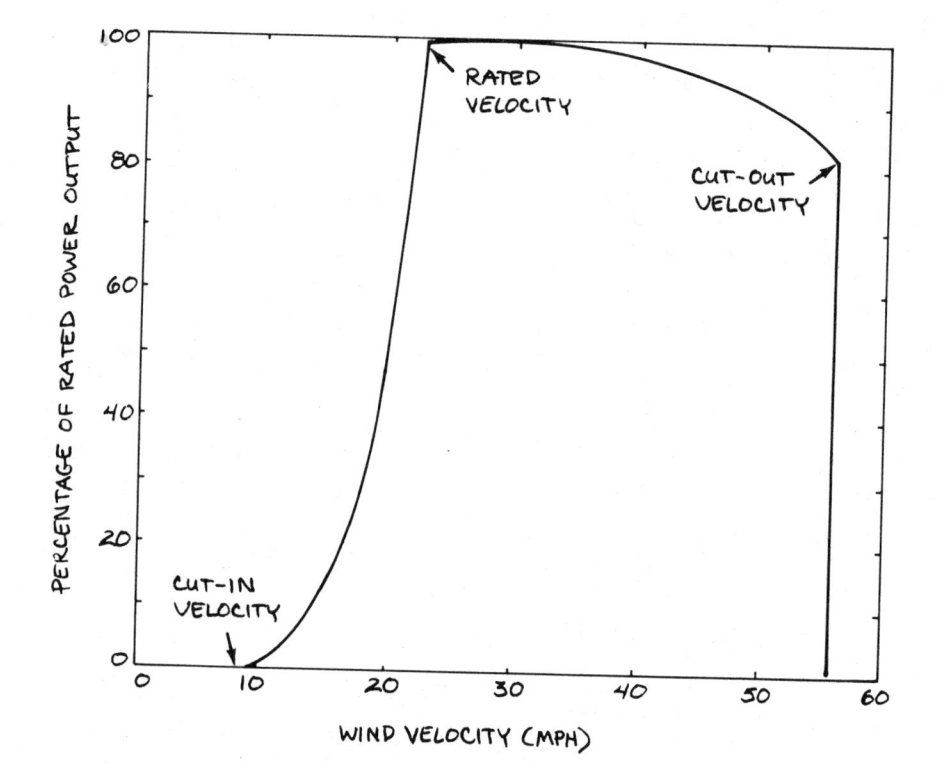

FIGURE 5-19: TYPICAL HORIZONTAL AXIS WIND MACHINE BEHAVIOR

supplied electricity to make up any additional power needs you may have. This device is attractive because, if your wind machine is producing more power than you require at a given moment, the synchronous inverter will automatically send the excess to the utility. Some wind machines are equipped with devices called AC induction generators which are controlled by AC electricity from the utility and thus require no inverter. Such systems may be more cost-effective and less complex than a DC wind machine hooked to an inverter.

Wind Conditions at Your Site

The most important factor for a wind system is, of course, the availability of wind. (See table 5-17, which shows mean wind velocities and the distribution of those velocities for a number of cities in the United States.) The higher the average wind speed, the better, because the energy content of the wind is equal to the cube of the velocity. For example, a doubling of wind velocity represents an eightfold jump in available wind energy. An increase in average wind speed from 12 mph to 13 mph, which is only an 8 percent increase, results in a 27 percent increase in wind energy (see figure 5-20). This fact means that, at any site, the relative proportion of high and low wind speeds is important in determining how much energy can be removed from the wind. A site where the wind velocity is 8 mph half of the time and 16 mph the other half—giving an average wind speed of 12 mph—will actually offer about 2.5 times as much wind energy as a site with a relatively constant wind speed of 12 mph.

Another important factor is the height

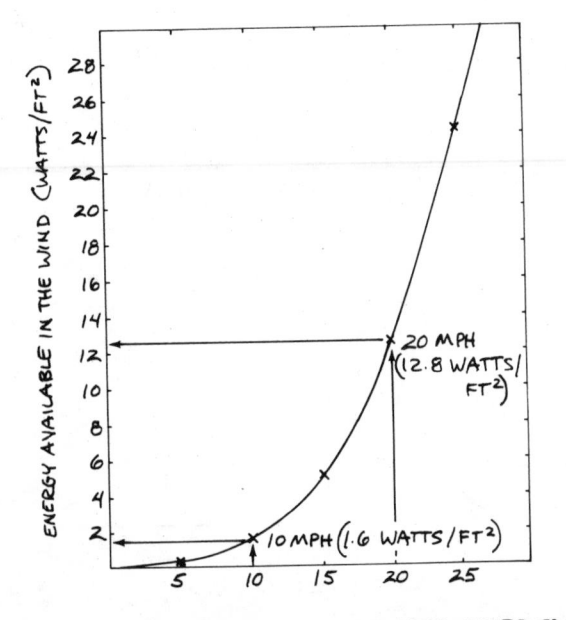

FIGURE 5-20: INCREASE IN WIND ENERGY AS WIND SPEED INCREASES

Table 5-17
Annual Frequency of Occurrence and Mean Wind Velocities for Selected Cities in the United States

Wind Velocity Range (mph)	0-3	4-7	8-12	13-18	19-24	25-31	32-38	Mean Wind Velocity* (mph)
City	Frequency of Occurrence (hrs per yr)							
Albuquerque	1,489	3,154	2,278	1,139	438	88	...	9.8
Atlanta	1,139	2,102	3,154	1,840	526	88	...	9.7
Baltimore	613	2,102	3,416	1,927	526	175	...	10.4
Boston	263	1,052	2,891	3,066	1,052	350	88	13.3
Buffalo	438	1,489	2,978	2,365	1,139	263	88	12.4
Chicago	613	2,278	3,154	2,190	438	88	...	10.6
Cleveland	613	1,577	3,066	2,540	788	175	...	11.6
Dallas	788	1,840	2,803	2,453	788	88	...	11.0
Denver	964	2,365	2,978	1,927	438	175	...	10.0
Detroit	701	2,015	3,241	2,278	438	88	...	10.3
Hartford	1,139	2,278	2,803	1,927	526	88	...	9.8
Houston	526	1,577	3,154	2,453	876	175	...	11.8
Kansas City	788	2,540	3,066	2,015	438	88	...	9.8
Los Angeles	2,453	2,891	2,365	964	88	...	...	6.8
Minneapolis	1,402	1,840	2,978	2,453	788	175	...	11.2
New York City	526	1,314	2,628	2,716	1,051	350	88	12.9
Oklahoma City	175	964	2,978	2,978	1,139	526	88	14.0
Omaha	1,051	1,489	2,450	2,453	964	263	...	11.6
Philadelphia	964	2,365	3,066	1,840	438	88	...	9.6
Phoenix	3,329	3,154	1,752	438	88	...	...	5.4
Pittsburgh	1,051	2,278	2,978	1,927	350	88	...	9.4
Portland	2,453	2,365	2,190	1,402	350	88	...	7.7
Providence	964	1,752	2,803	2,453	613	175	...	10.7
St. Louis	876	2,540	3,154	1,840	263	88	...	9.3
Salt Lake City	1,051	2,891	3,154	1,226	350	88	...	8.7
San Francisco	1,402	1,840	2,278	1,927	964	263	...	10.6
Seattle/Tacoma	1,139	1,402	3,066	2,278	701	175	...	10.7

SOURCES: J. W. Reed, *Wind Power Climatology of the United States* (Albuquerque, N.M.: Sandia Laboratories, 1975); G. Sullivan, *Wind Power for Your Home* (New York: Cornerstone Library, 1978).

*The mean velocity is the same as the average.

Table 5-18
Instrumentation Required for a Wind Site Assessment

Manual Data Collection
 Logbook for recording wind velocities, directions, dates, times, etc.
 Wind gauge (measures wind velocity)
 Weather vane (measures wind direction)

Automated Data Collection
 Recording anemometer (measures wind velocity and direction)
 Optional: magnetic tape recorder for later analysis of weather data by computer

NOTE: Instruments should be mounted on a tower of wind machine height, if possible. A telephone pole can also be used. Data taken near trees or at rooftop level are likely to be biased by wind shadow effects. Sampling can be less frequent if site data can be correlated with nearby weather station data.

of the wind machine. Average wind speed at a 30-foot height is almost 30 percent greater than at 5 feet. At 60 feet, the increase over 5 feet is 43 percent, and at 90 feet it is 51 percent faster (see table 5-19). This means that because of the relationship between available energy and the cube of the wind velocity, raising the height of the wind machine can greatly increase the amount of energy available to the machine.

In order to determine the suitability of your site, you need to measure the range of wind velocities and wind directions and the frequency of different conditions. This can be done with an *anemometer*, or wind gauge. Anemometers small enough to be hand-held cost about $10; more sophisticated devices that automatically record wind velocity and direction at preset time intervals cost from $150 to $300. Before purchasing a wind machine, wind speeds and directions should be monitored for at least one year. If you feel that a year is too long to wait, you can record wind velocities and directions at your site once or twice daily for a month and then compare your readings with those from a nearby weather station. If readings correlate closely between your site and the weather station, you can then use the annual information from the station to determine your local wind characteristics. This method cannot be considered highly accurate unless you are located in the same general area as the weather station. (Table 5-18 gives a list of instrumentation necessary for doing a complete site wind assessment.)

Wind Turbulence

You will also need to evaluate wind turbulence at your site. Turbulence is uneven wind flow caused by nearby obstructions such as hills, trees and buildings. Too much turbulence is unacceptable, because it exposes a wind machine to rapidly changing, perhaps dangerous, stresses that can damage it. To determine the degree of turbulence at your site, take an ordinary kite, remove the tail and in its place fasten a string with ribbons at least 3 feet long attached at 10-foot intervals (see figure 5-21). Once the kite is in

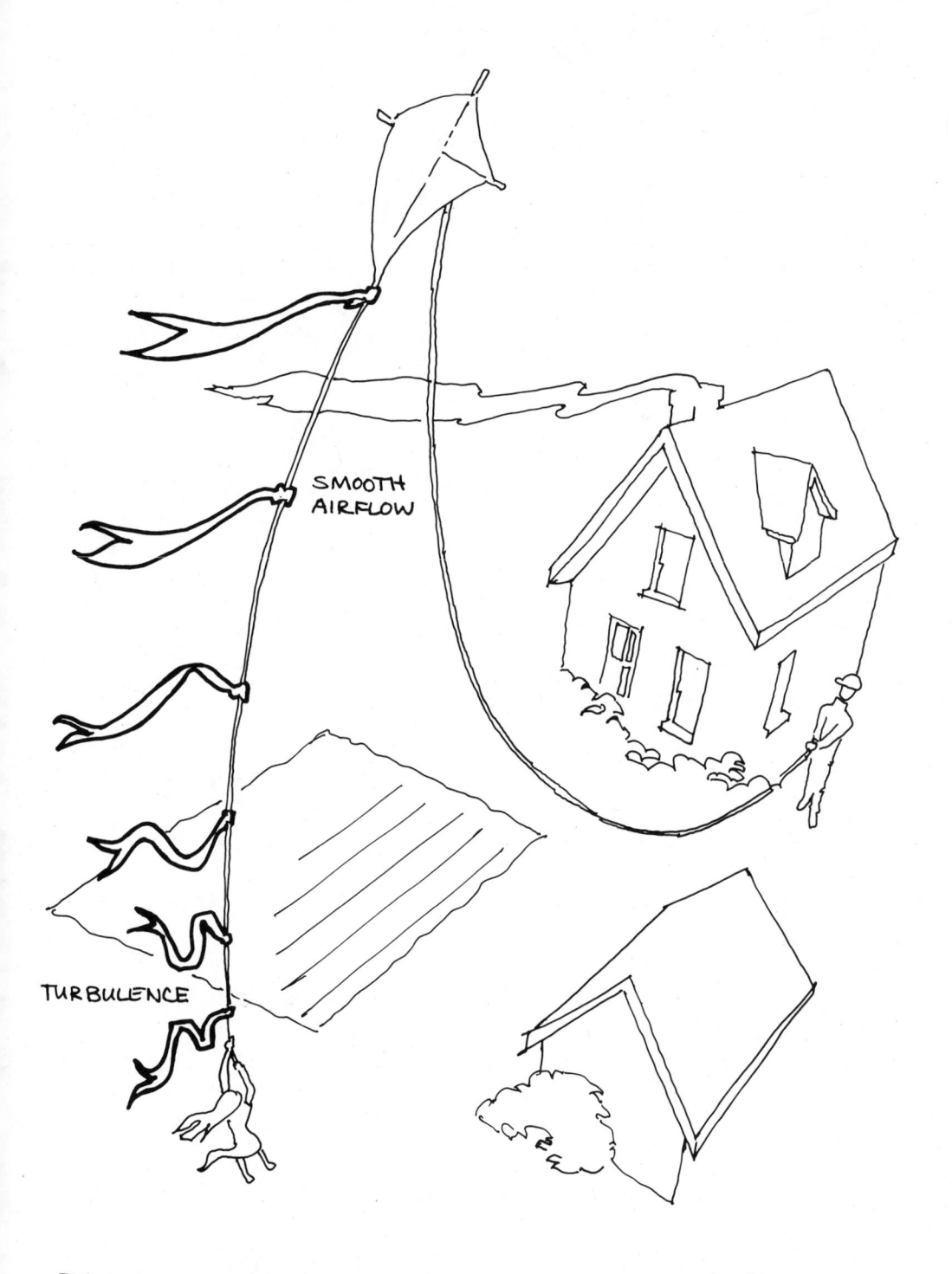

SMOOTH
AIRFLOW

TURBULENCE

FIGURE 5-21: DETERMINING WIND TURBULENCE WITH A KITE

the air, the end of the ribboned string should be held firmly on the ground in a vertical position. The behavior of the ribbons in the wind will indicate the degree and height of turbulence. Smoothly streaming ribbons indicate little turbulence while furling and unfurling ribbons indicate much turbulence.

One way to avoid turbulence is to increase tower height. Generally speaking, a wind machine should be raised at least 25 to 30 feet higher than any obstructions within a 300- to 500-foot radius. If the obstruction is a wide one, it may be necessary to go even higher, because the downwind turbulence generated by the obstruction can be extensive in both width and height. More preferable than raising the tower height is siting as far upwind or downwind as possible from nearby obstructions, space permitting, because raising the tower height can be very expensive. Even more turbulence can be expected if the machine is to be erected on a building roof because updrafts and downdrafts occur along the sides of the building. It is important to note that because tower heights are often restricted by local zoning ordin-

ances, it may be impossible to increase the height of a roof-sited machine greatly. However, if your building is surrounded by other buildings of comparable height, winds will pretty much behave as though they are passing over a level surface. Nearby obstructions can sometimes be helpful because closely spaced buildings occasionally create a wind tunnel, funneling winds into a narrow gap and creating an area of higher wind velocity. You may be able to use such increased wind velocities to great advantage if the effect is a fairly constant one. Remember, however, that such a site is likely to be highly turbulent at times and that excessive turbulence can diminish the performance of or even destroy a wind machine.

Environmental Hazards

Also important to wind machine siting are the frequency and severity of potentially damaging or destructive weather conditions such as wind shear (a change in wind direction or speed over a distance smaller than the rotor diameter), lightning, high winds, tor-

Table 5-19
Increase in Wind Speed and Wind Energy as Tower Height Is Increased
(assuming an average of 10 mph at 20 feet)

Tower Height (ft)	Wind Speed (mph)	Wind Energy (watts/ft^2)
20	10.0	1.6
40	14.0	4.4
60	17.0	7.6
80	19.5	11.5
100	21.7	17.0

nadoes, large hail, heavy snow, icing and salt spray. Wind energy systems are usually designed to withstand extreme weather conditions, but extraordinary stresses can cause deformation or cracking of the blades, injury to the generator or damage to the tower. It is important that the tower be installed to be as invulnerable as possible to extreme conditions and, in the event of an accident, to inflict a minimal amount of damage to nearby people and structures.   .

Extremely destructive weather conditions occur almost everywhere in the United States. The one environmental hazard for which good protection exists is lightning, and all wind machines should be installed with such protection. Wind shear is fairly common in densely built-up areas. If it is present it should be detected during your turbulence survey. Heavy snow and icing in the northern regions of the United States can impose unacceptable weights on blades, but little can be done to guard against this. In areas close to oceans, salt spray is of special concern because of its corrosiveness. Regular cleaning and maintenance can minimize such corrosion. One of the more serious environmental threats to wind machines is heavy winds. Tornado velocities can easily exceed 150 mph and, short of taking a wind machine down, there is little that can be done to guard against such winds.

The Economics of Wind Power

After determining whether your site is a suitable one for a wind machine, you must determine whether a wind machine is economically justifiable for your needs. The economics of wind power are dependent upon the following: your annual electricity consumption (in kilowatt-hours) and costs, the annual production of electricity that could be realized by a wind machine at your site, your highest electricity demand at any one time (in kilowatts), the cost of the wind system and the future cost of electric power from your utility.

First, you must determine how much electricity you (or the residents of your building) consume. A year's worth of electricity bills should indicate your annual and average monthly electricity requirements. To determine the cost of electricity simply divide a monthly bill by the number of kilowatt-hours that month. This will give you an average cost per KWH for electricity.

Table 5-20 shows estimates for the annual and average monthly consumption of electricity in the four prototype buildings. For comparison, the average national consumption in a typical single-family home is around 400 to 500 KWH per month. This translates to a monthly cost of $24 to $45 at costs of 6¢ to 9¢ per KWH. In an apartment building in which consumption for all units is measured on a single electricity meter, monthly costs will, of course, be much greater. However, because of present electricity rate structures which charge less per KWH as you use more electricity (a policy that discourages conservation), the cost per KWH of electricity will be less. These *declining block rates* are likely to be eliminated within the next several years (see "Energy Pricing and Rate Structures," chapter 7) to create an incentive for using less power.

Next, you must determine what size machine you require. You should realize that the cost of power from a machine sized to meet your individual needs will

Table 5-20
Estimated Electricity Consumption for Building Prototypes

Appliance/System	Annual Energy Consumption for One Appliance (KWH)	Estimated Number of Appliances/Systems per Building			
		Row House	Triplex	Single-Family	Apartment
Air conditioner	450	1-2	1-2	1	8-12
Clothes dryer	900	1	1	0-1	3-4
Dishwasher	250	...	...	0-1	...
Freezer	900	0-1	0-1	0-1	0-4
Hot water or warm air heating system (pumps or fans)	250	3-4	3	1	6-12
Lighting system	1,000	3-4	3	1	18-24
Range	1,200	...	...	0-1	...
Refrigerator	1,125	3-4	3	1	24
TV	200	3-4	3	2	24
Water heater	3,400	...	...	0-1	...
Miscellaneous	930	3-4	3	1	18-24
Totals (KHW per year)	Low High	11,865 16,720	11,865 13,215	4,155 10,805	74,340 95,520
Monthly consumption (KWH per month)	Low High	247-330 348-464	330 367	345 900	258 per unit 332 per unit

NOTE: This table represents an estimate of the numbers and types of appliances in each building type, the average energy consumption per appliance and the total estimated electrical consumption of each building type. They do not appear unreasonable in view of the fact that the average monthly consumption in urban areas averages around 300 to 500 KWH. The national average is about 500 KWH per month, but all-electric homes consume a good deal more.

be higher than that from a machine which can serve several families, a building or a whole block.

This is so because the cost of a machine does not increase at the same rate as its power output. All else being equal, therefore, the cost of electricity from a large machine (10 kilowatts) will be less than from a small machine (2 kilowatts).

It is also important to recognize that

the productivity of a wind machine of a given size will vary considerably from site to site, season to season and year to year. Its annual output can only be roughly estimated. Table 5-21 shows the estimated annual output of a 1-kilowatt machine under wind conditions characteristic of a number of American cities. For example, a 1-kilowatt machine at a site in Boston will generate about 2,679 kilowatt-hours

Table 5-21
Annual Energy Output of a 1-Kilowatt Wind Machine in Selected Cities in the United States

Wind Velocity Range (mph)	8-12	13-18	19-24	25-31+	
% of Rated Capacity	8.8	32.7	93.5	100	
City	Kilowatt-hours Output				Annual Total (KWH)
Baltimore	307	630	492	175	1,598
Boston	254	1,002	984	438	2,679
Buffalo	262	974	1,065	351	2,652
Chicago	278	716	410	88	1,491
Cleveland	270	831	737	175	2,012
Detroit	285	745	410	88	1,528
Houston	278	802	819	175	2,074
New York City	231	888	983	238	2,540
Oklahoma City	262	974	1,065	614	2,915
Omaha	216	802	901	263	2,182
Philadelphia	270	602	410	88	1,370
San Francisco	200	630	901	263	1,994
Seattle/Tacoma	270	745	655	175	1,845

NOTE: Wind machine outputs are calculated by taking the four wind-velocity ranges shown above (all greater than cut-in velocity) and the number of hours per year that each wind range occurs, and assuming a cubic increase in wind machine output from a cut-in velocity of 8 mph to a rated velocity of 23 mph. For a 1-kilowatt machine, the number of kilowatt-hours generated over a particular wind range is equal to the percentage of rated capacity times the number of hours that range occurs (as listed in table 5-17). Machines of rated capacity greater than 1 kilowatt are assumed to have outputs that are simple multiples of the output of the 1-kilowatt machine.

of electricity per year, or roughly 30 percent of the maximum possible output of the machine. The monthly output will, of course, vary depending upon wind conditions. Table 5-22 shows how much power a wind machine produces at varying wind speeds. Remember that the rated output of a wind machine is the power produced at a specific wind velocity, usually around 22 or 23 mph.

Peak Demand

Another quantity that you must calculate (which also influences the machine size) is the highest or peak electrical demand, the maximum amount of power you will require at any one time. You can calculate this *peak demand* by adding together the power ratings of all those electrical appliances that you might operate

Table 5-22
Power Output of a Wind Machine
(in kilowatts)

Rated Power Output	Wind Velocity (mph)			
	10	12	14	22.5
0.5	0.04	0.08	0.12	0.50
1.0	0.09	0.15	0.24	1.00
1.5	0.13	0.22	0.36	1.50
2.0	0.18	0.30	0.48	2.00
5.0	0.44	0.76	1.20	5.00
8.0	0.70	1.22	1.93	8.00
10.0	0.88	1.52	2.41	10.00

NOTE: Calculations assume cut-in velocity of 8 mph and cubic increase in output with wind velocity between cut-in and rated wind speeds.

Table 5-23
Estimates of Peak Demands in Building Prototypes

Appliance/System	Power Rating (kw)	Estimated Number of Appliances/Systems per Building			
		Row House	Triplex	Single-Family	Apartment
Air conditioner	1.00	2	2	1	10
Freezer	0.50	1	1	1	4
Lighting system	0.40	4	3	1	18
Range	1.00	...	...	1	...
Refrigerator	0.25	4	3	1	24
TV	0.12	4	3	2	24
Water heater	1.40	...	...	1	...
Miscellaneous	0.50	4	3	1	18
Peak hourly demand (kw)		7.60	6.32	5.20	37.20
Avg hourly demand (kw)		1.35-1.90	1.35-1.50	0.48-1.23	8.50-10.90

Table 5-24
Power Ratings of Some Electrical Appliances
(in kilowatts)

Appliance	Power (kW)
Air conditioner (window unit)*	0.8-1.6
Clock*	0.001-0.01
Clothes dryer*	4.6-5.0
Dishwasher*	1.2
Fan*	0.25-0.37
Food freezer	0.30-0.80
Furnace, oil burner*	0.10-0.30
Grill	0.65-1.30
Heater, portable	0.66-2.00
Hot plate	0.50-1.65
Iron	1.10
Radio, table*	0.04-0.10
Record player*	0.08-0.10
Refrigerator*	0.20-0.30
Refrigerator/Freezer*	0.20-0.30
Refrigerator, frost-free*	0.36
Skillet	1.00-1.35
TV, black & white*	0.06-0.20
TV, color*	0.20-0.35
Toaster	1.15
Vacuum cleaner*	0.60
Water heater	1.20-1.70

SOURCE: Rocky Flats Wind Systems Program, *Is the Wind a Practical Source of Energy for You?* (Golden, Colo: Rockwell International Energy Systems Group, n.d.).

*Requires alternating current (AC).

simultaneously. (Table 5-23 gives estimates of the peak and average demands for each of the four building types, and table 5-24 lists the typical power ratings of a variety of electrical appliances.) Your peak demand figure will indicate the required power rating of a wind machine to meet your needs.

Choosing a machine that is large enough to meet your peak demand, rather than some average demand, will enable you to be almost completely independent of your electric utility. This might seem desirable, but it can also cause problems because you might end up producing large quantities of excess power which, as we will explain later, could end up costing you money. It is generally better to buy a machine sized to meet only a portion of your needs and to purchase additional power from the utility. As an example, let's say your peak demand is 2.75 kilowatts. You could meet this demand by installing a 3-kilowatt machine, which would only generate 3 kilowatts at the rated wind velocity, a rather infrequent occurrence. Or, you could purchase a 5-kilowatt machine, which would probably meet your peak at all times but would also generate large amounts of excess power that would cost you money unless you were able to obtain a fair price from the utility for the electricity. Or, you could install a smaller, 2-kilowatt machine that could meet some of your needs, with the remainder being drawn from your utility.

If you find that your peak demand is much greater than your average demand, it probably makes sense either to reduce your peak demand by using your appliances more wisely and to size the machine accordingly, or to supplement wind-generated power with more power from the utility. To calculate your average demand, divide your monthly electricity consumption (in KWH) by the number of hours in a month. There are 720 hours in a 30-day month and 744 in a 31-day month.

System Cost

Obviously, the larger the power rating of the machine, the more expensive it will be. Representative costs of wind machines are shown in table 5-25. These costs for the most part exclude the tower, necessary additional components or installation. You must also consider the cost of financing, which at times can be prohibitive because of high interest rates. It is safe to say that with additional components and installation, the cost of the system will increase by at least 50 percent and with financing by another 100 percent over the lifetime of the system. Therefore, if wind conditions are not too favorable, if prevailing interest rates are high and if no tax credits or tax deductions are available, electricity from a wind energy system can prove quite costly.

Table 5-25
Representative Costs of Commercially Available Wind Machines

Rated Output (kw)	Price Range (wind generator only) ($)
0.5	2,000- 2,500
1.0	2,900- 3,300
1.5	3,000- 3,900
2.0	3,600- 4,800
5.0	5,000- 8,000 (estimate)
8.0	6,000-10,000 (estimate)
10.0	6,200-17,000

SOURCE: The staff of Solar Age, "Wind Product Supplement," Solar Age, February 1980.

On the other hand, in some cities, under more favorable conditions, a 10-kilowatt wind generator can produce electricity at costs competitive with utility-supplied power. Representative electricity costs, shown in table 5-26, have been calculated after taking federal solar tax credits and deducting the cost of interest payments on the loan from your income tax. The interest rate on the loan described in the table is 14 percent. Note that as the size of the wind machine increases, the cost of electricity drops. This is due to the fact that, although the cost of a machine increases with size, it does not increase as rapidly as the size.

Table 5-26
Cost of Electricity from Wind Machines in Selected Cities in the United States

Rated Output*	1 kw	2 kw	5 kw	10 kw
Capital Cost($)†	3,600.00	4,800.00	8,000.00	11,000.00
Annual Cost($/yr)‡	588.27	842.70	1,321.16	1,819.72
City§	Cost of Electricity (¢/KWH) ‖			
Baltimore ($0.07)	36.8	26.4	16.5	11.4
Boston ($0.07)	22.0	15.7	9.9	6.9
Buffalo ($0.07)	22.2	15.9	10.0	6.9
Chicago ($0.06)	39.5	28.3	17.7	12.2
Cleveland ($0.07)	29.2	20.9	13.3	9.0
Detroit ($0.07)	38.5	27.6	17.3	11.9
Houston ($0.04)	28.4	20.3	12.7	8.8
New York City ($0.10)	23.2	16.6	10.4	7.2
Oklahoma City ($0.04)	20.2	14.5	9.1	6.2
Omaha ($0.04)	27.0	19.3	12.1	8.3
Philadelphia ($0.07)	42.9	30.8	19.3	13.3
San Francisco ($0.05)	29.5	21.1	13.3	9.1
Seattle/Tacoma ($0.02)	31.9	22.8	14.3	9.9

*Annual output for 1-kilowatt machine taken from table 5-21. Outputs of larger machines are simple multiples of output of 1-kilowatt machine.

†Capital cost after 40% federal tax credit is taken. No state tax credits are assumed. Initial capital costs are: 1 kilowatt, $6,000; 2 kilowatt, $8,000; 5 kilowatt, $12,000; 10 kilowatt, $15,000. Cost assumed to include installation.

‡Assumes 20-year, 14% loan. Payments are calculated on loan after deduction of tax credit. Also assumes that the equivalent of 30% of the annual loan payment is treated as a tax credit. Operating and maintenance costs are: 1 kilowatt, $50/yr; 2 kilowatt, $75/yr; 5 kilowatt, $125/yr; 10 kilowatt, $175/yr.

§Number in parentheses represents estimate of cost per kilowatt-hour of utility-supplied electricity.

‖Cost of electricity during first five years only.

The Future Cost of Electricity

Another variable in the wind power economics equation is the future cost of utility-supplied electricity. Since 1974, electricity prices in the United States have risen at an annual rate of about 10 percent, generally somewhat faster than the inflation rate. Electricity prices are likely to increase at an even faster rate in the coming years as oil becomes more expensive and as more costly new generating plants are introduced into utility rate structures. Paradoxically, efforts to implement energy conservation measures in order to limit the need for new generating capacity may also help increase the cost of power by reducing overall demand for electricity. In order to maintain their accustomed profitability, utilities may find it necessary to ask for larger rate increases. If we assume that electricity prices continue to increase by 10 percent each year over the next 20 years (a rate that may be less than inflation and somewhat conservative), average electricity prices in the United States will go from the present cost of about 6¢ per kilowatt-hour to about 15.6¢ per KWH in 1990. Comparison of these figures with those in table 5-26 shows that in some cities in the United States, wind-generated electricity may be competitive with utility-supplied power within the next 5 years. Furthermore, a wind machine, purchased and installed today, will not be seriously affected by inflation, and because its energy source is free for the taking, it will provide electricity at a relatively constant cost over the next two decades. Under these conditions, the economics of small wind machines look moderately good. Machines purchased in the future will be more expensive than they are today, and it will take a longer time for their generating costs to become competitive with utility-supplied electricity. As the price of electricity rises, the payback for a wind machine purchased now becomes shorter. Table 5-27 shows paybacks for 1-, 2-, 5- and 10-kilowatt wind machines sited in various cities. The table shows that the rates of return on investment on large machines are fairly good in some cities and poor in others. You should remember, however, that as machine output goes up, electricity costs go down. A larger machine will usually represent a more favorable investment than a small one. You should also be aware that state solar tax credits can also bring the costs of wind machine electricity down and can shorten the payback time.

Legal and Institutional Barriers

There are few laws or precedents for dealing with urban wind machines, and a number of potential obstacles and pitfalls could deter installation. These include the following: resistance by utilities to wind machine installation and connection into the power grid, local zoning ordinances, building codes and state regulations that might restrict or forbid wind machine installation, damage liabilities of the owner, the absence of definitive wind access rights and potential for negative reactions from neighboring homeowners.

Because wind machines often operate in direct competition with electric utilities, power company attitudes toward wind energy systems have not always been favorable. Among the actions that utilities have taken with regard to wind machines are refusing

Table 5-27
Paybacks for Wind Machines in Selected Cities in the United States
(in years)

City	Rated Output			
	1 kw	2 kw	5 kw	10 kw
Boston	25	20	15	10
Buffalo	26	20	15	10
New York City	21	16	11	7
Oklahoma City	32	27	21	16
Philadelphia	34	29	24	19
San Francisco	34	29	24	18
Seattle/Tacoma	47	42	37	32

to supply power, charging unreasonably high rates for back-up power when the wind is not blowing hard enough to generate electricity, refusing to purchase power and paying unreasonably low rates for surplus power. Some utilities have also insisted on unfair legal responsibilities on the part of wind machine owners. Fortunately, power companies are required by federal law (Public Utilities Regulatory Policy Act, 1978, Sections 201 and 202 and National Energy Conservation Policy Act, 1978) to purchase surplus power from small power producers for a fair price, although they are not required to pay the full retail price. Utilities must also supply back-up power, but they are allowed to charge higher-than-retail rates (which are charged to all independent power producers, not just wind machine users) by arguing that a wind energy system will reduce a customer's total demand for electricity while leaving his or her peak demand unchanged. That is, the customer will generally purchase less electricity from the utility but, when the wind is not blowing, may still require the generating capacity to meet his or her peak demand, which may or may not occur during the utility's peak demand period. As a result, the argument continues, additional generating capacity must be maintained to meet this occasional peak demand. In order to be fair to those who do not have wind generators, the cost of paying for and maintaining this additional capacity must be figured into the higher electricity rates of wind machine users.

The price that the utility is willing to pay for surplus power can also seriously affect wind power economics. Until now, most power companies have been willing to pay only the deferred cost of power, which is the cost of the conventional power saved. If the conventional power saved is generated by burning fossil fuels (which are very costly), the savings may be significant. On the other hand, for nuclear plants, whose fuel cost is fairly low, the deferred cost is low. In either case, the price paid to the wind machine user will be less than the retail price of electricity. If the wind system is sized to meet only a portion of your demand, little or no excess power will be generated, and the low utility purchase price for power should not matter. If, however, the wind machine frequently produces surplus power, the low utility purchase price can have a marked effect on the net cost of power from the wind generator. This comes about because you as a wind machine owner will not be fully reimbursed for the complete cost of generating the excess power in the first place. You will also be forced to pay higher retail prices than those paid by the average consumer when buying the power back from the utility. Table

5-28 shows the effect of wind machine electricity costs in such a situation. As the table shows, the cost penalty associated with the installation of an oversized machine can be enormous and can seriously affect the economic viability of the system.

Additional costs may arise to the wind machine user as a result of the need for equipment to make the wind-system utility grid interconnection. If you have a DC wind system and wish to sell excess power to a utility, you will be required to install a synchronous inverter, costing from $800 to $2,500, depending upon the size of the wind machine. In addition, you may be required to pay a charge for a second electricity meter to monitor the amount of power fed by the machine into the grid. A synchronous inverter will not be required if the machine has an AC induction generator with a 60-cycle, 120-volt AC output.

Table 5-28
Added Cost of Wind Machine Electricity if Surplus Is Sold to a Utility and Later Repurchased
(based on a 10-kilowatt machine installed in representative cities)

| City | Added Cost of Electricity* | | |
	10% surplus (¢/KWH)	25% surplus (¢/KWH)	50% surplus (¢/KWH)
Boston	1.0	2.6	5.2
New York City	1.2	3.0	6.1
Oklahoma City	0.8	2.0	4.1
San Francisco	1.2	2.9	5.6

*Assumes that 10%, 25% and 50% of wind machine's output is sold to utility as surplus and later repurchased to fill consumer's demand. Also assumes that surplus is sold at half of retail price and repurchased at retail price from utility.

Product Standards and Liabilities for Wind-Powered Systems

Until now, uniform product standards have not been developed for wind machines. On the one hand, this lack of standards has benefited the fledgling wind industry by spurring innovative designs. On the other hand, the absence of standards leaves potential wind machine consumers with no means to assess the quality, durability, performance and output of a given system. As more experience with wind energy systems is accumulated, poorly designed and inefficient systems will fail in the marketplace, but in the meantime, the potential buyer has little to go on besides the claims of the manufacturer, the reassurances of the seller and the experience of the few people who have purchased and operated particular brands. Some manufacturers offer warranties, but their reliability is not yet clear. If you decide to purchase a wind machine, get the best warranty you can. Wind machines are often sold with only a 90-day guarantee. Be sure to ask for the names of other buyers of the same machine because they may be able to advise you on performance and quality.

The lack of product standards may also cause problems in obtaining financing and insurance for a wind system. Banks may refuse to loan money for a wind machine if they have no way of judging its reliability. The bank, after all, does not want to end up owning a useless wind generator. Insurance companies, unable to make a reliable determination of the hazards and risks associated with a particular wind machine, may refuse to grant coverage.

Access Rights and Effects on Neighbors

There are currently no legal access rights to the wind and as wind machines become more common, various legal prescriptions will have to be developed that define their rights and access. This problem is discussed further in chapter 7.

Complicating matters further, wind machines may have some effect on neighboring houses and areas. A large wind machine may affect television reception in nearby homes, although this has not been observed with small, residential-scale machines and should not occur with nonmetallic blades. The tower may have a small effect on reception, however. Wind machines may also be noisy if defective or poorly maintained, although general experience suggests this to be a very infrequent problem. Finally, wind machines are large and visible objects. Neighbors might object to a wind machine for purely aesthetic reasons or out of fear of possible hazards. You should be sure to consult your neighbors on their attitudes toward a wind machine installation, if for no other reason than to avoid lawsuits or a potentially rancorous situation. Until such time as laws allowing and protecting urban wind machine installations are in effect, the prospective wind machine owner should be prepared for problems such as these.

Zoning, ordinances, building codes and municipal and state regulations may affect wind machines in various ways by imposing installation controls, establishing height restrictions and setting allowed distances from property lot lines. They are discussed in detail in chapter 7.

Calculating Simple Paybacks and Returns on Investment for Conservation, Solar and Wind Energy Improvements

You may wish to calculate simple paybacks and returns on investments for combinations of energy improvements. The following charts allow you to calculate simple paybacks and returns for a wide range of improvement costs and fuel costs.

Assume, for example, that you have paid $750 for an energy conservation improvement that saves 100 gallons of oil per year. Also, assume that the cost of oil is $1.20 per gallon. First, calculate the cost of the improvement for each gallon of oil saved *during the first year*:

On figure 5-22 locate the curve along the right-hand side labeled "$7.50." Follow the curve to the point at which it meets the vertical line for oil at $1.20 per gallon. Then, follow a straight line to the left-hand side of the chart. In this case, the payback is a bit more than six years, and the return on investment is around 16 percent. The appropriate lines on this chart demonstrate this example. Figure 5-23 and figure 5-24 can be used in the same way.

$$\frac{\text{Cost of energy improvement}}{\text{gallons saved per year}} = \frac{\$750}{100 \text{ gal/yr}} =$$

$$\$7.50/\text{gal/yr}$$

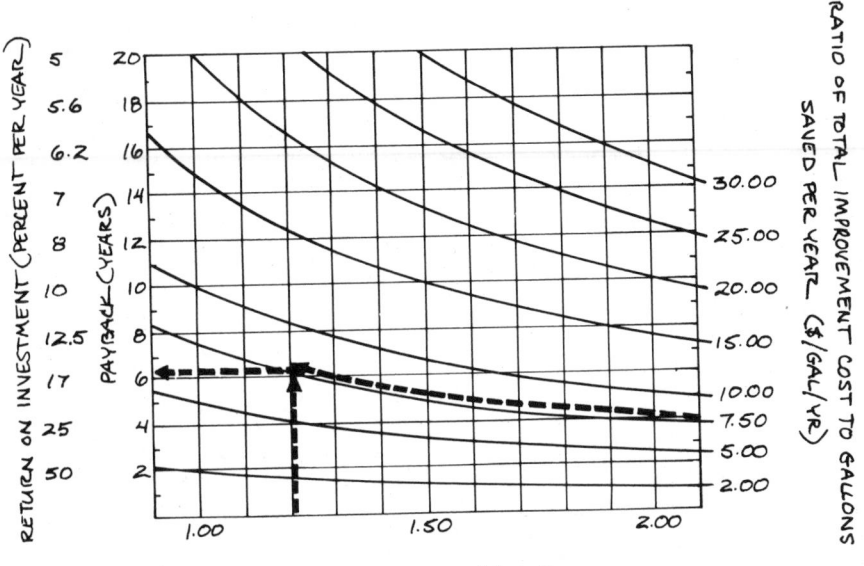

FIGURE 5-22: INVESTMENT PAYBACK CHART (FUEL OIL)

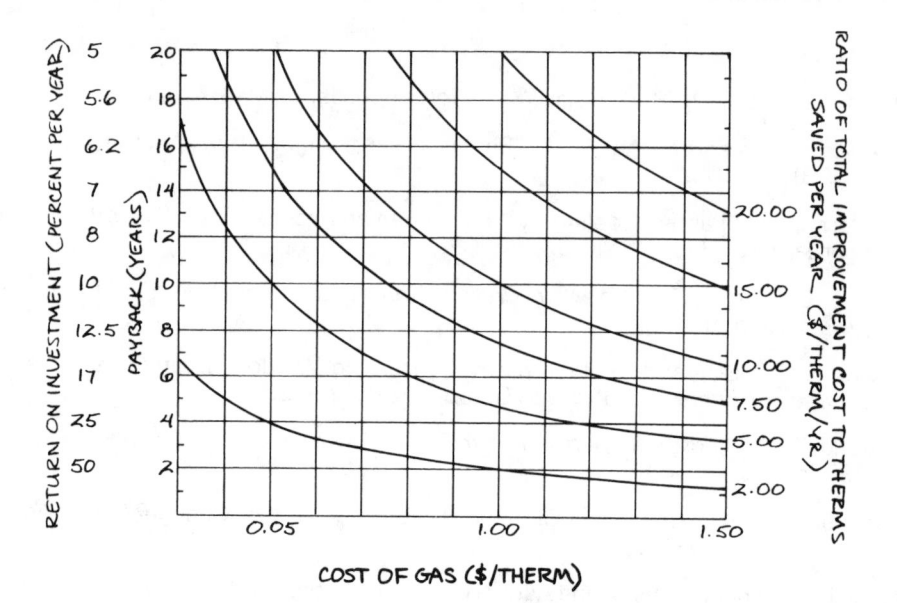

COST OF GAS ($/THERM)

FIGURE 5-23: INVESTMENT PAYBACK CHART (NATURAL GAS)

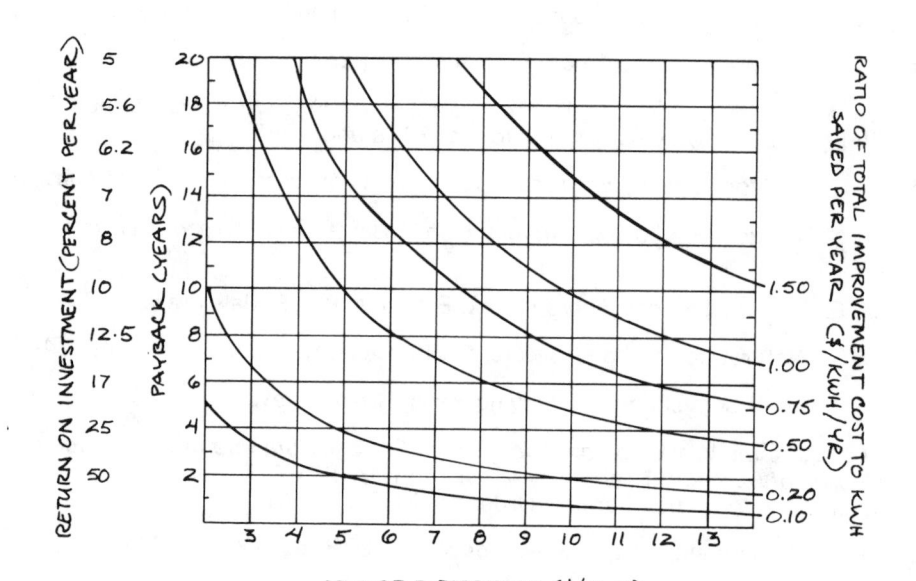

COST OF ELECTRICITY (¢/KWH)

FIGURE 5-24: INVESTMENT PAYBACK CHART (ELECTRICITY)

For Further Reference

Passive Solar

Adams, J. "An Easy-to-Build Solar Wall." *Solar Age*, January 1979, pp. 26-28.

Anderson, B. N., and Michal, C. J., "Passive Solar Design." *Annual Review of Energy 3* (1978): 57-100.

Bahadori, M. N. "Passive Cooling Systems in Iranian Architecture." *Scientific American*, February 1978, pp. 144-52.

Barnaby, C. S. et al. *Solar for Your Present Home*. Sacramento: The California Energy Commission, n.d.

Buesing, J. "Dollars and Sense: Energy Conserving Window Treatments." *Alternative Sources of Energy*, September/October 1980, pp. 26-31.

Diamond, S., and Lorris, P. S. *It's in Your Power*. New York: Rawson, Wade Publishers, 1978.

Eissenberg, D. M., and Wyman, C. "What's in Store for Phase Change?" *Solar Age*, May 1980, pp. 12-16.

Energy Department of the Maine Audubon Society. *Passive Solar Greenhouse*. Falmouth: Maine Audubon Society, 1978.

Farallones Institute. *The Integral Urban House*. San Francisco: Sierra Club Books, 1979.

Kando, P. F. "Eutectic Salts." *Solar Age*, April 1978, pp. 15-16, 40.

Keller, S. F. "Retrofitting Warehouses." *Solar Age*, September 1978, pp. 28-30.

Langdon, W. K. *Movable Insulation*. Emmaus, Pa.: Rodale Press, 1980.

Lawrence Berkeley Laboratory. *Windows for Energy Efficient Buildings*, nos. 1 and 2. Berkeley, Calif.: University of California, 1979 and 1980.

Leckie, J. et al. *More Other Homes and Garbage*. San Francisco: Sierra Club Books, 1981.

Marlboro College. *Marlboro Conference on Passive Solar Greenhouses*. Marlboro, Vt., 1978.

Mazria, E. *The Passive Solar Energy Book*. Emmaus, Pa.: Rodale Press, 1979.

Morris, S. "Natural Convection Collectors." *Solar Age*, September 1978, pp. 24-27.

_____. "Natural Convection: No Moving Parts!" *Solar Age*, January 1979, pp. 38-42.

Proceedings of the Workshop on Solar Energy Storage Subsystems for the Heating and Cooling of Buildings, Charlottesville, Va., April 16-18, 1975. Springfield, Va.: National Technical Information Service. Order no. NSF-RA-N-75-041.

Riordan, M. "Thermal Storage." *Solar Age*, April 1978, pp. 12-17, 19-20.

Sayigh, A. A. M., ed. *Solar Applications in Buildings*. New York: Academic Press, 1979.

Scully, D.; Prowler, D.; and Anderson, B. *The Fuel Savers*. Harrisville, N.H.: Total Environmental Action, 1976.

Shurcliff, W. A. *Thermal Shutters and Shades*. Andover, Mass.: Brick House Publishing Co., 1980.

Steadman, P. *Energy, Environment and Buildings*. Cambridge: Cambridge University Press, 1975.

Strauss, H. "Passive Cooling Where It's Hot and Humid." *Alternative Sources of Energy*, July/August 1980, pp. 24-27.

Temple, P. and Kohler, J. "Glazing Choices." *Solar Age*, April 1979, pp. 25-29.

Trombe, F. et al. "Concrete Walls to Collect and Hold Heat." *Solar Age*, August 1977, pp. 13-19, 35.

Wing, C. *From the Walls In*. Boston: Little, Brown & Company. 1979.

Yanda, W., and Fisher, R. *The Food and Heat Producing Solar Greenhouse*. Santa Fe: John Muir Publications, 1977.

Yellott, J. I. "Passive Solar Heating and Cooling Systems." *ASHRAE Journal*, January 1978, pp. 60-67.

Active Solar

AIA Research Corp. *Solar Dwelling Design Concepts*. Washington, D.C.: U.S. Government Printing Office, 1976. Order no. 023-000-00334-1.

Barnaby, C. S. et al. *Solar for Your Present Home*. Sacramento: The California Energy Commission, n.d.

Burt Hill Kosar Rittleman Associates. *Minimum Energy Dwelling (MED) Handbook*. Springfield, Va.: National Technical Information Service, 1977. Order no. SAN 1198-1.

Diamond, S., and Lorris, P. S. *It's in Your Power*. New York: Rawson, Wade Publishers, 1978.

Farallones Institute. *The Integral Urban House*. San Francisco: Sierra Club Books, 1979.

Graham, B. J. "Evacuated Tube Collectors." *Solar Age*, November 1979, pp. 12-17.

Haworth, W. L., and Mattiler, J. M. "A Challenge for Warranties." *Solar Age*, May 1979, pp. 30-35.

Kohler, J., and Temple, P. "The Fundamentals of Site-Built Collector Design." *Solar Age*, July 1980, pp. 12-17.

Leckie, J. et al. *More Other Homes and Garbage*. San Francisco: Sierra Club Books, 1981.

Meinel, A. B., and Meinel, M. P. *Applied Solar Energy*. Reading, Mass: Addison-Wesley Publishing Co., 1976.

Northeast Solar Energy Center. *Five Award-Winning Solar Designs for Multi-family Housing*. Boston, 1980.

Office of Technology Assessment. *Application of Solar Energy of Today's Energy Needs*, vols. 1 and 2. Washington, D.C.: U.S. Government Printing Office, 1978. Order no. 052-003-00539-5 (vol. 1); 052-003-00608-1 (vol. 2).

Reno, V. "Controls Demystified." *Solar Age*, July 1980, pp. 32-36.

Solar Energy Applications Laboratory, Colorado State University. *Solar Heating and Cooling of Residential Buildings: Sizing, Installation and Operation of Systems.* Washington, D.C.: U.S. Government Printing Office, 1977. Order no. 003-011-0008-7

Solar Domestic Hot Water

Lewis, D. C. "SDHW Systems: What Experts Look For." *Solar Age*, February 1979, pp. 16-20.

Schwolsky, R.; Williams, J.; and Ross, A. "DHW System Checklist." *Solar Age*, May 1980, pp. 43-46.

_____. "Nuts & Bolts of Installation." *Solar Age*, March 1980, pp. 41-50.

_____. "Weatherproofing Domestic Hot Water Systems." *Solar Age*, April 1980, pp. 27-33.

Smith, R. O. and Associates. "Report Summary of Performance Problems of 100 Residential Solar Water Heaters Installed by New England Electric Company Subsidiaries in 1976 and 1977." Newton Highlands, Mass, October 1977. Order no. BNL 419 929-S.

Solar Cells

Kendall, H. W., and Nadis, S. N., eds. *Energy Strategies: Toward a Solar Future.* Cambridge, Mass.: Ballinger Publishing Company, 1980.

Law, S. R., and Bottaro, D. *Institutional Issues Associated with Residential P.V. Applications: A Bibliography.* Cambridge, Mass: MIT Energy Lab, 1979. Working Paper MIT-EL-79-050WP.

Maycock, P. D., and Stirewalt, E. N. *Photovoltaics: Sunlight to Electricity in One Step.* Andover, Mass.: Brick House Publishing Co., 1981.

Office of Technology Assessment. *Application of Solar Technology to Today's Energy Needs,* vols. 1 and 2. Washington, D.C.: U.S. Government Printing Office, 1978. Order no. 052-003-00539-5 (vol. 1); 052-003-00608-1 (vol. 2).

Wind Power

Cheremisinoff, N. P. *Fundamentals of Wind Energy.* Ann Arbor: Ann Arbor Science, 1978.

Clews, H. *Electric Power from the Wind.* Norwich, Vt., 1974.

Energy Task Force. *Windmill Power for City People.* Washington, D.C.: U.S. Government Printing Office, 1977. Order no. 059-000-00001-2.

Enertech Corp. *Planning a Wind-Powered Generating System.* Norwich, Vt., 1977.

Golding, E. W. *The Generation of Electricity by Wind Power.* London: E. & F. N. Spon, 1976.

Leckie, J. et al. *More Other Homes and Garbage.* San Francisco: Sierra Club Books, 1981.

Marier, D. *Wind Power for the Homeowner*. Emmaus, Pa.: Rodale Press, 1981.

Massachusetts Executive Office of Energy Resources. *Wind Report*. Boston, 1978.

Park, J. *The Wind Power Book*. Palo Alto: Cheshire Books, 1981.

Park, J., and Schwind, D. *Wind Power for Farms, Homes and Small Industry*. Springfield, Va.: National Technical Information Service, 1978. Order no. RFP-2841-1270-78-40.

Putnam, P. C. *Power from the Wind*. New York: Van Nostrand Reinhold Co., 1948.

Sullivan, G. *Wind Power for Your Home*. New York: Cornerstone Library, 1978.

Wegley, H. L. et al. *A Siting Handbook for Small Wind Energy Conversion Systems*. Springfield, Va.: National Technical Information Service, 1978. Order no. PNL-2521.

Large-Scale Energy-Conserving Systems

The energy-conserving technologies described in previous chapters generally apply to single buildings. There are, however, other technologies that are well suited for application to groups of buildings. They include *cogeneration, district heating* and *seasonal energy storage systems.*

Cogeneration is the simultaneous production of both space heat and electricity from an electrical generating system. District heating involves supplying heat and possibly domestic hot water from a central production facility to a cluster of residential, commercial and office buildings through a network of water mains. A seasonal energy storage system is designed to store heat or coolth during one season, when it is not needed, for use during another season.

In order to be cost-effective, these three technologies are usually applied to groups of buildings, although cogeneration and seasonal energy storage systems may be sized for small-scale applications. These technologies are not mutually exclusive. District heating, for example, may involve cogeneration or summer storage of solar energy for winter space heating. All three of these technologies share significant potential for energy conservation for city blocks, neighborhoods and whole towns.

COGENERATION

In electric power plants, electricity is produced from steam that turns power-generating turbines. Typically, much of the heat contained in the steam after it condenses is released to the environment. In industrial plants, steam or process heat is often produced for production purposes, but the mechanical energy in the steam that could turn generating turbines is wasted. Clearly, there would be less waste if these two processes were combined. Cogeneration does this by combining the production of heat and the generation of electricity to yield a total conversion efficiency that can be much higher than that of either process occurring separately. Also, the economic and work values of the energy produced by congeneration can be very high. As the costs of fossil fuels and electricity continue to increase, cogeneration will become more attractive from both an economic and an energy conservation viewpoint. Cogeneration is already available for large-scale applications and should become widely available on a small scale within the next five years.

Fossil-Fueled Cogeneration Systems

The *gas turbine* uses hot, high-pressure gases produced by burning oil or natural gas to drive a generating turbine. The hot exhaust gases are used to create steam in a boiler system. About 62 percent of the fuel's energy is removed in this process, but the efficiency can be raised to as high as 90 percent if the system is properly designed.

The *steam turbine* uses high-pressure steam produced in a boiler (from burning fossil fuels or urban waste) to generate electricity. The resulting low-pressure steam (from which the mechanical energy has been removed) is then used for heating purposes. Total efficiency for this process can be as high as 85 percent, and the system is cheap, clean and reliable. However, it is limited in the amount of electricity it can produce.

The *diesel engine* drives a generator directly. Waste heat is then recovered either from the water-filled cooling jacket placed around the engine or from the exhaust gases. This heat can in turn be used to heat water or to produce steam. Diesels often have lower efficiencies than either gas or steam turbines, but with cogeneration they can have total conversion efficiencies as high as 90 percent. They are also capable of generating more electricity than gas or steam turbines for a given amount of heat and are the most appropriate for small-scale applications. One potentially serious problem with some diesel cogeneration systems is air pollution because high levels of nitrogen oxides, carbon monoxide and hydrocarbons are emitted. New, large diesel engines are, however, generally much cleaner than those produced 10 to 15 years ago.

Marine diesel engines, a long-existing technology, can be adapted for cogeneration and are at least as reliable as conventional cogenerators. They are also highly resistant to corrosion, which may be an advantage under some circumstances.

Fluidized bed combustion is a new technology that burns coal in a non-polluting, efficient manner and is capable of producing both electricity and heat. This technology uses a mixture of finely crushed coal and limestone that rides on a stream of air. This allows the coal to be burned at temperatures lower than those of conventional coal burners, thereby reducing unwanted nitrogen oxide production. The limestone absorbs sulfur from the coal, which decreases unwanted sulfur dioxide production. This allows the use of high-sulfur coal, which makes up much of this country's new coal supplies. Cogeneration using this technology should become commercially available within the near future.

Other types of cogeneration systems are likely to be developed in the future. While most systems now use fossil fuels, it is also possible for them to use renewable fuel sources such as wood, wood gas or methane from sewage and garbage. Completely new designs such as household solar systems that combine photovoltaic cells for power and solar collectors for heat are on the horizon.

Cogeneration systems share several problems. Because they are industrial installations, they aren't always welcome neighbors in residential zones. Because gas, diesel and steam cogeneration systems are still dependent on fossil fuels, they are subject to fuel shortages and increasing fuel costs. For short-term shortages, such as those that might be caused by an international fuel crisis, an emergency supply of fuel can be kept in storage. If long-term shortages of fossil fuels become a reality, cogeneration systems may have difficulty in obtaining fuel supplies. Having

a system that can be converted from a conventional fossil fuel to a coal- or renewable-derived fuel will be advantageous. Fluidized bed combustion offers protection against supply shortages but not against coal strikes or fuel price increases. Other multi-fuel systems offer similar, if limited, protection. Cogeneration systems using renewable fuels are, for the most part, not susceptible to these problems.

Another way to ensure a steady fuel supply for a cogeneration system is to use a waste product. Sun-Diamond in Stockton, California, is the world's largest walnut-processing plant. It turns millions of walnut shells into electricity for the plant and for nearby homes. The walnut shells are the fuel that produces steam to drive a turbine generator. The resulting low-pressure steam is then used to heat as well as to refrigerate the plant. The Sun-Diamond cogeneration system produces about 32 million KWH of electricity per year, but it only uses 12 million. It sells the surplus power to Pacific Gas and Electric Company. If the supply of walnut shells runs out, which is unlikely, almond hulls, wood chips and similar materials could fuel the system.

Residential-Scale Cogeneration

Small-scale, or residential-scale, cogeneration units are those capable of serving one or several families and are usually in the 5- to 20-kilowatt range. Cogeneration systems are generally described in terms of electrical generating capacity, rather than heat output. Virtually the only cogeneration system of this size currently available is one developed by the Fiat Motor Company. It is called the TOTEM (Total

Energy Module) and costs about $6,000 to $10,000. Based upon a standard, four-cylinder automobile engine, the TOTEM unit most commonly burns natural gas, although it can be adapted to a variety of other fuels, including alcohol. It has a heat recovery efficiency of about 66 to 70 percent and an electrical generating efficiency of about 26 percent. The heating efficiency is comparable to a conventional heating system, but since TOTEM also generates electricity its total efficiency is about 92 to 96 percent. The unit can produce 15 kilowatts of electrical power and heat a 4- to 10-unit apartment building. One problem with TOTEM is that it must undergo major maintenance after every 3,000 hours of operation, or about once every one to two years. This procedure is much like overhauling an automobile engine and costs a good deal more than a conventional furnace tune-up. However, the sytem is cost-effective even with this overhaul requirement.

In the system, space heat is the primary product, and electricity is the by-product. Since the demand for electricity at any one time may be less than the amount being generated, the excess electricity must be sold to a utility. Conversely, when electricity demand is greater than what is being supplied by the system, additional back-up power must be available. As in the case of wind-generated electricity, neither the prices utilities pay for excess power nor their rates for back-up electricity have traditionally been favorable to cogenerators. Many cogenerators have chosen to be completely independent of electric utilities as a result. Like a wind system, a cogeneration system that must sell much excess electricity, only to buy it back at a later time, may suffer economically.

Nonetheless, at the present time, the economics of a system like TOTEM appear generally favorable when compared to most other space-heating systems (see table 6-1).

The next five years should see the development of diesel cogenerators as small as 4 kilowatts, costing not much more than conventional heating systems and capable of operating 10,000 to 15,000 hours between overhauls. Widespread use of these small units, however, will depend upon reduction of noise and toxic emissions.

Community-Scale Cogeneration

Large cogeneration units, which have had a long and successful operating history, are quite common and more durable than small-scale units. Units that can produce 50 to 100 kilowatts, can heat multi-dwelling apartment buildings and are fueled by natural gas or diesel fuel presently cost about $400 to $600 per kilowatt when installed. Units of 200 to 2,000 kilowatts that operate on residual fuel oil or diesel fuel are also available and are suitable for large apartment buildings or small district heating systems. They also cost about $400 to $600 per kilowatt when installed. Larger systems (3,000 or more kilowatts) have an installed cost of around $600 to $1,000 per kilowatt, depending upon fuel source. Table 6-2 contains a comparison of heat and electricity costs for large-scale systems.

These large systems operate at around 35 percent electrical conversion efficiency and 45 percent heat conversion efficiency, which means that 80 percent of the energy in the fuel is converted to heat or electricity. They are generally sized for their space-heating capability

Table 6-1
Comparison of Costs of Space Heat from Small-Scale Cogeneration and Conventional Heating Systems
(in dollars per million Btu)

System*	Climate Zone†			
	I	II	III	IV
Electric resistance‡	21.08	21.85	21.40	21.18
Gas central §	10.45	9.08	8.28	7.87
Oil central ‖	15.14	14.24	13.43	13.03
TOTEM#	13.64	6.86	4.60	3.46
TOTEM & heat pump**	21.89	9.20	4.23	1.75

*Systems are assumed to have the following costs: electric resistance, $1,000; gas system, $2,000; oil system, $2,000; TOTEM, $8,000; TOTEM and heat pump, $13,000. Operating and maintenance costs are assumed to be 1% of capital costs or $60/year, whichever is greater. System is paid for by 10-year, 14% loan. Systems are installed in buildings with the following seasonal heating loads: Zone I, 89.6 million Btu; Zone II, 179.2 million Btu; Zone III, 269 million Btu; Zone IV, 358.3 million Btu. Oil is assumed to cost $1.25/gallon; natural gas, $0.50/therm; electricity, $0.07/kilowatt hour.

†Energy cost is calculated by dividing total annual cost by heat supplied to building. Annual cost is sum of loan payments, fuel costs and operating and maintenance costs, minus any credits for electricity generated.

‡Standard electric resistance system with efficiency of 100%.

§Standard gas-fired system with seasonal efficiency of 70%.

‖Standard oil-fired system with seasonal efficiency of 70%.

#TOTEM is sized to meet seasonal heating requirement; surplus electricity is sold to utility. Value of electricity generated is deducted from total annual cost.

**TOTEM supplies 57% of heat load; electricity is used to operate a water source heat pump with a COP=2.5 in order to meet remainder of space-heating requirements. Value of electricity generated by TOTEM and used by heat pump is deducted from total annual cost.

rather than for the electric power they produce. Since it is inefficient for these units to cycle on and off frequently or to run at less than full capacity, it is common for a number of smaller units to be installed rather than one large unit. Then, only units that are necessary are operated at one time. Often, as in the case with small cogeneration units, more heat is supplied than can be used, so these systems may also include some heat storage components.

Large-scale cogeneration can provide district heating in different ways (see figure 6-1). Cogeneration units could be installed to provide process or space heat to an industrial, commercial or apartment complex and electricity to both the cogenerator and nearby homes. The heat would be provided directly or stored until needed. The heat from a cogeneration unit could also be used as a heat pump source, perhaps with electricity from the unit powering the

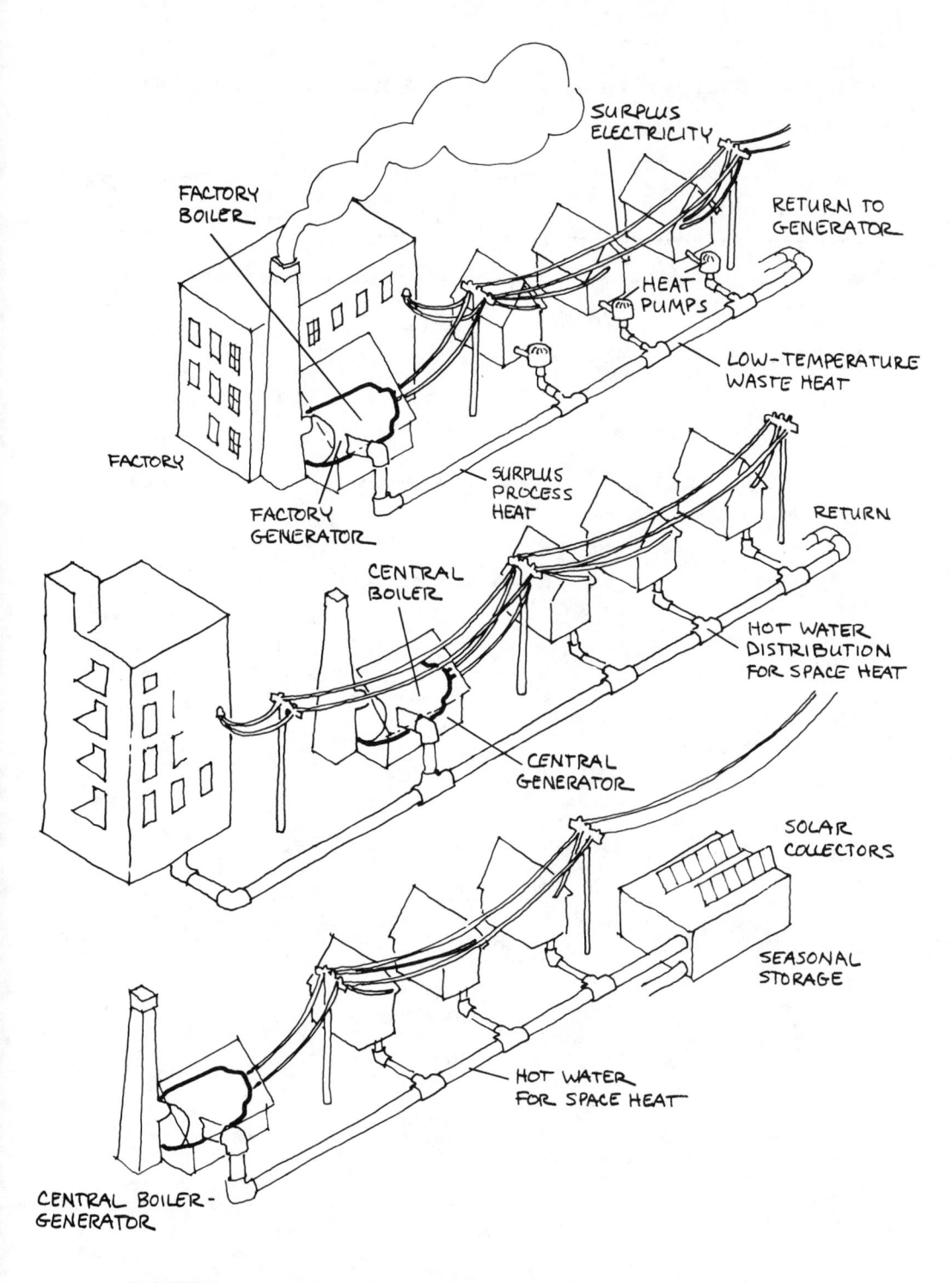

FIGURE 6-1: COGENERATION AND DISTRICT HEATING

Table 6-2
Comparison of Costs of Large-Scale Cogeneration with a Conventional Gas-Fueled System

System and Zone*,†	Capacity		Electricity Generated (KWH/yr)	Energy Cost ($/10⁶Btu)
	Btu/hr	Kilowatts		
(Cogeneration system supplying 80% of building heat requirements)‡				
Zone I	376,800	60	82,590	9.06
Zone II	484,800	78	164,935	5.97
Zone III	700,000	112	248,050	5.70
Zone IV	864,000	138	327,300	5.34
(Cogeneration system supplying 40% of building heat requirements and using heat pumps to supply remaining 40%)§				
Zone I	188,400	30	41,295	9.25
Zone II	242,400	39	82,468	5.66
Zone III	350,000	56	124,025	5.42
Zone IV	432,000	69	163,650	5.01
(Gas-fueled system with seasonal efficiency of 75%, supplying 80% of building heat requirements)‖				
Zone I	376,800	. . .	. . .	14.46
Zone II	484,800	. . .	. . .	10.67
Zone III	700,000	. . .	. . .	9.39
Zone IV	864,000	. . .	. . .	8.79

*Design-day building heat requirements are: Zone I, 471,000 Btu/hr; Zone II, 606,000 Btu/hr; Zone III, 875,000 Btu/hr; Zone IV, 1,080,000 Btu/hr. Seasonal heating requirements are: Zone I, 646 million Btu; Zone II, 1,290 million Btu; Zone III, 1,940 million Btu; Zone IV, 2,560 million Btu.

†Cogeneration systems are assumed to cost $600/kilowatt, installed; heat pumps, $1,000 per 10,000 Btu/hr of capacity; gas system, Zone I, $30,000; Zone II, $35,000; Zone III, $40,000; Zone IV, $45,000. Systems are financed by 10-year, 14% loan. Energy cost is annual cost of system divided by space heat supplied to building; annual cost is sum of annual loan payments, operation and maintenance, and fuel costs.

‡System supplies 80% of building design-day heat load with heat recovery efficiency of 55% and electrical generating efficiency of 30%. Energy cost includes credit for generated electricity.

§System supplies 40% of building design-day heat load with heat recovery efficiency of 55%, electrical generating efficiency of 40%. COP=2.5 water source heat pumps supply remaining 40% of heat load. Energy cost includes credit for generated electricity.

‖Standard gas-fired boiler with seasonal efficiency of 75%, supplying 80% of building design-day heat load.

heat pumps. Widespread use of cogeneration systems would benefit electric utilities by operating during periods of peak electrical demand, thereby lessening the need for new and expensive peak generation facilities.

Is Cogeneration for You?

Utility attitudes and policies do not generally favor small power producers. Electric utilities are in the business of generating and selling electricity. They tend to view small power producers as competitors and have established rate structures that tend to discourage independent power generation. In the long run, cogeneration will benefit the utilities because it will reduce the need for expensive new generating capacity, but at the present time electric utilities have more generating capacity than they can use and therefore have little reason to encourage cogeneration. The Public Utilities Regulatory Policy Act (PURPA), which may not cover certain diesel engines, requires utilities to buy surplus power from and to supply back-up power to small power producers and cogenerators at nondiscriminatory and "fair" rates. Unfortunately, this does not necessarily mean they must be cooperative.

Safety. Cogeneration equipment must be safely connected to the utility grid. Utilities frequently object to independent power generation by arguing that safety hazards might threaten workers if independent systems continue to operate during system-wide black-outs. Such problems can be avoided by installation of appropriate, standard safety equipment at the cogeneration site.

Flexibility. It's a good idea to invest in a cogeneration system that can use different fuels, since the availability of natural gas, residual fuel oil, heating oil, diesel fuel and gasoline for cogeneration is not assured over the next several decades. The potential for conversion to an alternate fuel source such as coal liquids or wood gas should be an important factor in choosing a system.

Cost-effectiveness. Large-scale systems may be more cost-effective and preferable to smaller ones. If a good system is properly sized and installed it will cost less per unit of energy produced, cause less air pollution per unit of power and be less expensive to maintain per unit of power than small cogeneration units. In some cases, though, it may cost less to install a small, on-site cogeneration system than to install the distribution system required for a large cogenerator. Cost-effectiveness is highly dependent upon the particular situation. Cogeneration must, of course, compare favorably with the cost of conventional heat and electricity supplies in order to make economic sense. When determining the cost of owning and operating a cogeneration system, you must include regulatory, maintenance, repair and fuel costs. Because of the maintenance requirements, purchase of a small cogeneration unit may not be advisable for all applications even though the economics are favorable for some systems. For larger installations, the regulatory requirements may be difficult and expensive to meet.

Access to other facilities. If there is a potential cogenerator in your neighborhood such as a factory, it might make sense to discuss the possibilities for such a system with the managers or owners of the facility. Factories often require heat and electricity but not in equal amounts. Any surpluses could be sent into the neighborhood.

Sizing the System

While choosing an appropriate-size cogeneration system might seem fairly straightforward, the actual sizing of the system, particularly for small cogenerators, can be complicated. The relatively high cost per kilowatt of power from small systems makes it imperative that a number of factors be considered. First of all, is heat more important than electricity, or is the opposite the case? If heat is more important, the system should be sized to meet all or a sizable portion of the design-day heating requirements. In sizing the system, it is also important to keep the amount of excess electricity low. Otherwise, if power has to be occasionally purchased from or frequently sold to the utility, the economics can become unfavorable. For very small systems (less than 15 to 20 kilowatts) there is little that can be done to avoid this, and it may simply be necessary to accept these problems. For larger systems, you can install several small units rather than one large one. If only one cogeneration unit is to be installed, electricity will be generated only intermittently on all but the coldest days (when there is a continuous demand for heat) and then at levels in excess of demand. On the other hand, if multiple, smaller units are in place, at

Table 6-3
Sizing a Cogeneration System to a Residential Building

	Climate Zone			
	I	II	III	IV
Approximate building design-day heat load (Btu/day)	612,500	787,500	1,137,500	1,400,000
Required hourly heat capacity (Btu/hr)	25,520	32,810	47,400	58,330
Generating capacity (kw)	2.95	3.80	5.50	6.70
Kilowatt-hours generated on design day	71	91	132	161
Kilowatt-hours generated on average day	35	45	66	80
Kilowatt-hours generated during heating season	4,040	8,080	12,120	16,160

NOTE: System is assumed to have efficiency of 66% for heat conversion and 26% for electrical generation. Generation capacity is therefore determined by heating requirements of the building. The building is 2,500 square feet with a 17,500 Btu/DD heat load.

least one of the units can be operating continuously, providing electricity at all times.

If some of the electricity generated is used for space heating, the system can be downsized by about one-third. If the electricity is used to power water source heat pumps, an even smaller system is required. Table 6-3 shows the cogeneration capacity necessary to meet the heating loads of the four building prototypes.

Environmental and Regulatory Aspects of Cogeneration

Depending upon the type of system and its size, location, ownership and purpose, a cogeneration unit may or may not fall under the provisions of one or more of the many environmental and regulatory acts that cover power generation and industrial installations. Furthermore, because coverage by these various laws seems to be constantly changing, it is virtually impossible to ensure that what is written in one month will be valid the next. Nonetheless, there are a variety of regulations that could cover cogeneration.

The regulation of residential systems is fairly unclear, but it appears that household-size systems (5 to 100 kilowatts) are likely to be exempt from all environmental regulations except local building and zoning codes.

Larger systems with a capacity in the area of some 500 to 2,500 kilowatts and capable of serving 400 to 2,000 homes appear to be exempt from most, but not all, regulations. In all likelihood, they must satisfy the provisions of the Clean Air Act of 1970. To meet these provisions, a system must comply with emission limits for five pollutants:

nitrogen oxides, sulfur dioxide, small suspended particulates in the air, carbon monoxide and photochemical oxidants (chemicals found in smog). State regulations may also apply to small cogeneration systems; you should check with the environmental department or agency for your state.

Systems with a generating capacity of 75,000 kilowatts or less appear to be exempt from almost all federal regulations governing power generation; however, experience with residential-scale cogeneration in most states is limited, and the provisions of PURPA (which apply to small power generators) have yet to be tested. If a system is more than 50 percent utility-owned, it is subject to a broad range of statutory regulations governing commercial power generation. These include state and federal laws and regulations covering electricity prices and sales, facility siting, fuel use and environmental compliance. Other regulations that may affect small cogeneration systems include those governing noise pollution, water discharge and solid waste disposal.

Large cogeneration systems are generally defined as those with outputs greater than 2,500 kilowatts. Such systems must comply with the provisions of the Clean Air Act. Systems larger than about 75,000 kilowatts, or that sell 25,000 kilowatts or one-third of their generating capacity, whichever is greater, must also comply with the Environmental Protection Agency's New Stationary Sources Performance Standards for Electric Utility Steam Generating Units, which sets pollutant limits for new power plants. Large systems may also be subject to the Powerplant and Industrial Fuel Use Act of 1978 which restricts new oil and gas burning and encourages coal consumption. In densely populated areas, a large cogeneration system may be required to comply with other emission standards, to install pollution control technology, to obtain an emission offset (intended to maintain air quality in certain areas by requiring that pollution levels not rise above certain limits), to meet noise pollution standards and to obtain water and air discharge and solid waste disposal permits.

As this list of regulations suggests, installing and starting up a large cogeneration system can be a difficult proposition. Encouraging cogeneration while maintaining environmental and neighborhood quality and keeping nearby residents happy is a problem that, to date, has not been satisfactorily resolved.

DISTRICT HEATING

District heating involves supplying hot water for space heating and possibly domestic hot water from a central production facility to residential or commercial buildings by means of a large-scale network of water distribution mains. District heating networks are common in Europe, serving large portions of the populations of various countries. For example, in Sweden, 25 percent of the population is served by district heating; in Denmark, 33 percent; in the Soviet Union, 55 percent; in Iceland, 50 percent. In the United States, by contrast, district heating serves only about 1 percent of the population, and this through antiquated steam supply systems.

The advantage of district heat is that it replaces relatively inefficient home heating systems with a more efficient, centralized boiler or cogeneration

system, thereby offering the potential for significant energy savings. Although some heat is lost during distribution of hot water, energy savings are still substantial. Furthermore, district heating systems can use the waste heat from electric generation and industrial plants that would ordinarily be released to the air or to nearby water supplies. For this reason, some observers have suggested that district heating could save as much as one billion barrels of oil per year in the United States, which is about one-half of our total oil imports from foreign suppliers. Not all district heating systems can be considered equally attractive, however. Some people have suggested that systems be built using the waste heat from nuclear power plants.

Heat Production

A district heating system has three main parts: a heat and possibly an electric generating facility, a heat distribution system and a heat delivery system. Heat production is by far the most variable component of a district heating system. Most existing systems use either a centralized boiler or a cogeneration system to produce heat. Large, centralized oil-fired boilers can remove as much as 90 percent of the energy contained in the fuel. Cogeneration systems may have a total heat and electricity efficiency approaching this. One benefit of district heating is that large boilers can use residual fuel oil, which costs less than the distillate fuel oil used in homes and which produces less sulfur dioxide pollution. In a number of European cities, waste heat from fossil fuel electric power plants is used for district heating purposes, with an overall energy use efficiency of 85 percent. (These power plants were not originally constructed as cogenerating units.)

A number of other heat sources for district heating systems are also possible. Waste heat from industrial process plants can supply heat to nearby residential areas. Geothermal sources can also provide heat for district heating systems, which is done in Iceland and Boise, Idaho. District heating based on renewable heat sources such as solar, wood, urban waste incineration and wind is possible; such a system is being planned for Uppsala, Sweden. In fact, the use of solar energy to heat enormous tanks of water for district heating purposes is one of the more promising and interesting concepts that is being studied.

Distribution

Hot water is the most practical choice for distributing heat, because it is economical and can be transported over long distances with little heat loss. This allows larger distribution networks to serve more people. Steam heat distribution systems, in contrast, can only serve high-density urban core regions. For example, the largest steam system in the United States, that of New York's Consolidated Edison Company, serves only a small fraction of Manhattan Island.

The distribution system normally consists of an intricate network of pipes running beneath the service area. Large pipes, or mains, carry 200 to 250°F supply water under pressure from the production facility to delivery points,

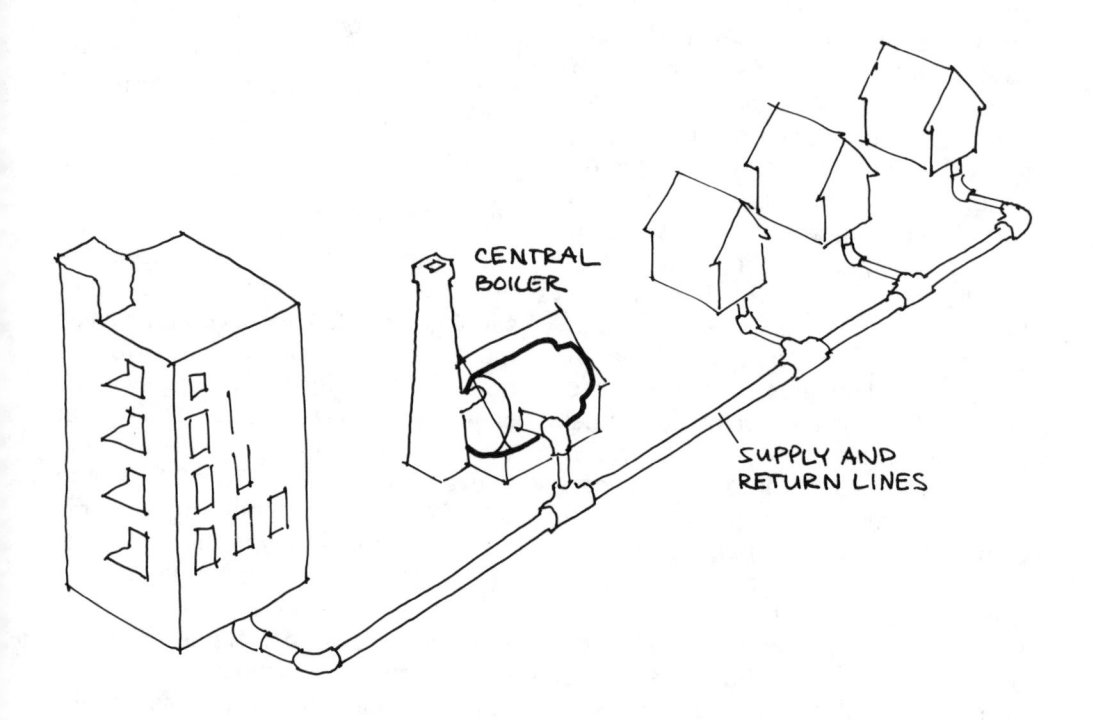

CENTRAL BOILER

SUPPLY AND RETURN LINES

FIGURE 6-2: DISTRICT HEATING

while other mains return the cooler, "used" water at 120°F to the central facility. (Water boils at 212°F, but if it is kept under high pressure, its boiling point is raised.) Smaller mains branch off from the large ones into neighborhoods and down streets. Each home is served by a supply line and a return line. The distribution mains are made of welded steel pipe buried in insulated trenches or culverts.

The installation of the distribution system is by far the most costly part of a district heating system, because it requires extensive excavation and disruption of streets. The immense cost of this excavation is one of the major arguments against district heating. In Europe, many district heating systems were installed during the rebuilding that followed World War II. Newer systems built since then have required extensive excavation, but the cost has not been considered burdensome. These costs might be reduced with the use of new types of pipes, insulating materials and excavation techniques now being developed in Sweden. Plastic piping (in long rolls) is laid in plastic insulation and placed in narrow trenches. Using these techniques, hundreds of feet of this pipe can be laid in the time it would take to install much shorter lengths of conventional steel pipe. The use of plastic pipe eliminates problems due to ground shifting that have plagued steel pipe systems. However, the use of plastic pipe can increase the amount of oxygen in the system, which causes corrosion of metal radiators. Conventional radiators may have to be replaced by plastic ones. The cost of a plastic pipe distribution system is estimated to be slightly more than half that of a conventional steel one.

An alternative way to reduce the expense and disruption caused by excavating streets is to run the smaller supply and return mains through the basements of adjacent buildings. Excavation would then be mostly limited to yards and driveways, rather than streets. This approach would also eliminate most of the need to run supply mains from the street into buildings.

Delivery

The delivery system for district heating is fairly conventional. In many instances, hot water delivery systems can be run directly to existing radiators with no alterations unless conventional radiators are replaced by plastic ones. District heating systems supplying lower temperature water (less than 100°F) require boosting which can be accomplished with heat pumps installed in each building. In general, except for the absence of a boiler or furnace, district heat-supplied space heating is indistinguishable from conventional space heating.

The Cost of District Heating

District heating systems are usually quite expensive. A system capable of serving 100,000 homes could cost as much as $1 billion or $10,000 per home. (Conventional heating systems cost from $3,000 to $4,000, heat pumps from $4,000 to $5,000 and active solar systems from $9,000 to $18,000.) District heating systems could be financed by municipal bonds at low interest rates, to be repaid over a 30- to 40-year period. In that case, the annual cost per home could be competitive with or less than that of conventional heating systems. On the positive side is the fact that district heating systems using waste heat from industrial and generating plants are less susceptible to fuel price increases than conventional systems. District heating systems that used renewable energy sources would be unaffected either by price or availability of fossil fuels. If district heating were to serve half the population of the United States, as some suggest, the cost of service could run about $300 billion, or $20 billion per year over 15 years. This cost, while very great, should be compared with the $30 billion per year that the district heating would save in energy costs.

Considerations before Investing in District Heating

If district heating is so great, why hasn't it taken the nation by storm? Until recent years, of course, district heating was not cost-effective in comparison with conventional home heating systems. Also, because a district heating system would presumably be constructed and operated by a local governmental agency and thereby infringe upon an area traditionally reserved for private enterprise, there has been considerable resistance to such arrangements. District heating could drive home heating suppliers, furnace manufacturers and heating equipment suppliers out of business, and the installation phase of the system could be quite disruptive. Finally, because all existing systems in the United States use steam, it has generally been assumed that new systems would also have to use steam, which is incapable of supplying heat to large numbers of people.

On the other hand, the energy conservation potential of district heating is quite large and the positive effects upon air quality would be significant. Because the cost of the system would be known upon completion and because additional fuel charges would increase more slowly than the actual price of the fuel, you would end up paying a monthly bill that would rise only at a slow rate. Furthermore, construction of the system would generate large numbers of new jobs, both skilled and unskilled, in the inner cities.

Planning for District Heating

If district heating sounds like a promising heat supply system for your neighborhood or community, there are several things you should do and know that may prove helpful in speeding up its development. First, you should survey your neighborhood or community for industrial facilities, hospitals, power plants and other sources of waste heat. One way to do this is to look for cooling towers or clouds of steam. This waste heat could be a useful heat supply source for a district heating system. Second, you should try to find out if any new facilities that produce significant amounts of waste heat are planned for your area. If they are, inquire about what will be done with the waste heat from the facility and press for consideration of a plant cogeneration system. Third, if no such plants exist or are planned, you should study your area for large open spaces that might be used for solar collection and hot water storage facilities. A district solar heating system could be an attractive possibility. Fourth, find out whether any major street repairs are planned for your area. That would be a good time to press for district heat, since the first water mains could be installed at relatively low cost. Fifth, you should survey your neighborhood to determine the condition of individual heating systems. For example, if you live in an area of mostly triple deckers and each apartment requires a new heating system at a cost of $3,000, this adds up to $9,000 per building, not much less than the cost of district heat. Homes should, of course, be weatherized before any consideration is given to installing a new heating system. Finally, you must collect financial and design information. This should be done in cooperation with a consultant or other expert in district heating.

SEASONAL ENERGY STORAGE SYSTEMS

A seasonal energy storage system is designed to store natural heat or coolth during one season, when it is not needed, for use during another season. These systems have a large energy storage component, although the modes of collection and distribution of the heat or coolth vary. While fairly complex technologically, these systems have significant energy-conserving potential because they collect essentially "free" heat or coolth when they are in abundance and save them until required. Furthermore, the only energy consumed is that needed to run the various parts of the system. Three such systems are discussed here: the *Annual Cycle Energy System*; the *Integrated Community Energy System*; and the *Annual Storage Solar District Heating*. The first two systems can provide both heating and cooling; the third, heating only. Many of the systems described here are still in the early stages of development.

Annual Cycle Energy System

The Annual Cycle Energy System (ACES) has two basic parts: a very large insulated storage tank of water and a heating-only heat pump. The tank contains coils of pipe filled with brine (salt water) warmed by the water in the tank. The brine, in turn, circulates through a heat exchanger and transfers its heat to the heat pump refrigerant. During the heating season, heat is removed from the water tank (which stays at a temperature near 32°F) by the brine and transferred to the building at a useful temperature of 100 to 130°F. The system may also be used to provide domestic hot water.

As heat is removed from the tank, the temperature of the water falls below the freezing point and ice begins to form on the brine circulation coils. By the end of the heating season, ice fills the entire tank. (If the winter is a particularly cold one, the tank may become ice-filled before the end of the heating season. In this case, it is necessary to supply some extra low-temperature heat, perhaps from a small solar system, to melt some of the ice.) This ice can then be used during the summer to provide chilled water for air conditioning. As long as ice remains in the tank, the only power required for cooling is that needed for operation of a small circulator pump and a fan. The annual coefficient of performance (the ratio of heat produced to energy consumed) for heating and cooling of such a system can be as much as 3.5. For the cooling cycle alone, the COP can be as high as 12. Accounting for efficiency losses at the power plant, such a system delivers about 120 percent of the energy in the original power plant fuel.

DISTRICT SCALE ACES

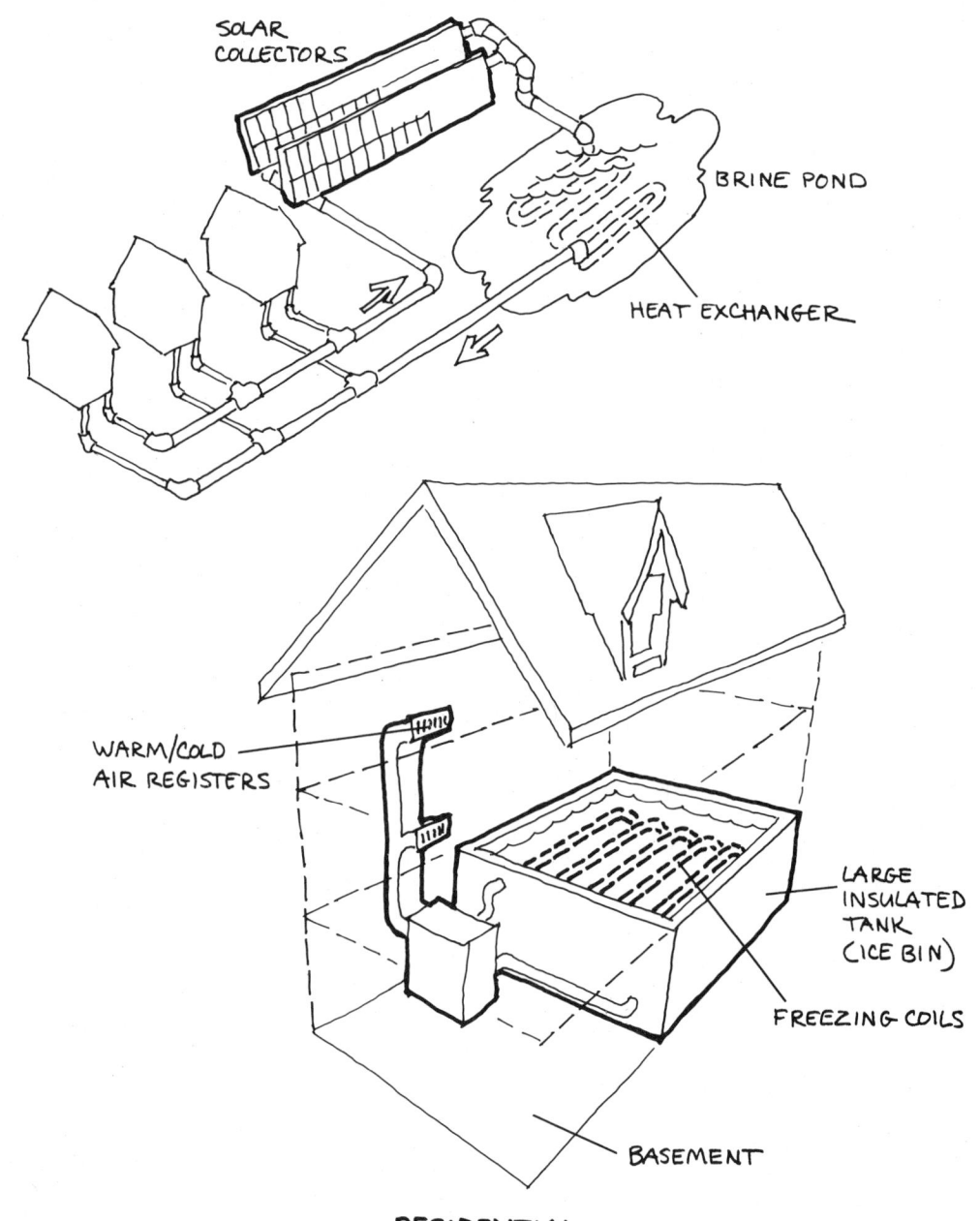

SOLAR COLLECTORS

BRINE POND

HEAT EXCHANGER

WARM/COLD AIR REGISTERS

LARGE INSULATED TANK (ICE BIN)

FREEZING COILS

BASEMENT

RESIDENTIAL SCALE ACES

FIGURE 6-3: ANNUAL CYCLE ENERGY SYSTEMS (ACES)

An ACES is generally sized to meet the summer cooling requirements, rather than the winter heating load, of a building. In order to meet the total heating requirements of a building, an ACES is best suited for climates in which the heat provided to the building from the tank during the winter roughly equals the heat to be removed from the building for cooling and transferred back into the tank during the summer. This condition occurs in areas where winter and summer climates are not too extreme, such as Maryland and Virginia.

At the present time, the ACES is not available commercially, but it is being tested in several experimental installations. Data gathered from these buildings show that for heating, the system requires only about 44 to 51 percent (COP=2.27) of the electricity consumed in a similar house with conventional electric resistance heating. It is somewhat more efficient than a conventional air-to-air heat pump system, because the heat source is maintained at a constant, known temperature to increase heat pump efficiency. (See Heat Pumps in chapter 3.) In a moderately cold climate (6,000 degree-days), an ACES would use about 25 percent less electricity than a conventional heat pump with a coefficient of performance of 1.5.

ACES Economics

The first cost of an ACES is much higher than that for conventional home heating and cooling systems, primarily because of the added cost of the storage tank. The estimated cost of an ACES for a home is about $10,000 to $13,000. Energy savings from such a system in a house with electric resistance backup would be about $1,000 per year, which indicates a 10- to 13-year payback. These costs are based on systems constructed from off-the-shelf components, rather than complete commercial systems.

Integrated Community Energy System

Integrated Community Energy System (ICES) is a generic term for a class of district heating (and cooling) systems that uses heat pumps to collect and concentrate energy. The use of heat pumps, rather than fossil-fueled boilers, is advantageous because "free" heat that would otherwise be lost can be removed from sources such as industrial and commercial boiler waste heat, groundwater, lakes and reservoirs or solar and geothermal sources to be used for space heating purposes. Even though heat pumps must consume high-quality energy (electricity, oil or gas) in order to operate, they use these sources of energy more efficiently than ordinary centralized boilers.

All ICESs have three major components: heat pumps, a large heat source which may also act as heat storage and a distribution system. Beyond this, ICES designs can vary in a number of ways in terms of heat pump size, operation modes and type of heat source.

ICES Components

The heat pump portion of an ICES can be placed in a system in three different ways. Based upon this, the system is either centralized, distributed or cascaded. In the *centralized system*, one or more large heat pumps function

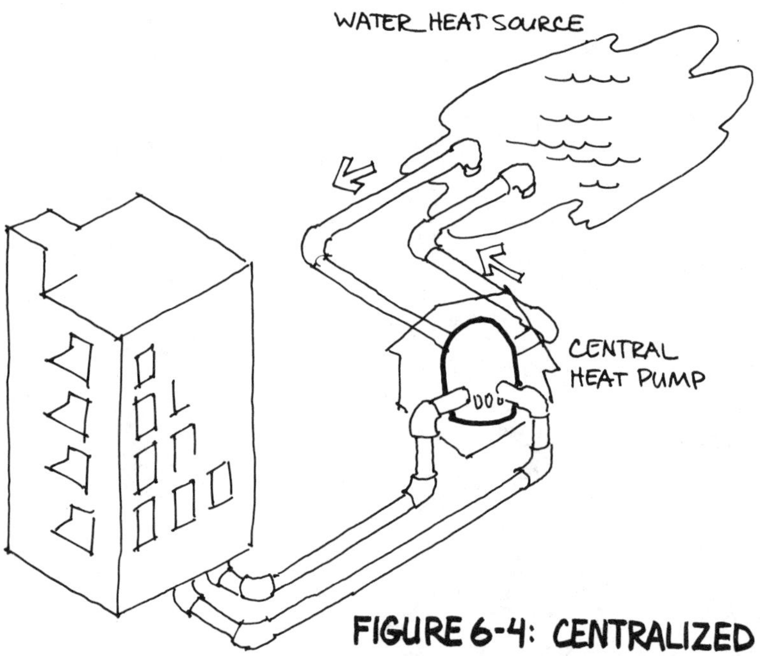

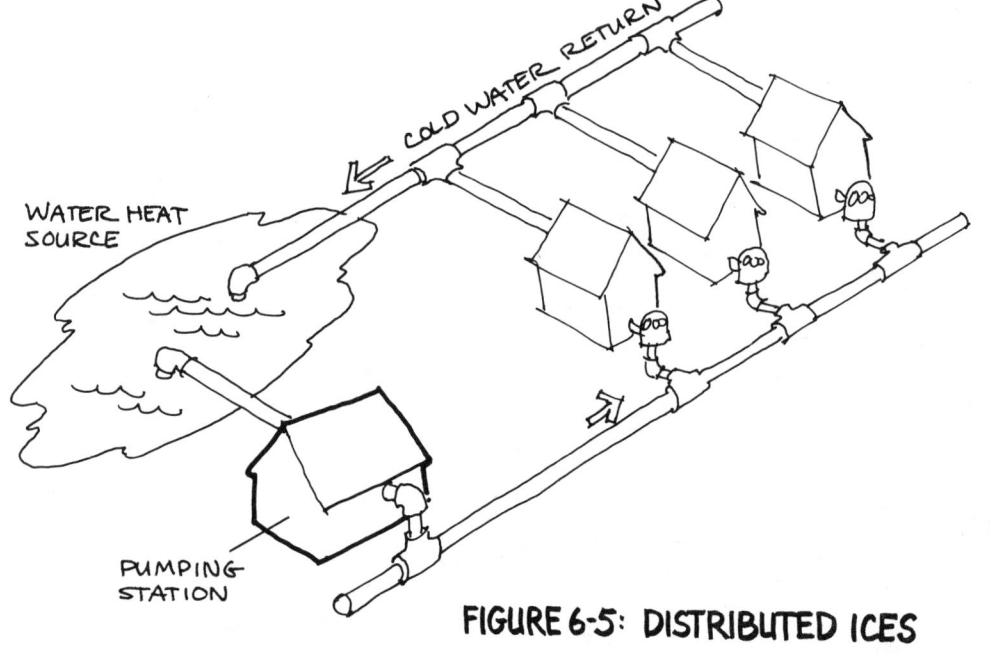

WATER HEAT SOURCE

CENTRAL HEAT PUMP

FIGURE 6-4: CENTRALIZED INTEGRATED COMMUNITY ENERGY SYSTEMS (ICES)

COLD WATER RETURN

WATER HEAT SOURCE

PUMPING STATION

FIGURE 6-5: DISTRIBUTED ICES

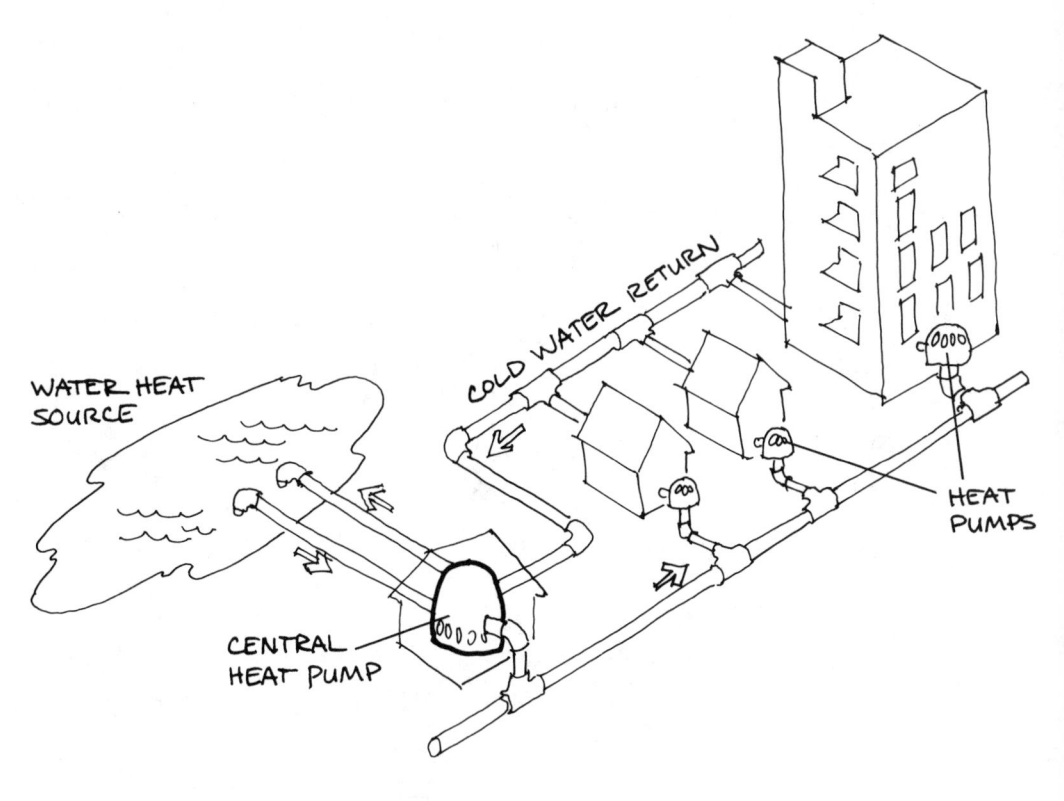

FIGURE 6-6: CASCADED ICES

in a manner very similar to the centralized boiler of an ordinary district heating system. The heat pumps are located in a central facility, and they remove heat directly from a heat source. This heat is then used to warm distribution water, which is then pumped to individual buildings.

In the *distributed system*, small heat pumps are located in each building. Water from the heat source is sent directly to an individual heat pump. Heat removed from the distribution water is then used to warm the building. Some heat pumps may be located in

large office buildings and may also provide cooling. In this case, the heat from the office building is rejected into the warm side of the distribution system.

The *cascaded system* has both centralized and individual heat pumps. The central heat pump removes low temperature heat from the primary source and adds it to the distribution water, which is sent to individual buildings. Heat pumps in the buildings then use this distribution water as a secondary heat source. This system is used when the primary source water is too corrosive (such as salt water) or contaminated (such as sewage) to be piped to the individual buildings.

Distribution. The distribution system of an ICES is identical to that of a conventional district heating system. Each ICES generally has warm water supply and cool water return mains. Systems that supply both heating and cooling at the same time (for example, when both houses and office buildings are serviced) may have independent distribution systems for hot and cold water. Distributed systems using groundwater as a heat source may have only a distribution water supply line; the return water can be dumped directly into a sewer. It is important to note that cascaded and distributed ICESs have separate heating distribution systems for each building and therefore differ slightly from conventional district heating systems.

Heat source. The heat source is generally the most variable part of an ICES. Virtually all such systems use a large body of water for this purpose because water provides a concentrated source of heat, usually at a fairly constant temperature. Depending on what is accessible and the severity of winter climate, the heat source can be a lake or reservoir, a fabricated underground storage tank, an aquifer (an underground river or lake), solar-heated water, sewage and waste water, geothermal energy or waste heat from large industrial or commercial facilities. For those ICESs that serve both small and large buildings, surplus internal heat from the large buildings can be used to provide source heat to smaller ones. ICESs designed for areas with moderate winter temperatures may use air as a heat source.

Heat storage. Not all ICESs are actually seasonal storage systems. Those that depend upon waste heat or geo-

thermal energy, for example, are operated on the assumption that the heat source will be available year-round. Systems using lakes or reservoirs depend on the natural collection of heat by these water sources during the year. The solar energy-based system collects heat during the summer as does the system using a fabricated storage tank. An ICES supplying simultaneous heating and cooling may have short-term hot and cold stores in which to keep extra heat or coolth that might be produced during the course of a day. An aquifer can sometimes be used to store warm water, too, if groundwater flow is sufficiently slow.

ICES Operation

The operation of an ICES depends primarily upon two factors: the nature of the heat source and whether the system is centralized, distributed or cascaded. If the system is a centralized one with a *surface water* (lake or reservoir) heat source, heat is removed directly from the source and transferred to points of use through the distribution system. Otherwise, source water is pumped directly to buildings for use by individual heat pumps. Use of this type of heat source is possible only in warmer climates or with very large sources, because freezing would otherwise render it useless.

An *aquifer* heat source is suitable for a distributed ICES, because the extent of the distribution system can be physically limited by the nearness of the source. All that is required are two wells—one for groundwater supply and the other for groundwater return. The groundwater is distributed to individual heat pumps and then returned to the aquifer. If the aquifer is sufficiently large, pairs of wells can be drilled at conveniently spaced intervals throughout the community. In a groundwater system that supplies both heating and cooling, the aquifer can be used to store heat or coolth. If warm water is injected and withdrawn only from the upper part of the aquifer and cool water is injected and withdrawn only from the lower part, the slow movement of water in the aquifer and the difference in temperature will help to keep the warm and cool layers separated.

Geothermal sources need not be very hot to be used as a heat pump source. In fact, many geothermal sources of water have a temperature range of 70 to 90°F. Such sources exist in many parts of the United States. An ICES using a geothermal source would probably have to be cascaded, with a large centralized heat pump and additional heat pumps in each building. This would be necessary because mineral deposits from geothermal sources could block pipes and pumping equipment. Heat is removed from the source water and added to the distribution water by the central heat pump. The individual heat pumps then use the distribution water as a source of heat for individual buildings.

Solar energy can be used to warm heat pump source water. In this system solar collectors are mounted on a large, fabricated, insulated water tank in which the warmed water is stored. Most of the heat is collected in the summer for use during the winter. In the winter, the hot water can be used directly for space heating until it is too cool (about 85 to 90°F). The remaining heat can be removed and concentrated by a centralized heat pump.

An ICES using a large fabricated *tank of water* as a heat source can

operate as a community-scale ACES. The water in the tank is at a temperature slightly higher than 32°F. During the winter a centralized heat pump removes heat from the tank, causing the formation of ice. This ice is then used for summertime air conditioning or for winter cooling of large buildings.

Sewage and wastewater heat sources are generally not too much colder than the buildings from which they come. A cascaded ICES can remove heat from waste water and transfer it to the distribution system which then acts as a secondary heat source for heat pumps in individual buildings.

Waste heat is often rejected into the environment by industrial facilities in the form of hot water. This hot water can be used directly by heat pumps in a centralized ICES.

If a number of large buildings are included in an ICES, the *internal heat* generated in these buildings can be used to provide residential space heating in either a centralized, distributed or cascaded system.

Advantages of ICES

ICESs have a number of advantages over conventional district heating systems or individual building heating systems. An ICES will commonly serve business, commercial and residential districts. Because the peak heating and cooling demands of these different sectors do not occur at the same time of the day, a single moderately sized system can meet the varying peaks of these different sectors. If an ICES contains a short-term heat storage component, such as a water tank, the system can operate continuously and at a steady level around the clock with peak heat demand requirements being drawn from storage. Even though most

ICES have a primary heat source, such as a lake or aquifer, the presence of large buildings that require cooling when others must be heated provides an additional source of heat. This increases system reliability.

Conventional home heating systems burn fossil fuels at very high temperatures just to heat water to 120°F. District heating systems do the same. In both cases, when the hot water cools off to 90°F or less, it is no longer warm enough to supply heating. This remaining heat is eventually lost to the environment. The advantage of an ICES is that it can recover this low-temperature heat that would otherwise be wasted. This helps to increase system efficiency. Large-scale heat pumps in ICESs can be engine- or turbine-driven, using fossil fuels directly in place of electricity. This allows reductions in electricity consumption and more efficient use of fossil fuels.

ICES Economics

Studies of ICESs have generally found them to be economically competitive with conventional heating systems such as furnaces and/or boilers in individual buildings or district heating systems using fossil fuels. Because all ICES designs are fairly preliminary, exact cost comparisons do not exist. However, it has been estimated that the capital costs would be a good deal higher than those of conventional systems. ICESs have lower energy requirements than conventional systems, however, because "free" environmental energy is substituted for the burning of fossil fuels. In some ICESs, electricity consumption is greater than in conventional systems lacking heat pumps, but total consumption of all forms of energy is lower.

No ICESs has been constructed anywhere in the world, although a number of systems have been designed for specific communities and developments. If district heating is developed in the United States, it is more than likely that many of these systems will be central heat pump ICESs.

Annual Storage Solar District Heating

Both ACESs and ICESs are dependent upon heat pumps and rather elaborate storage systems, and both require significant amounts of energy to operate them. An annual storage solar district heating system, on the other hand, would be less complex and could offer the possibility of supplying almost all of a community's annual space heating requirements with minimal input of nonrenewable energy. Such systems cannot provide air conditioning so they are mainly suited to northern climates. They can work in these regions because over the course of a year even very northerly locations such as Canada receive as much sunlight per square foot as Saudi Arabia. The problem is that most of the sunlight falls in the summer when it is not needed for heating. That is the beauty of annual solar storage: the system collects heat in the summer for use during the winter.

An annual storage solar district heating system has three components: a heat store, a collecting area and a distribution system. The storage can be either an insulated earth pit or a below-ground concrete tank. Both have insulated concrete covers and are filled with water. Collectors are mounted on the cover of the storage tank and are rotated during the day so they always face the sun.

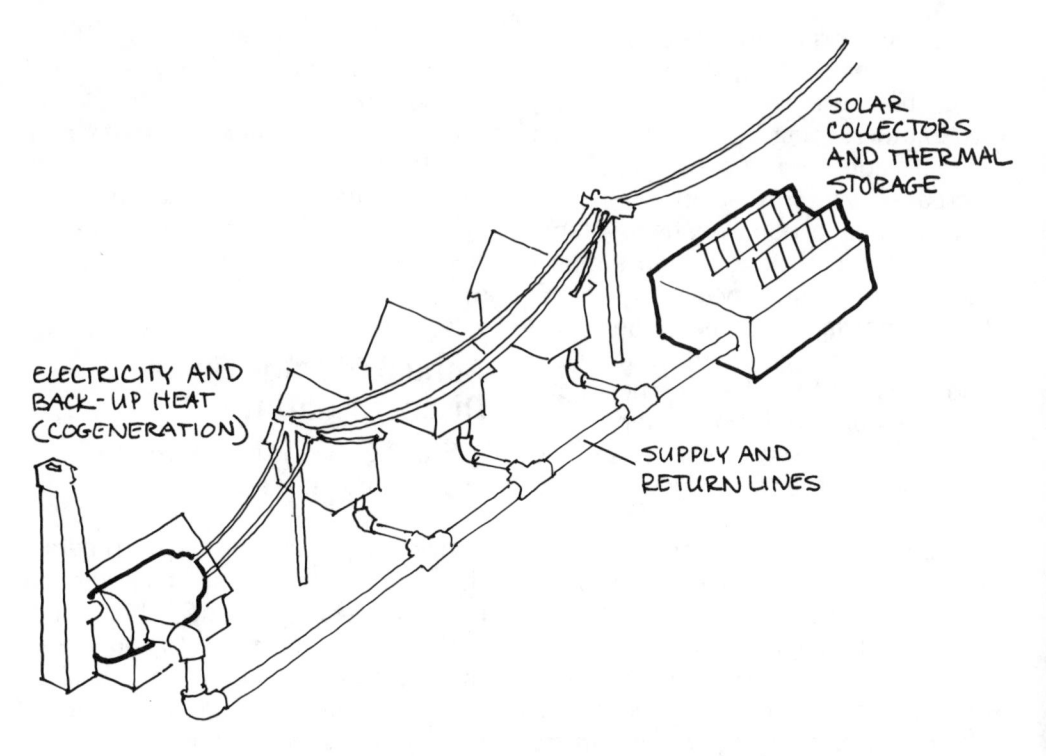

FIGURE 6-7: ANNUAL STORAGE SOLAR SYSTEM

During the summer, the collecting system heats water for storage and possibly for domestic hot water. During the winter, the collecting system also heats water that is used directly for heating purposes. Whenever additional heat is required, the hot water stored in the tank or pit is used. Water is removed from the topmost layers of the storage tank. The cooler return (used) water is then either pumped back through the collectors or into the bottom of the storage tank. With enough collector area and storage, it is possible to provide a community with 100 percent solar heating.

The distribution network for this system is identical to that described for district heating, and it therefore has similar cost problems that may be partially resolved through the use of the same cost-reducing techniques being applied to district heating in Sweden and elsewhere. Annual storage solar district heating systems have been installed in a 30-unit apartment building in Ontario, Canada, and at various locations in Sweden. The two existing Swedish systems serve groups of 50 and 500 homes, and larger systems serving 2,000 to 10,000 homes are being planned. All of these systems operate in latitudes far to the north of virtually all American cities.

Specifications and costs for an annual storage solar district heating system capable of serving 1,000 homes are shown in table 6-4. Such a system would supply 90 percent of the annual heating requirements for the homes in the community. Depending upon the climate zone, required collector area per house would range from 71 to 284 square feet, and the cost per house would range from $8,600 to $16,700. These costs would exceed those of conventional heating systems, but the total system cost might be reduced if residential heat loads were lessened through increased levels of weatherization and the addition of passive solar features.

The system considered here would heat water to temperatures between 90 and 170°F. If heat pumps were to be used, however, the system could heat water to lower temperatures (which would then be boosted by the heat pumps), and the cost of collectors could be reduced. The financing of such a

Table 6-4
Annual Storage Solar District Heating System

	Climate Zone			
	I	II	III	IV
Annual heat load per house (Btu)	15,000,000	30,000,000	45,000,000	60,000,000
Collector area per house (ft²)*	71	142	210	284
Storage requirements per house (gal)†	19,880	37,050	51,355	63,555
Cost per house ($)‡				
Storage @ $0.075/gal	1,500	2,800	3,800	4,800
Collectors @ $22.50/ft²	1,600	3,200	4,700	6,400
Distribution	3,500	3,500	3,500	3,500
Miscellaneous	2,000	2,000	2,000	2,000
Total per house	8,600	11,500	14,000	16,700
Storage tank surface area in ft² (assuming 35-ft depth)	75,715	141,140	195,715	242,000
Storage tank radius (ft)	155	212	250	278

SOURCES: System costs are from P. Margen, "Central Plants for Annual Heat Storage," *Solar Age*, October 1978. Conversation with Jack Gleason, MITRE-New England Sustainable Energy Project/District Heating Report, 1980.

NOTE: This table is based on sizing and costs for an annual storage solar district heating system serving 1,000 homes of 1,500 ft² each with 90% of their annual heating requirements. Houses have heat loads of 5 Btu/ft²/DD.

*Collectors assumed to have average 50% collection efficiency.

†Storage and distribution losses assumed to be 25%.

‡Water storage over 80°F temperature range.

system could be done through issuance of municipal bonds, which could then be repaid through monthly bills to the consumers of the system's energy. Since the lifetime of the system would far exceed that of conventional heating systems, perhaps reaching 50 to 60 years, repayment of the bonds could extend over a far longer period than is customary with solar systems, and monthly payments could be reduced accordingly.

Solar district heating offers a number of advantages over conventional single-residence active systems. First, by placing the collectors in a set-aside open area, problems with ensuring access to the sun do not arise and future problems can be avoided. Second, heat storage capacity is not constrained by space limitations in any one building; instead, the storage tank can be as large as necessary. Third, because the system is equipped with annual storage, collection is not critically dependent upon day-to-day weather conditions and can continue all through the summer. Finally, because the system is community-based, financing arrangements can be very creative.

For Further Reference

Cogeneration

Beer, M. "Fiat Markets 15-kw Cogenerator." *Canadian Renewable Energy News*, July 1980, p. 3.

Commoner, B. *The Politics of Energy*. New York: Alfred A. Knopf, 1979.

Energy and Minerals Division of the General Accounting Office. *Industrial Cogeneration—What It Is, How It Works, Its Potential*. Washington, D.C.: General Accounting Office, 1980. Order no. EMD 80-7.

Governor's Commission on Cogeneration. *Cogeneration: Its Benefits to New England*. Boston: Commonwealth of Massachusetts, 1978.

Iceman, L., and Staples, D. M. "Industrial Cogeneration: Problems and Promise." *Energy* 4 (1979):101-17.

Lornell, Randi. "A PURPA Primer." *Solar Law Reporter*. 3(1981):31-65.

Parisi, A. J. "Fiat Will Sell 2-in-1 Energy Generator." *New York Times*, 30 August 1980.

Thermo Electron Corp. *Summary Assessment of Electricity Cogeneration in New England*. Waltham, Mass., 1977.

Was, C. S., and Golay, M. W. "Cogeneration—An Energy Alternative for the U.S.?" *Energy* 4 (1979): 1023-31.

Williams, R. H. "Industrial Cogeneration." In *Annual Review of Energy*, vol. 3. Edited by Jack M. Hollander. Palo Alto, Calif.: Annual Reviews Inc., 1978.

District Heating

Geller, H. S. *Thermal Distribution Systems and Residential District Heating.* Princeton, N.J.: The Center for Energy and Environmental Studies, Princeton University, 1980. Report PU/CEES No. 97.

Gleason, J. *"Efficient Fossil and Solar District Heating System.* Golden, Colo.: Solar Energy Research Institute, forthcoming.

Karkheck, J., and Powell, J. R. "District Heat—A Major Step toward U.S. Energy Self-Sufficiency." *Energy* 5 (1980):285-93.

Karkheck, J.; Powell, J. R.; and Beardsworth, E. "Prospects for District Heating in the United States." *Science,* 11 March 1977, pp. 948-55.

Margen, P. "New Types of Hot Water Distribution Systems for Low-Density Heat Areas." *Building Systems Design,* December 1978/January 1979, pp. 23-44.

ACES

Baxter, V. D. *ACES: Final Performance Report, December 1978 through September 15, 1980.* Springfield, Va.: National Technical Information Service, 1980. Order no. ORNL/CON-64.

Biehl, R. A. "Annual Cycle Energy System: A Hybrid Heat Pump Cycle." *ASHRAE Journal.* July 1977, pp. 20-24.

Eissenberg, D. M. "Annual Cycle Storage for Building Heating and Cooling." In *Proceedings of Solar Energy Storage Options,* edited by M. B. McCarthy. Springfield, Va.: National Technical Information Service, 1979. Order no. CONF-790328-P3.

Fischer, H. C. "Annual Cycle Energy System (ACES) for Residential and Commercial Buildings." In *Proceedings of the Workshop on Solar Energy Storage Subsystems for the Heating and Cooling of Buildings, April 16-18, 1975. Charlottesville, Va.,* edited by Lembit U. Lilleleht; J. Taylor Beard; and F. Anthony Iachetta. Springfield, Va.: National Technical Information Service, 1975. Order no. PB252449.

Miller, R. S. "A Performance and Economic Evaluation of Annual Cycle Energy Storage (ACES)" *Energy* 5 (1980):183-90.

Minturn, R. E. et al. *ACES 1979: Capabilities and Potential.* Draft. Oak Ridge, Tenn.: Oak Ridge National Laboratory Energy Division, n.d. Order no. ORNL/CON-48.

Research Program on the Economic Feasibility and Commercialization Potential of ACES. Final Report. Rockville, Md.: NAHB Research Foundation, 1979.

ICES

Calm, J. M. *Heat-Pump-Centered Integrated Community Energy Systems.* Argonne, Ill.: Argonne National Laboratory, 1980. Order no. ANL/CNSV-7.

Calm, J. M., and Sapenzia, G. R. "Site and Source Energy Savings of District Heating and Cooling with Heat Pumps." Paper presented to the Fifth Annual Heat Pump Technology Conference; April 1979. Oklahoma State University, Stillwater, Okla. Argonne, Ill.: Argonne National Laboratory.

Calm, J. M., and Werden, R. G. "Heat Pump Systems for District Heating and Cooling." *ASHRAE Journal*, September 1979, pp. 54-58.

Powell, W. R. "The Case for CASES." *Environment*, July/August 1978, pp. 14-20, 40-41.

Solar Annual Storage District Heating

Beard, J. T. et al. *Annual Collection and Storage of Solar Energy for the Heating of Buildings*. Report #3. Charlottesville, Va.: University of Virginia, 1978. Order no. ORO-5136-78/1.

Givoni, B. "Underground Longterm Storage of Solar Energy—An Overview." *Solar Energy* 19 (1977):617-23.

Hooper, F. C. "Annual Storage." *Solar Age*, April 1978, pp. 16-17.

Margen, P. "Central Plants for Annual Heat Storage." *Solar Age*, October 1978, pp. 22-26.

Meyer, C. F., and Todd, D. K. "Conserving Energy with Heat Storage Wells," *Environmental Science and Technology*, June 1973, pp. 512-16.

Office of Technology Assessment. *Application of Solar Energy of Today's Energy Needs*, vols. 1 and 2. Washington, D.C.: U.S. Government Printing Office, 1978. Order no. 052-003-00539-5 (vol. 1); 052-003-00608-1 (vol. 2).

Zangrando, F., and Bryant, H. C. "A Salt Gradient Solar Pond." *Solar Age*, April 1978, pp. 21, 32-36.

7

Institutional Issues:

Lifting the Barriers to Energy Conservation

Can you protect your building from being blocked from sunlight and wind? Do you have the right to stop the construction of a building that would cut off your solar access? Are there certain building codes and zoning regulations in your town that restrict energy improvements and additions to your building? With rental housing, who is really responsible for making energy improvements? Why do utility rate structures still charge less for more consumption, when they should be charging more to encourage conservation? These questions point up some of the institutional issues that affect energy conservation and renewable energy development. They can't really be answered with technology in the way that a wasteful heating system can be improved. These are political issues, and depending on the way they are or aren't resolved they can pose significant barriers or open the doors to widespread conservation and renewable energy development. This chapter discusses these issues and presents some ideas toward ensuring that this development is not hindered but rather promoted by society's laws, rules and regulations.

RIGHTS OF ACCESS TO RENEWABLE ENERGY

In suburban and rural areas, there generally are few problems of solar access because most single-family dwellings have some exposure to direct sunlight. In urban and more densely populated residential areas, however, the situation is quite different. In a place like Manhattan, for example, it is clear that there will be some sites with no hope of ever gaining solar access and other sites whose access may have to be foregone because of the enormous development value of adjoining land. To date there have been very few legal cases that specifically concern the right of access to sunlight or wind. Most cases concerning renewables involve property rights, zoning violations or deed restrictions. Any attempts to legislate solar and wind access laws frequently run into intense opposition from landowners whose property rights might be infringed upon. In other words, there is not likely to be a solution to the problem that will satisfy everyone.

What remedies and recourse exist if you want to ensure a continuing supply of energy for your solar or wind system? Before answering this question, three points must be made. First, while solar access is well established as an issue, there has still been little discussion of how to ensure wind access, in part because there have been few attempts to install wind machines in urban locations and also because wind energy is much less predictable than solar. Second, although solar access or sun rights laws and solar zoning ordinances have been enacted in some states and communities, they are not common.

Third, the prospect of enactment of an all-encompassing solar access legislation seems fairly remote at the present time, if only because much of the initial enthusiasm for such laws failed to take into account negative reactions of property owners.

Solar Access Solutions

Perhaps the best-known response to the access question is the traditional English Doctrine of Ancient Lights, which is intended to provide a building resident with sufficient light to read without "grumbling." However, the Doctrine was long ago rejected by the courts of all fifty states as inconsistent with the needs of a growing industrial society.

The do nothing, or ad hoc, approach is one under which you build only where there is a reasonably good chance that your access will not be obstructed in the foreseeable future. This approach offers no legal recourse if access is obstructed, but it does allow you to establish a first right, so to speak, to the energy resource at your site should access legislation ever be enacted.

An *easement* is a right-of-way across someone else's property. It's generally paid for and guaranteed by a written contract. In the same way, a solar easement would guarantee perpetual access, preventing your neighbor from ever obstructing your sunlight. The easement remains in force even if the property is sold. An easement can be quite expensive to purchase, particularly if it will restrict vertical building rights above an urban lot or building.

With a *legislated solar easement,* written law would forbid landowners from doing anything to obstruct solar or wind access to a neighboring

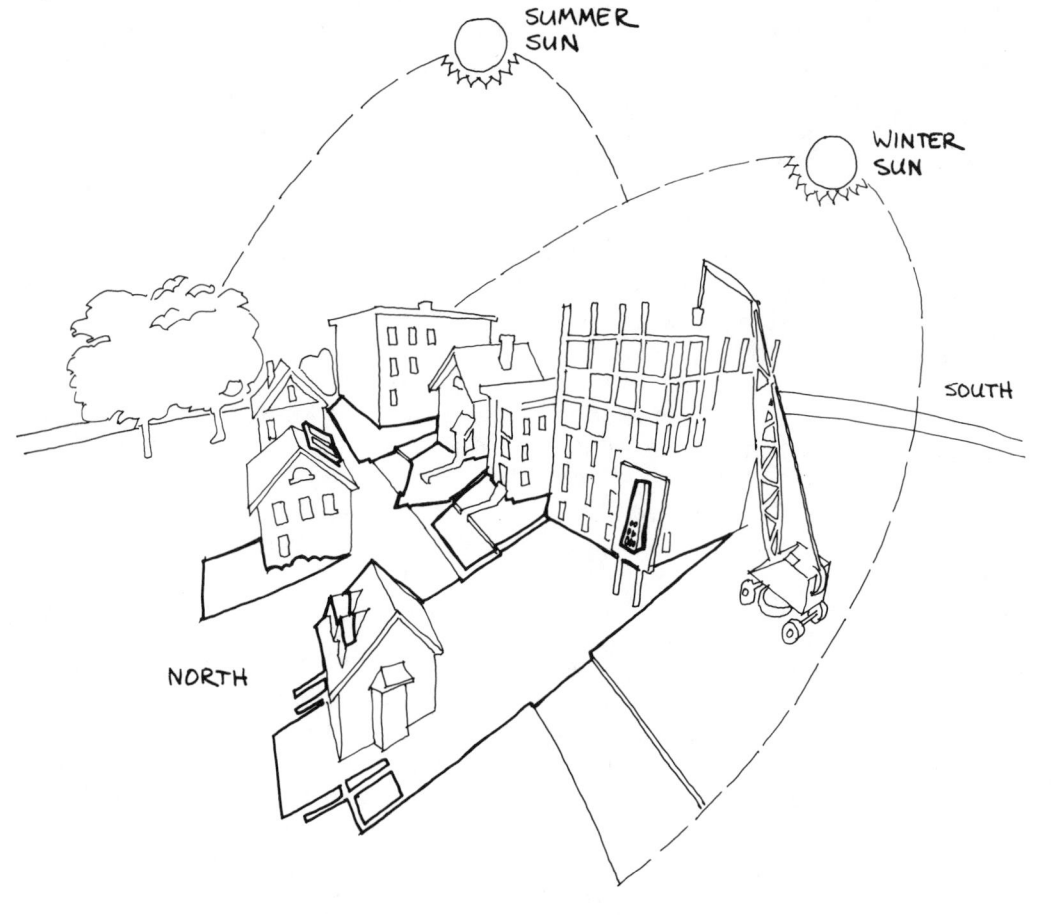

FIGURE 7-1: SOLAR ACCESS

property. Such legislation would probably not be retroactive. A barrier to a law like this is that it does constitute a significant infringement upon traditional property rights and could possibly be interpreted as a taking without just compensation and, therefore, unconstitutional.

Under a *solar permit* system, you as a builder or homeowner would register with the local government your intention to install a solar or wind system which

would put neighboring landowners on notice. A hearing would be held to allow neighbors to object to the proposed permit, and the hearing board would then decide whether or not to issue the permit. In cases with no objections, there would be no problem in obtaining the permit. This grants access rights to the permittee for the lifetime of the system. In situations where the permit is refused, however, the person seeking the permit does not receive guaranteed access and might have to resort to the courts in order to obtain it.

Use of *existing zoning ordinances* is probably the least intrusive means of protecting solar access. Height restrictions, limits on how close to property lines buildings can be erected (setbacks) and controls on trees and shrubs can protect access. Unfortunately, such ordinances are not retroactive and cannot regain access that was previously lost. Along these lines, the concept of "solar envelopes" has been proposed, in which the floors of a building above a certain height must be set back from the lower floors in order to allow sunlight to penetrate to ground level. Once again, this approach would affect only new buildings and additions to existing ones.

The *sun rights* approach operates on the theory that a solar resource is much like a water resource and that the first to use the resource should possess perpetual rights to it. New Mexico has passed a sun rights law that gives continuing use of sunlight to whomever uses it first. This approach is not popular because it restricts the rights of neighboring landowners to develop their potentially valuable property.

Transferable development rights restrict the development of land next to a site with a solar system if access will be blocked. At the same time they also transfer development rights to another site of equal value whose development might otherwise be restricted by zoning regulations. A potential problem here is finding a site to which development rights can be transferred. In commercial districts, land of value equal to the restricted property may be difficult to locate.

Land use planning ensures that streets and buildings are laid out and oriented in such a way as to provide equal and guaranteed solar access to all residents in a community. Of course this approach would only apply to areas being newly developed or redeveloped.

Finally, if all else fails, there is *litigation* in the courts. The prospects for winning a solar access case are not good at present. The few recent cases that have claimed such a right have been unsuccessful. Some observers feel, however, that while there is no automatic "right to light" in the United States, there might be some grounds for a "right to energy," but this concept has not been tested in the courts. A major drawback to this approach is that litigation can be very expensive.

Wind Access Solutions

Many approaches to guaranteeing solar access are also applicable to wind power. The wind, however, is much more variable over small areas than sunlight and its patterns can be changed, increased or diminished by nearby construction. Furthermore, while it is quite easy to point to the building or tree that is shading your solar collector, it would be much more difficult to identify the building(s) that have cut off your wind supply, particularly if they are some distance away. Indeed, given the fact that obstructions can affect wind patterns over a large area, it may be impossible to protect your access to the wind completely. At the present time there are no legal access rights to the wind, but as wind machines become more common, various legal prescriptions will have to be developed that define access rights.

The free air space around a wind machine in every direction should be a radius equal to about ten rotor diameters in order to ensure the proper operation of the machine. In urban areas it may be virtually impossible to meet this criterion, and you may find it necessary to increase tower height in order to get above nearby obstructions. Unfortunately, there is a cost penalty associated with increasing tower height and there may be zoning restrictions ruling out this course of action.

Protecting Your Access to Solar and Wind

Clearly, the protection of solar and wind access in urban areas can be a sticky problem. Many of the proposals previously described are either inappropriate for urban areas or are unlikely to be implemented because of questions regarding infringement of other property rights. Given this situation, what should you do to protect your access to the sun or wind if you have or plan to install a renewable energy system?

Carefully survey your neighborhood and nearby areas for redevelopment potential. Unless your neighborhood is undergoing rehabilitation, it is not very likely that existing access will be eliminated in the future. If, however, you live in or near a commercial district that is undergoing redevelopment, large buildings may be erected that could block your access or disrupt wind patterns. If possible, find out if there are any plans for such construction and how they might affect your site.

Carefully survey surrounding lots for growing vegetation and ask your neighbors whether they plan any major building additions. Any young trees growing near the site of your renewable energy system could eventually grow to block out access. Now is the time to ask your neighbor(s) to move or remove such trees. Also, make sure that none of your neighbors plans to make an addition to his or her house that could block your access or disrupt wind patterns. If such plans do exist, you may have to find another site or abandon your plans for renewable energy.

Study your local zoning ordinances and building codes. Determine what, if any, height restrictions and setback requirements exist in your community. If the height limit is 40 feet and your neighbor's house is 35 feet high, additional floors cannot be added. For a solar system, this can ensure access. For wind, however, height restrictions may restrict construction of the tower for a wind machine, unless you can get a variance. Finally, be sure that your system complies with all relevant regulations and building codes.

Carefully keep a record of your activities and your system's performance. This is the most important thing you can do to protect your access, since records may be necessary at some time in the future in order to substantiate your precedence at a site. First, photograph your site before and after system installation. If you are installing a solar system, be sure to take photos several times during a midwinter day in order to provide a record of your solar access at the site at the time of installation. Second, keep records of your energy consumption (oil, gas, electricity) before and after system installation so you can document energy savings. Third, make detailed plans of the system design and placement, and if possible, have them notarized. Fourth, be sure not to lose your building permit. Finally, keep careful records of system performance from year to year and of the status of your site and neighboring lots. The rationale behind these activities is to have complete documentation for your system in the event that solar or wind access laws or ordinances are passed. You may have to go before a hearing board or to court to prove your prior right to access at a particular site.

BUILDING CODES AND ZONING ORDINANCES

Any major energy conservation improvements made to your home or renewable energy systems installed on your property will probably have to conform to the building codes and zoning ordinances of the city and/or state in which you live. At first encounter, you may feel that these laws serve only to obstruct attempts to conserve energy, for it is not inconceivable that your improvements or additions will violate one or another obscure provision of the codes. Since the penalty for violation of these regulations can be severe (if you get caught), it is good practice to be familiar with the relevant portions of the codes and ordinances before you begin to build or modify.

There are several sets of ordinances of interest here. In some places, each set is independent; in others, they are combined. Your local building inspector can tell you which regulations apply in your area. The *building code,* while generally used to cover all other codes, specifies how and of what materials a building must be constructed. This code generally applies to new construction and major modifications to existing buildings. The *fire prevention code,* which may be a subsection of the building code, covers potential fire hazards in a building. The *mechanical code* deals with heating and cooling systems and their installation and operation. The *plumbing code* covers plumbing. The *electrical code* governs the installation of electrical fixtures and equipment. The *housing code* sets requirements for water supply, heating, cooling, light, ventilation and sanitary facilities in multi-family housing. This code may also cover rehabilitation of an existing building.

The *energy conservation code,* if one exists in your city or state, sets standards for energy use and the installation of energy conservation measures in homes. Finally, *zoning ordinances* regulate the types, locations and heights, among other things, of structures erected in particular neighborhoods. Some or all of these codes may apply in your area.

The exact nature of these various codes can vary greatly from state to state and city to city. At least 20 states have state-promulgated codes, and of these some are very strict, others lenient. The remaining states leave the entire matter to individual cities and towns. Consult your local building inspector to get a clear picture of the regulations governing buildings. Compliance with the codes is generally monitored by this inspector and, if one exists, the state agency responsible for development of the code. The latter should have a staff capable of answering questions concerning the code. Most local building inspectors rely, at least in part, on the certification of a registered architect or engineer that building designs or changes meet the provisions of the local or state code. While such certification may be costly, it can be of great help in obtaining approval for your project.

Building Codes

Depending upon where you live, it is likely that your building code is patterned upon one of the three model codes developed by building code groups: the *Basic Building Code* (most common in the northeastern and midwestern United States), the *Uniform Building Code* (West and Southwest) and the *Southern Building Code* (South). These codes are only advisory and do not have the force of law unless adopted by a political jurisdiction. They differ in specifics but have similar general provisions. Another code that may apply in your area is the *One- and Two-Family Dwelling Code,* which has been adopted by all three code groups and covers single- and two-family residences.

Compliance with the building code is usually necessary if you make any major modifications or additions to your house or building. Such changes generally require a building permit; if you do not have such a permit, you may be liable for a fine, or you may have to remove the changes made to the building. For example, if you rip out walls in order to install insulation or if you add a solar greenhouse, you will probably need a permit, and you will have to prepare plans of the modifications in order to get the permit. In some areas, you are allowed to draw up these plans, while in others, a registered architect or engineer must draw them. Installation of solar collectors on a building roof or erection of a tower for a wind machine will have to meet structural design and strength specifications set forth by the code, and a permit will probably be required. You do not, however, need a permit in order to install fiberglass batts in your attic or make similar, non-structural changes to your building.

Fire Prevention Code

The fire prevention code is often, but not always, a subsection of the local building code. This code (or the relevant subsections of the building code) covers such items as the placement and construction of fireplaces, wood stoves and heating and cooling systems and the use and placement of certain materials that may be flammable, such as insulation. Woodstoves must generally be placed specified distances from plaster lath or paneled walls, or lesser distances from masonry walls and footings. Flue pipe penetrations through walls must be fireproof, and the flue must vent a certain height above the roof line. Insulation installed in attics must not be placed over recessed lighting fixtures or electrical junction boxes or around flue pipes. If placed in an occupied space, insulation must be covered by a fire-resistant wall material such as gypsum board.

Mechanical Code

The mechanical code is frequently part of the building code. It governs the installation of heating and cooling systems in a building. The code specifies sizing requirements for the system, which means the system must supply a certain percentage of the building heating load on the design-day. Also covered are conversion of oil and gas burners, repairs or modifications to existing systems and installation of new furnaces or boilers, active solar systems, heat pumps or cogeneration systems. The code also covers pumps, shut-offs, pressure-limiting devices, flues, dampers, combustion air inlets, controls and distribution systems. The mechanical portions of a wind energy system might be covered by this code.

Automatic flue dampers and pilotless ignition systems may require approval by either Underwriters Laboratories or the American Gas Association under this code.

Plumbing Code

Any heating or cooling system incorporating liquid heat collection and distribution comes under this code. This would naturally include solar space and domestic water heating systems and modifications to an existing distribution system. One problem that may arise involves installation of new systems because some codes require that a licensed plumber do any plumbing. Most plumbers are not yet familiar with active solar system details and, as might be expected, the preponderance of problems with such systems are caused by incorrect installation of plumbing.

Electrical Code

The electrical code (which may be part of the building code) deals with all electrical wiring, fixtures and installations. Any rewiring or new wiring resulting from installation of a wind, solar cell or cogeneration system, or necessitated by energy conservation modifications or caused by changes in heating, cooling or lighting systems will fall under these regulations. Some codes require placement of electrical outlets every few feet. Many codes allow electrical work to be done only by a licensed electrician; others, however, allow do-it-yourself work.

Housing Code

Energy-related issues covered by the housing code (which may be incorporated in the building code) include heating and cooling supply, ventilation and lighting. Many codes mandate that a residential building be equipped with a heating system capable of maintaining a 70°F temperature at a height of three feet above the floor. Thus, if your home is substantially heated by renewable energy systems capable of providing heat only to 65°F, you may be required to have a large back-up system even if you personally have no need for it.

Housing codes usually require that habitable rooms (living room, dining room, kitchen, bedrooms) be equipped with operable windows in order to provide both daylighting and ventilation. (Such windows are also required by the fire code as possible escape routes.) If you eliminate north-facing windows or install a greenhouse that encloses windows opening into a habitable room, you may be technically in violation of the code. The housing code may also require a certain ventilation rate in a building; this can conflict with attempts to reduce infiltration. Finally, some codes set minimum lighting standards for the common spaces of large buildings, and delamping efforts may conflict with this provision of the code.

Energy Conservation Code

Of increasing importance in recent years are energy conservation codes that set standards for energy consumption in buildings. These include codes passed by states and cities, the Minimum Property Standards of the Department of Housing and Urban Development, Standard 90-75 of the American Society of Heating, Refrigerating and Air-Conditioning Engineers (ASHRAE) and the Building Energy Performance Standards of the Department of Energy. These codes generally cover three areas of building performance: heating and cooling loads, lighting and hot water use. Some codes may also cover appliances.

Energy conservation codes are generally classified in one of three categories: *prescriptive, true budget* or *component performance*. A prescriptive code describes in great detail the construction features required in a building such as the thickness of insulation, the area of windows or the layers of glazing. A true budget code sets energy consumption achievement standards that must be met by a building (number of Btu's consumed by a building of a certain size under design-day weather conditions). Finally, a component performance code sets minimum R-values (or U values) for doors, windows, walls and so on.

Cooling loads are specified in the same way as heat loads in energy conservation codes. Lighting is specified in terms of intensities allowed for different tasks and building areas. A lighting audit may be required (see chapter 4). Water use is often specified in terms of allowed flow rates through taps. Finally, energy efficiency standards may be specified for certain heating and cooling appliances.

A growing number of states and cities (such as Massachusetts, Florida, Ohio and Davis, California) have passed energy conservation codes or incorporated such provisions into existing codes. While specifics vary from place to place, such a code might typically accomplish the following: establish required minimum heat loss limits through a building shell; set minimum efficiencies for air conditioners, water heaters, heat pumps and other heating appliances; control installation of electric resistance heating; set maximum limits on water flow rates through showers and faucets; set minimum ventilation rates; establish maximum amounts by which heating and cooling systems may be sized in excess of a building's calculated design heating and cooling load; set maximum allowed lighting levels (in watts per square foot of floor area) for exterior or interior general lighting.

The Housing and Urban Development (HUD) Minimum Property Standards (MPS) is a general building code that covers all housing (one, two or multiple-family) constructed with federal financing assistance. Buildings built or improved with loans from HUD, the Federal Housing Administration, the Veterans' Administration and other federal urban renewal programs, must comply with the MPS. The energy conservation provisions of the MPS use the component method of compliance.

The ASHRAE 90-75 (Energy Conservation in New Building Design) is a set of prescriptive energy conservation standards developed by ASHRAE in cooperation with various federal agencies, building design associations and other groups in the building industry. The standard has generally been accepted by the industry as a reasonable energy conservation code for new buildings. (ASHRAE 90-75 does not, however, cover existing buildings.) The standards include provisions covering heating, ventilation and air conditioning systems and equipment, water heating, electrical distribution systems, lighting, renewable energy systems and solar heat gain through windows. Under the provisions of the Energy Policy and Conservation Act (1979), a number of states have adopted ASHRAE 90-75 as the technical basis for energy conservation standards for buildings. Consequently, this standard may, at some point, be applied to existing buildings.

The Building Energy Performance Standards (BEPS) code is an energy performance code developed under Congressional order by the Department of Energy, working in cooperation with various national laboratories, associations and consulting groups. The provisions of BEPS cover heating and cooling systems and equipment, ventilation, lighting and domestic hot water use. BEPS was originally planned as a mandatory building code, much like the gasoline mileage standards set for new cars, intended to go into effect in 1981. Resistance from the building industry and the general distaste with federal regulations has changed BEPS into, at best, a voluntary standard. Because BEPS is stricter than ASHRAE 90-75, most builders are more likely to follow the latter.

Changing Building Codes

What do you do if your energy conservation measures come into conflict with the code? The best approach is to try and make your modifications conform to the relevant code provisions, or persuade your local building inspector that your changes do, indeed, conform to the code. Failing this, it is possible to change building code provisions. In some states, codes can be changed only by the state agency in charge of building codes. In this case, changes must be pushed through this agency, and this requires knowledge of the art of political persuasion, a topic beyond the scope of this book. Other states allow cities to modify the code, in which case changes must be affected through the city council. If the provision to be changed is a relatively unimportant one, or if you can persuade

fellow citizens of the need to change the code, it should be possible to get the city council to act (unless you are trying to change a provision with obvious safety issues in mind). If this approach is unsuccessful, it is possible to go to court over the issue, but this can be expensive. In the final analysis, compliance with the code is the best way to avoid problems.

Zoning Ordinances

Ordinarily, zoning ordinances do not affect energy conservation improvements. The only likely conflicts are with regard to historic preservation, building appearance and building height regulations. Renewable energy installations may run afoul of zoning ordinances regulating height, appearance and property line setbacks. Cogeneration units may be regarded as industrial installations and therefore unsuitable for residential zones. Since ordinances and their enforcement vary with locality, it is important in all cases to consult your local zoning and building department before proceeding with any improvements.

Historic Preservation Ordinances

Some sections of older cities in the United States are historic districts, and many buildings throughout these cities may be registered historic structures. Since zoning ordinances applicable to such districts or buildings generally prohibit changes to the building exterior, conflicts with energy conservation improvements may occur. An example of this is windows that may not be altered in exterior appearance. In such a case, their thermal performance cannot be improved unless an identical replacement unit of good thermal quality can be provided. This is possible but may be quite expensive. Another solution is to improve the windows from the inside of the structure.

Historic zoning ordinances generally prohibit new structures from being built on a roof and will rule out building-mounted solar collectors and wind machines. You can get around these restrictions by placing the collector or wind machine elsewhere. For example, solar collectors might be mounted on an isolated rack on the building grounds.

How Zoning Ordinances May Affect Solar, Wind and Cogeneration Systems

Very few cities in the United States have passed any zoning ordinances concerned specifically with solar energy systems, although such systems may be affected by ordinances applying to buildings in general. For example, a solar system could be affected if mounted on a building whose height is already very close to the limit for that particular zone. Also, a ground-mounted system must conform to setback ordinances. This might be a problem where lots are small in relation to the building. In addition, in some condominium and housing developments, ordinances known as *restrictive covenants* may prohibit changes to the building facade or forbid alteration of landscaping. These can make it impossible to install any kind of renewable system.

Wind systems are usually affected to a much greater degree by zoning ordinances. In commercial districts, height restrictions and setback requirements, if they exist at all, are usually quite liberal. In residential areas,

however, the opposite is likely to be true, and wind systems may be subject to severe zoning restrictions. Experience with wind systems in urban areas is so limited that it is very difficult to generalize, but such systems are usually subject to restrictions imposed by zoning ordinances, building codes and state regulations.

Zoning ordinances generally impose limitations on tower height (if they allow tower installation at all) and establish required property setbacks. In some communities, towers are considered accessories to the main building on a lot and are therefore not subject to height or setback limitations at all. Building codes place restrictions on electrical work and set specifications for tower strength and stability, but it is up to local zoning officials and building inspectors to decide how to interpret and apply existing regulations to wind energy systems. Each state may have regulations that also apply to tower construction and/or wind systems. In particular, the regulations for the Residential Conservation Service may contain certain restrictions on wind site assessment that could have the effect of eliminating all prospective urban wind energy sites. If you are thinking about installing a wind machine in an urban area, be sure to consult your local building inspector, zoning department and state building code commission before proceeding. This will prevent your having to take the system down if you have violated some provision of one code or another.

Cogeneration systems may be considered industrial installations if they serve more than one building and may therefore be forbidden in residential zones under certain zoning regulations. In addition, air quality ordinances may also affect a cogeneration installation (see chapter 6).

Changing Zoning Ordinances

There are three ways to deal with respective zoning ordinances. The first is to obtain a *variance*, that is, a ruling by the local zoning appeals board that your system constitutes an exception to existing regulations. The second is to have the zoning ordinance overturned by your local municipal governing board. The third, when all else fails, is to go to the courts. We recommend trying to comply with zoning ordinances unless it is absolutely impossible, because obtaining a variance or a new ordinance can be time-consuming and frustrating. In addition, litigation is often not attractive because it can be expensive and quickly eliminate any savings you might otherwise realize from your renewable energy system.

TENANT-LANDLORD RELATIONS

A serious institutional problem is that of developing cooperation between tenants and landlords relative to making energy conservation improvements. Problems arise for many reasons: conflicting financial interests, lack of incentives to invest in energy-saving measures, tenant fear of rent increases or landlord fear of rental income decreases, lack of knowledge about the most cost-effective measures with the shortest payback periods, or concern about obtaining financing and being ripped off by unscrupulous contractors. No one has come up with a perfect solution to the tenant-landlord problem, but this section reviews some of the problems and offers some approaches to resolving them.

There are several sources of energy-related conflicts: If the tenant pays the heating bill, the landlord has no incentive to spend money for energy conservation, and since the tenant doesn't own the building, he or she has little incentive to make an investment that will increase the building's value and, possibly, the rent. If the landlord pays the heating bill, he or she can pass fuel cost increases directly on to the tenant. Thus the landlord has no incentive to conserve. In addition, since nothing the tenant can do will prevent rent increases, he or she also has no incentive to conserve. The low vacancy rates in most cities make tenants reluctant to agitate for energy-conservation measures for fear of being evicted. Low vacancies also make it easy for landlords to find new tenants.

Rent control or stabilization laws may make it difficult for the landlord to pass along the cost of energy conservation measures but easy to pass along increased fuel costs. Improving the building may increase property values, creating an added tax burden on the landlord. There is no guide for tenants and landlords to determine which energy conservation measures should have priority, which are most cost-effective and have the fastest payback and who is to benefit from the measures. The landlord who owns a few buildings as an investment may have little or no experience in property management, in obtaining financing for property improvements, in finding reputable contractors or in ensuring that work is done properly and to specifications.

These are not, of course, the only

difficulties. There are bad tenants, bad landlords and bad relations. There are buildings about to be taken over by the city for nonpayment of taxes. There are buildings that are nothing but money-losing tax shelters. The high first costs of making energy improvements act as a deterrent. Some of these problems defy solution and will, unfortunately, continue to do so.

The important thing for tenants and landlords to recognize is that implementation of energy conservation

measures—both in terms of improvements and lifestyle changes—can have multiple benefits. Reduced fuel costs, for example, can reduce pressure on the landlord to increase rents or even lead to rent reductions. The improved monthly cash flow (through reduced expenditures on fuel) can free the funds of both tenants and landlords for other purposes. Energy improvements increase tenant comfort and could lead to improved tenant-landlord relations. Improvements to the property make

Table 7-1
Some Low-Cost/No-Cost Energy Conservation Measures for Apartments
(oil heat)

Measure	Fuel Savings (gal/yr)	Approximate Cost ($)	Simple Payback (months)*
Install drapes with valance (15 windows per apartment)	180	450	24
Install flow reducers on taps and showers	60	20	3
Install temporary plastic storm windows (15 windows per apartment)	330	35	1
Insulate water heater	40	20	5
Lower water heater temperature from 140°F to 120°F	44	0	immediate
Maintain proper heating distribution system operation	120	15	1
Set back thermostat 4°F day; 12°F night	194	0	immediate
0°F day; 12°F night	168	0	immediate
Caulk windows with rope caulk (15 windows per apartment)	230	20	1

Correction factors: For fuel savings, multiply by the correction factor; for payback, divide by the same factor. Zone I: 0.33; Zone II: 0.67; Zone III: 1.00; Zone IV: 1.33.

SOURCE: Massachusetts Executive Office of Energy Resources, *How to Cut Fuel Bills in Apartments* (Boston, 1978). The numbers in this table were updated.

NOTE: This table is based on an apartment with annual fuel consumption of 1,500 gallons before conservation measures.

*Assumes heating oil cost of $1.25 per gallon.

Table 7-2
Energy Conservation Economics for a 24-Unit Apartment Building
(oil heat)

Energy Conservation Measure	Estimated Monthly Fuel Savings/ Apartment Unit* (gal)	Do-It-Yourself Cost ($)	Payback in Yrs (@$1.25/ gal)	Total Cost per Unit ($)
Install storm windows (no weather stripping)	11.6	3,128	0.75	130
Install storm windows and weather stripping	12.8	3,898	0.83	162
Install weather stripping (no storm windows)	2.4	770	0.92	32
Insulate attic with R-19 fiberglass batts	7.0	880	0.33	37

Correction factors: For fuel savings, multiply by correction factor; for payback, divide by correction factor; for return on investment, multiply by correction factor. Zone I: 0.33; Zone II: 0.67; Zone III: 1.00; Zone IV: 1.33.

*Savings averaged over entire year.

†Assumes 5-year, 14% loan. Additional cost is monthly payback divided by 24 units.

the building more valuable and tenants more satisfied and more inclined to respect the property.

Although the interests of tenants and landlords might seem mutually exclusive, they need not be, and the place to begin is to determine where these interests coincide. For example, if the spending of a certain sum of money cuts fuel costs by more than the cost of the energy conservation measures, the benefits accrue to both parties. If the landlord pays for heating, the operating costs for the building will drop and this decrease can, perhaps, be passed along to the tenant. If the tenant pays for heating, the improvements will free income for other purposes, and the tenant may be

less likely to leave because of high fuel costs.

There are a number of low-cost/no-cost energy conservation measures that, if implemented, can dramatically cut fuel consumption in apartments, resulting in almost immediate financial benefits. No-cost measures include setting back your thermostat, lowering water heater temperature, keeping windows tightly closed, closing off unused rooms, keeping basement and closet doors closed and keeping shades and drapes closed. Low-cost measures including caulking, temporary plastic storm windows, heat reflectors behind radiators, water flow reducers on taps and shower heads, water heater insulation and insulation of pipes and ducts

Return on Investment (%)	Additional Monthly Cost per Unit† ($)	Contractor Cost ($)	Payback in Yrs (@ $1.25/gal)	Total Cost per Unit ($)	Return on Investment (%)	Additional Monthly Cost per Unit† ($)
133	3.23	7,360	1.8	307	57	7.14
120	3.78	11,625	2.5	484	40	11.28
109	0.75	4,265	4.9	178	20	4.13
300	0.85	3,620	1.4	152	70	3.51

have paybacks of much less than one year. Some of these measures and savings are shown in table 7-1.

It is also possible to invest in more costly conservation measures in apartment buildings such as storm windows, attic insulation and weather stripping. These measures have do-it-yourself paybacks of one year or less; even if done by a contractor, the most expensive measure still gives a return on investment of 20 percent, much greater than that realized by most financial investments.

Table 7-2 shows the economic benefits that can be realized by these conservation measures. For example, the monthly cost over five years for storm windows installed by a contractor is $7.14 per unit in a 24-unit apartment building; however, the fuel savings resulting from this measure are 11.6 gallons of heating oil ($14.50 at $1.25

per gallon) per unit per month. The savings will vary, depending upon climate zone and fuel. The net economic benefit is, therefore, $7.36 per unit per month ($14.50 minus $7.14) or $98.32 per year. As fuel prices increase, so will the net economic benefits. Furthermore, these savings will continue year after year, even after the loan has been paid off. What this means is that energy conservation measures, even fairly costly ones, can show immediate economic benefits.

How to Begin

If you're going to discuss energy conservation improvements with your landlord or tenants, be well prepared! Have a course of action mapped out. Document the potential savings resulting from conservation improvements by references to other, similar buildings in the neighborhood that have already been weatherized. Propose a cooperative effort. A tenants' cooperative could probably purchase materials for less than retail cost and could provide free labor. Such a co-op could also serve as a strong bargaining unit. Learn as much as you can about possible improvements. Not only will this help you in negotiations, it will also help you to make sure the job is being done correctly. The first thing to do is make low-cost/no-cost energy improvements. The immediate savings realized from these measures will provide incentives to go on to other, more costly improvements.

Tenants should be educated on how and where energy is wasted in apartments and how simple changes in life-style can reduce energy consumption. You may decide that fuel costs should be separated from rental payments when the landlord pays for heating in order to provide an incentive for tenants to reduce energy consumption. This may not be legal in all localities. Check with your city government.

The next step should be to identify improvements needed in your apartment building and to analyze them for potential fuel savings and costs. You can accomplish this using the energy survey in chapter 1 as a starter or by using another, more comprehensive, do-it-yourself audit. You can also hire a professional auditor which costs about $50 to $75 or ask your electric or gas utility to send an energy auditor which, under the Residential Conservation Service, should cost about $15.

After the audit decide which improvements are the most important in terms of comfort, fuel savings, cost and payback, and arrange to have these done first. Before doing any major work such as external rehabilitation or replacement of systems, be sure to consult your local building inspector about relevant building codes and zoning ordinances.

The labor for these improvements can either be contracted or do-it-yourself work. Doing the work yourself is, of course, much less expensive than contract labor but is time-consuming and subject to mistakes. Such efforts must be well organized and should be well thought out and agreed upon by tenants in advance. It pays to involve at least one individual who has had experience in making energy conservation improvements, even if this costs some money.

If the job is to be done by a contractor, you must find someone to do the necessary work. An unscrupulous or incompetent contractor may not only do a bad job but could also make off with substantial sums of money. The solution here is to locate contractors who come highly recommended and whose previous work can be checked out. In addition, self-education can be of great help. If tenants know what the contractor is supposed to be doing and how a particular energy conservation measure is supposed to look, they can monitor the progress of the work and notify the landlord of any apparent irregularities. Contractors who know they are being watched will be less likely to skimp on their work. In the end, both tenants and landlords will benefit from a job done correctly.

Obtaining financing for energy conservation improvements may be the most difficult part of the process. Home improvement loans often are not applicable to rental housing, and landlords may therefore have to remortgage their property at high interest rates. Tenants are relatively well insulated from this part of the process, but they can make a contribution here. Careful preparation is an essential element in applying for financing. Being able to show how energy conservation improvements can actually improve a landlord's and tenants' cash flows can be very helpful in making a loan application (see chapter 8).

None of these cooperative efforts will work if tenants and landlords refuse to work together or if the landlord has no interest in the welfare of his tenants, or vice versa. In such a case, there is little that can be done, and the tenant may have to live with the situation or move elsewhere. Until it becomes glaringly obvious, however, that working together is not possible, it pays to explore all potential avenues of cooperation. It may also be useful to locate an experienced, unbiased mediator who is capable of setting up a cooperative effort between tenants and their landlord.

ENERGY PRICING AND RATE STRUCTURES

For a long time, energy production costs were very low, and these production costs were reflected in the low energy prices charged to consumers, who therefore had little or no incentive to limit their energy consumption. Today, energy production costs are rising rapidly, but current energy prices still do not fully reflect these higher costs. Clearly, increased prices to the consumer will be an important and necessary incentive to encouraging conservation and the use of renewable fuels. Unfortunately, higher energy prices will also inflict a great hardship on the poor, and new energy pricing policies will somehow have to take account of this.

The costs of producing energy are continually rising because, as each unit of conventional energy is produced or removed from a source, it becomes more difficult and more costly to produce or remove the subsequent unit. Also, each refinery or transmission line or power plant is more expensive to build than the preceding one because of increased labor, material and financing costs. Energy has traditionally been priced to reflect not these increasing, or marginal, costs but, rather, the average cost of old, cheaper energy supplies or plants and new, expensive supplies or plants. A more realistic approach to energy pricing would reflect the marginal cost of new energy supplies. Under such a scheme, for example, energy supplies would be priced to reflect the cost of the most expensive supplies, such as oil at $35 a barrel, rather than the cost of production, which may be no more than $5 a barrel.

Besides providing an incentive to conserve, marginal pricing would enhance the economics of renewable energy systems. This is so because, while renewable energies are free, the equipment needed to capture them is costly. However, when compared to the marginal costs of conventional energy resources, renewable energies are often competitive. Indeed, the life-cycle costs (the total cost of equipment plus fuel over a system's lifetime) for many renewable energy systems are, even today and with present energy prices, less than or equal to those of conventional heating or electrical generating systems. Therefore, through marginal pricing of fossil fuels and electricity, consumers would be able to evaluate the economics of competing systems realistically. How can marginal costs be reflected in the energy rates charged to consumers? The most important step is the implementation of an energy-pricing system that makes each additional unit of energy consumed more expensive than the previous one.

Such pricing schemes could result in rapid and large increases in home energy bills, and because of this they are often strongly opposed by consumers. There is no reason, however, why energy bills should rise immediately under marginal pricing. Instead, energy consumption up to a certain average level could be charged at current or slightly higher rates while any energy use in excess of this level would be charged at the much higher marginal rate. Consumers would therefore not be penalized for moderate energy use but would pay much more for increased or excessive use. This would be the desired price signal. Such an approach would also limit the impact of higher energy prices on the poor, since low-income families generally do not consume large amounts of energy, although they devote a large proportion of their income to energy costs.

Rate Structures

The prices charged to consumers for electricity and natural gas reflect a good deal more than simply the cost of energy. These rates, generally established by the public utilities commission in each state, traditionally have been based on various factors in addition to fuel costs. These factors include the costs of building energy production and distribution facilities, the fair allocation of costs among different types of consumers based upon consumption amounts and patterns, allowance for a reasonable rate of return to the utility and recognition of rapid fuel price increases, primarily through fuel adjustment charges. In addition, rate structures have generally been designed to encourage consumption rather than conservation. Rate structures reflecting marginal costs will encourage conservation and limit the need for new peaking energy supply facilities.

Flat rates. If you buy oil or gasoline, you are used to paying a fixed price per gallon. This neither encourages nor discourages conservation as long as the price remains constant. Residential consumers of less than 500 to 600 KWH of electricity per month generally also pay flat rates. These rates usually reflect the average, rather than the marginal, cost of electricity. A flat rate also remains constant throughout the day (see figure 7-2). However, utilities may impose a set charge for service, and, therefore, small consumers (200 KWH or less) will have to pay more for electricity (see figure 7-3).

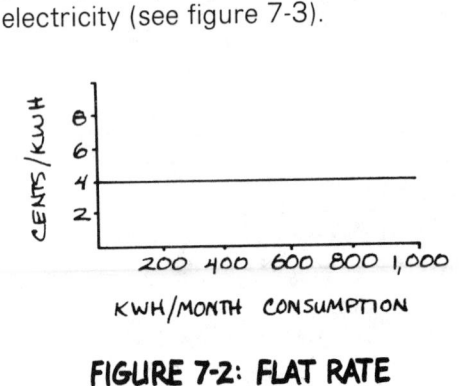

FIGURE 7-2: FLAT RATE

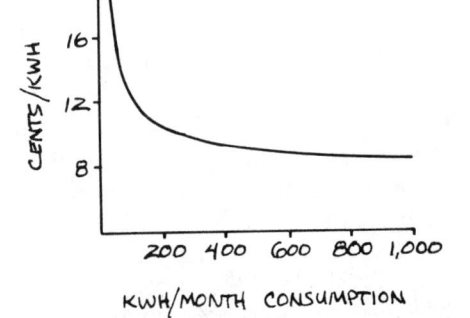

FIGURE 7-3: SET CHARGE RATE

Declining rates. Gas and electricity have traditionally been sold at declining rates, under which each therm or KWH is cheaper than the last, in order to encourage consumption. This is still the practice in many areas. Declining rates are constant throughout the day (see figure 7-4). Because it is difficult in practice to charge a little less for each successive KWH consumed, *declining block rates* are normally used. For very large consumers, therefore, electricity may be extremely cheap (see figure 7-5). Thus, for the first 800 KWH consumed, you would pay 8¢ per KWH, for the next 4,200 KWH, 5¢ per KWH and for anything above 5,000 KWH, 2¢ per KWH. The average cost of electricity for 10,000 KWH of consumption would be about 3.7¢.

Inverted rates. With inverted rates, each KWH consumed is more expensive than the previous one, reflecting the marginal costs of energy. This type of rate structure, therefore, encourages efforts to limit consumption. An inverted rate remains constant throughout the day (see figure 7-6).

Practically speaking, *increasing block rates* are more likely to be used (see figure 7-7). The first block of energy is relatively cheap, while each increasing block gets more expensive. Following the example in figure 7-6, the first 800 KWH would cost 8¢ each, the next 400 KWH would cost 12¢ per KWH and anything above 1,200 KWH would cost 16¢ per KWH.

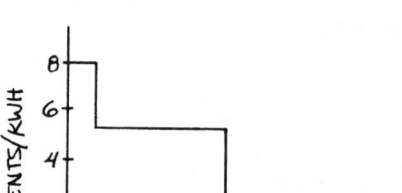

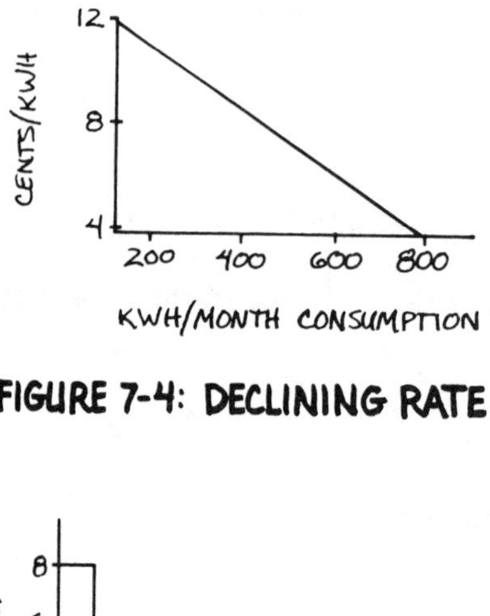

FIGURE 7-4: DECLINING RATE

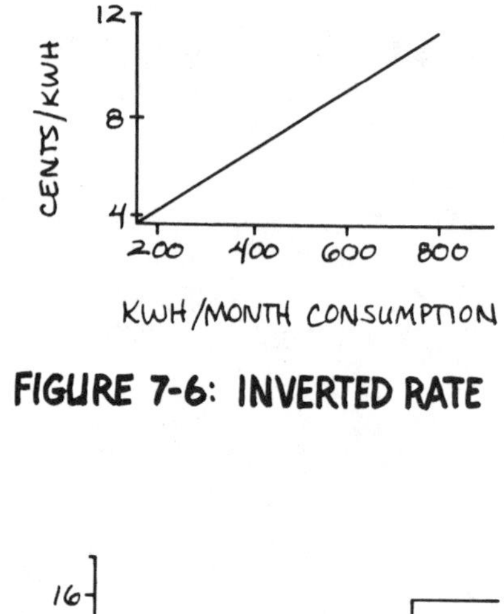

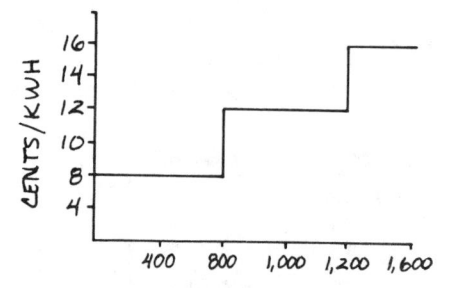

FIGURE 7-6: INVERTED RATE

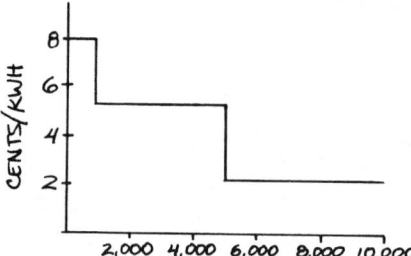

FIGURE 7-5: DECLINING BLOCK RATE

FIGURE 7-7: INCREASING BLOCK RATE

In order to moderate the impact of such rates upon the poor, *life line rates* have been devised in which the cost of an initial block of energy is low, with the cost increasing for increased consumption above the initial block (see figure 7-8). Here, the initial block is 400 KWH at a cost of 2¢ per KWH. Above that, electricity costs 12¢ per KWH.

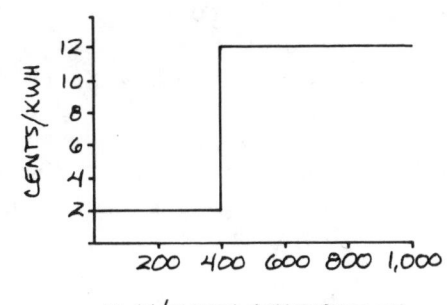

FIGURE 7-8: LIFE LINE RATE

Demand charges. To a certain extent, each consumer requires a certain storage, transmission and power plant capacity to supply energy at his or her rate of consumption. The gas or electric utility must, therefore, invest money to build this capacity as well as to purchase the fuel to meet that need. Some consumers have an almost constant need for energy while others have very high periodic or occasional consumption. In order to meet a customer's peak demand, especially if it is coincident in time with the utility's daily peak, the utility may impose an added charge on consumers to pay for the extra capacity. This also acts as an incentive to limit demand peaks. In a large apartment building with a single meter, the demand charge is typically one-third of the total bill. Thus, there may exist an incentive to reduce peak

demand through load management (see chapter 4). The demand charge is normally a constant dollar figure charged for each kilowatt of peak demand (for example, $4 per kilowatt), although the charge may increase as peak demand increases (see figures 7-9 and 7-10).

Time-of-use rates (TOURS). This type of rate includes a variable demand charge or energy price that changes with the time of day. Just as long-distance telephone charges vary during the day, time-of-use rates reflect changes in hourly energy demand (see figure 7-11). Such a rate provides an incentive to shift energy consumption to those

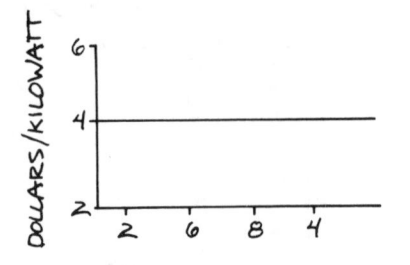

FIGURE 7-9: DEMAND CHARGE

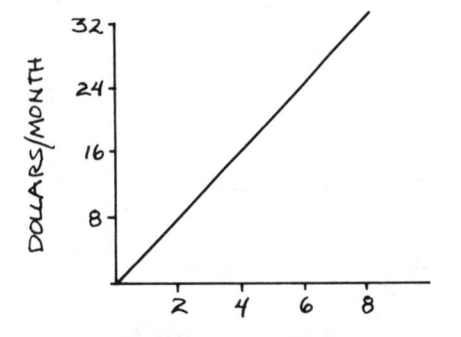

FIGURE 7-10: MONTHLY DEMAND COST

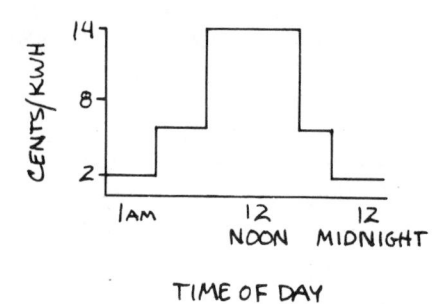

FIGURE 7-11: TIME-OF-USE RATE

times of the day when demand is low and, hence, energy is less expensive. (TOURS rates are also discussed in chapter 4.)

Interruptible rates. An extreme application of TOURS is a rate that allows the utility to cut off part or all of the customer's demand whenever the utility chooses. The utility is allowed to do this only a limited number of times per month and for a limited duration. In return, the consumer receives a major discount for what is clearly a lower-than-normal standard of service. In this way, the utility avoids the need to construct additional capacity to meet peak demand.

For Further Reference

Renewable Energy Access Rights

American Planning Association. *Protecting Solar Access for Residential Development.* Washington, D.C.: U.S. Government Printing Office, n.d. Order no. 023-000-00523-9.

_____. *Site Planning for Solar Access.* Washington, D.C.: U.S. Government Printing Office, n.d. Order no. 023-000-00545-0.

Environmental Law Institute. *Legal Barriers to Solar Heating and Cooling of Buildings.* Washington, D.C., 1977.

Hayes, G. B. *Solar Access Law.* Cambridge, Mass.: Ballinger Publishing Co., 1979.

Hayward, J. *Product Liability Laws and Their Impact on Solar Energy Development.* Boston: Northeast Solar Energy Center, 1979.

Katz, L. K., and Harwin, M. "The Dawn of a New Age." *Solar Age,* June 1976, pp. 8-10.

Kramer, S. F. *Solar Law.* New York: McGraw-Hill Book Co., 1978.

Matuson, J. L. "A Legislative Approach to Solar Access: Transferable Development Rights." *New England Law Review* 13:835-76.

Miller, A. S. "Legal Obstacles to Decentralized Solar Energy Technologies." *Solar Law Reporter* 1:595-612, 761-83.

Park, J. *The Wind Power Book.* Palo Alto, Calif.: Cheshire Books, 1980.

Park, J., and Schwind, D. *Wind Power for Farms, Homes and Small Industry.* Springfield, Va.: National Technical Information Service, 1978. Order no. RFP 2841-1270-78-4.

Ridgeway, J. *Energy-Efficient Community Planning.* Emmaus, Pa.: J.G. Press, 1980.

Taubenfeld, R. F., and Taubenfeld, H. J. *Barriers to the Use of Wind Energy Machines: The Present Legal/Regulatory Regime and a Preliminary Assessment of Some Legal/Political/Societal Problems.* Springfield, Va.: National Technical Information Service, 1976. Order no. PB-263-576.

Thomas, W. A. "The Legal Situation." *Solar Age,* June 1976, pp. 11-13, 31.

Wallenstein, A. R. *Barriers and Incentives to Solar Energy Development.* Boston: Northeast Solar Energy Center, 1978.

_____. ed. *Proceedings of a Workshop on Solar Access Legislation.* Boston: Northeast Solar Energy Center, 1979.

Building Codes and Zoning Restrictions

Hayes, G. B. *Solar Access Law.* Cambridge, Mass.: Ballinger Publishing Co., 1979.

Massachusetts State Building Code Commission. *An Illustrated Guide to the Massachusetts Energy Code for One- and Two-Family Dwellings.* Boston, 1978.

Miller, A. S. "Legal Obstacles to Decentralized Solar Energy Technologies." *Solar Law Reporter* 1:595-612, 761-83.

Park, J. *The Wind Power Book.* Palo Alto, Calif., Cheshire Books, 1980.

Ridgeway, J. *Energy-Efficient Community Planning.* Emmaus, Pa.: J.G. Press, 1980.

Thompson, C. P. *Building to Save Energy.* Cambridge, Mass.: Ballinger Publishing Co., 1980.

Vitale, E. *Building Regulations.* New York: Charles Scribner's Sons, 1979.

Tenant-Landlord Relations

Blumstein, C. et al. "Overcoming Social and Institutional Barriers to Energy Conservation." *Energy* 5 (1980):355.

Massachusetts Executive Office of Energy Resources. *How to Cut Fuel Bills in Apartments.* Boston, 1978.

Southwest Research and Information Center. *The Solar Renter's Handbook.* Albuquerque, N.M., 1977.

Financing Energy Conservation and Renewable Energy Systems

Once you have made the decision to invest in energy conservation or renewable energy for your home or building, you face what is probably the most difficult part of the process: finding the funds to pay for the project. Funds are available through normal channels such as banks and credit unions, but these sources can generally be tapped only by qualified individuals. In certain regions of the United States, utilities are initiating home weatherization financing programs, but many of these programs are still in the planning stage. Under the Carter administration, many energy-related financial aid programs were created to help both low- and middle-income families and communities. However, many of these programs are threatened with reductions or extinction by the budget-cutters in the Reagan administration. Their ultimate fate remains to be seen, however, because of the intricate nature of the federal budgeting process. Some may survive in one form or another or may be taken over by individual states.

FINANCING FOR INDIVIDUALS

Only a few government programs and initiatives exist to help people pay for urgently needed, but often expensive, energy repairs. The most valuable incentives, federal and state energy tax credits, do not really reach those who need help the most. Instead, they are used by upper-income families who are generally able to afford energy improvements but need the extra incentive to spur them to action. Middle-income families benefit from these tax credits but usually require some assistance in obtaining private financing. Few financial institutions offer loan programs geared specifically to energy conservation or renewables, and those that do may provide loans with prohibitive interest rates and terms so short as to make repayment burdensome. Low-income families fare the most poorly because, due to collateral and credit rating requirements, they cannot even turn to lending institutions. They must, instead, look to government programs for help.

While the situation may look bleak, there are some promising developments as far as financing of energy conservation and renewables are concerned. These opportunities can be expected to increase as energy prices continue to rise and energy supplies continue to diminish. For example, programs intended to assist urban groups, housing rehabilitation organizations, neighborhoods or communities may offer individuals alternative means of obtaining financing for conservation projects. Furthermore, several programs in the early stages of implementation may offer additional opportunities for funding. The Residential Conservation Service, utility financing of conservation and renewables and utility and private leasing of solar energy systems are some possibilities. As federal programs are cut back or cancelled, responsibilities formerly filled by the federal government will, in many instances, be taken up by individual states. Individual states might create their own financing assistance programs.

What follows is a summary of presently existing and proposed financing methods that include government and private programs. In evaluating these options and attempting to decide which make the most sense for you, take into account not just the specific financing terms such as interest rates and paybacks but also the procedural conveniences and obstacles, the degree of control you can or cannot exercise over a project and the time involved in processing an application. Programs or financing methods that at first seem expensive or unreasonable may in fact be better for you in other ways, while seemingly inexpensive options may have hidden costs in terms of time and independence.

Government Programs

Most government programs are aimed at specific (usually lower) income groups. While these programs may provide funds with little or no interest, the processing work usually takes much longer to complete than under conventional financing. Such programs can also place limitations on the flexibility of the work being done as well as on owner control over the work. In addition, federal programs are often run by municipal offices and are subject to local regulations.

The only federal program aimed

primarily at improving energy conservation, the Weatherization Program, can be hindered by bureaucratic requirements involving several different and sometimes conflicting government agencies. Tax relief incentives are available but tend to favor the wealthy who need them the least. Other federal programs are more directly aimed at improving housing stock or providing mortgages, but may have energy conservation and renewables components. The HUD 312 low-interest loan program may be a good source of funds but only for certain urban areas. You should check with your local housing office or the regional office of the Department of Housing and Urban Development for details. You should also be aware that most of these government programs are tentatively scheduled for reduction or termination of funding; however, new energy-related programs may be created as replacements.

If you are eligible for any government programs, you should get a copy of the program regulations and read them carefully before beginning the application process. It is important to understand the actual legal limitations of any program to which you apply.

Sources of information on government financing for individuals include the following:

Department of Energy (DOE)
1000 Independence Ave. SW
Washington, DC 20585
(202) 252-5000

Department of Housing and Urban Development (HUD)
451 7th St. SW
Washington, DC 20410
(202)755-6422 or 755-6270

Institute for Local Self-Reliance
1717 18th St. NW
Washington, DC 20009
(202) 232-4103

Internal Revenue Service (IRS)
1111 Constitution Ave. NW
Washington, DC 20224
(202) 566-5000 or 488-3100

Veterans Administration (VA)
941 N. Capitol St. NE
Washington, DC 20421
(202) 872-1151

DOE, HUD, the IRS and the VA also have regional offices.

Other sources include the following:

Your state energy office or energy extension service

Your state community development office

Local community development corporations or community action programs

Your city energy office or energy conservation program

Your gas or electric utility or home heating oil supplier

Your state public utilities commission

Your municipal housing office

Your local Community Action Agency (CAA)

The following states offer residential tax credits or deductions:

Energy conservation:

Alaska	Hawaii
Arkansas (tax deductions)	Oregon
California	Rhode Island
Colorado	Wisconsin

Renewables:

Alaska	Colorado
Arizona	Delaware
Arkansas	Hawaii
California	Idaho

Indiana
Kansas
Maine
Massachusetts
Michigan
Minnesota
Montana
New Mexico
North Carolina

North Dakota
Ohio
Oklahoma
Oregon
Rhode Island
South Carolina
Utah
Vermont

These states offer other types of tax incentives:

Illinois
Iowa
Nevada
New Hampshire
New Jersey

New York
South Dakota
Tennessee
Texas

Private Financing

Private financing is often much quicker than government financing and allows more flexibility and independence to the borrower. Such financing, however, is limited by the nature of the lending institution, which means that those most in need of financial assistance are often the least likely to qualify for loans.

The best place to start looking is at your local bank. Banks offer a variety of programs, some of which may be applicable to energy improvements. Second mortgages, which include provisions for energy conservation and renewables, may be offered, as may home improvement loans. Some banks may offer interest rate reductions on energy-related home improvement loans. Before you apply for any financing, it's a good idea to spend some time phoning local banks to compare interest rates, loan terms, collateral and credit requirements for second

mortgages and other possible financing options. You will find that these can vary widely from one bank to the next. You may also discover that some banks have limitations on the use of their money that they do not immediately mention. For example, a low-interest energy home improvement loan may be available only to those who are already mortgage customers of the bank.

There are sources of private financing other than banks. Credit unions often offer loans at generally favorable interest rates. You may also be able to borrow against a life insurance policy. If the policy is more than ten years old, the interest rates on such a loan can be very favorable. Finally, utilities may be getting into the financing of energy conservation and renewables.

Whatever course of action you take in applying for a loan, be sure you are prepared. First, you should take time to figure out what your monthly loan payments will be at different interest rates, compared with what you will be saving in energy costs. If the energy savings equal or exceed the monthly payments, the loan probably makes good sense. Table 8-2 shows monthly payments on a $1,000 loan with varying interest rates and terms. Second, you should be aware that loan officers are often reluctant to offer energy-related loans, particularly for renewable energy systems, because of doubts about reliability, safety and actual energy savings. After all, a bank has no desire to end up owning something that does not work. You should be prepared to demonstrate that you understand the energy conservation or renewable improvement for which you are seeking a loan, how it works, what it costs and what the payback will be. If you can show that the loan will aid in improving your cash flow, you can assure the loan officer that you will be a good credit risk. When all is said and done, if the loan officer is convinced that he or she is dealing with a well-informed individual, you are more likely to succeed in obtaining a loan.

It is important to note that all financing and assistance programs fluctuate in their availability over time. When money is scarce, as it is in a recession, banks often freeze their mortgage and loan programs. But don't give up. If you have decided to apply for private financing, it often helps to establish an account at one or more banks for a month or so before you apply, even if the amount deposited is small. You should not be discouraged if one bank turns down your loan application. If you can, try to find out why your application was turned down. Then, try another bank that has different criteria for lending money.

The National Energy Act (revised, 1980) allows utilities to finance residential energy conservation and renewable improvements provided that the utilities obtain their financing from local banks. Utilities are also allowed to provide actual energy improvements as long as independent subcontractors are used and unfair methods of competition are not used. Individuals served by a utility are eligible for this assistance. Utilities are not, however, required to offer these services.

Because it is less costly to conserve energy than it is to develop new energy supplies, there is some incentive for electric and gas utilities to get into the business of financing installation of energy conservation measures and renewable energy systems. Such a program has been underway for several years in Portland, Oregon, where the Portland Gas and Electric Company has been financing the installation of energy conservation measures in the all-electric homes of some of its customers. The utility is allowed to enter the cost of these improvements into the rate base since the savings from not building new generating capacity accrue to all customers. The homeowner is required to repay the cost of the conservation measures only if the home is sold.

In certain parts of California, the state Public Utility Commission is requiring gas and electric utilities to finance solar heating and domestic hot water systems with no-interest loans. This is called ZIP, for Zero Interest Plan. The logic here is that the cost of new energy supplies is so great that financing at no-interest terms will still represent a savings to the utility.

At the present time, utility financing is limited to California and parts of Washington and Oregon. In those states, electric utilities are offering interest-free loans to certain homeowners to install energy conservation improvements and renewable energy systems. The loan is repaid through a monthly charge on the electric bill. In some locales, utilities are simply financing these improvements, but not requiring repayment unless the house is sold. The cost of the improvements is figured into the utility's rate base.

Future Financing Programs for Individuals

A number of new, innovative ideas for financing of renewables are being developed in various parts of the United States, mostly in California. Among these programs are leasing of renewable energy systems, the Municipal Solar Utility, the Safe Energy Fund and the Sunny Mac, or Solar Secondary Loan Corporation.

Leasing. Under this program, utilities

or private companies would lease active solar heating and domestic hot water systems to homeowners. The system suppliers would provide, install and maintain the system and the homeowner would pay a monthly service fee for use of the system.

Municipal Solar Utility (MSU). The MSU is a nonprofit, municipal organi-zation that can raise funds through long-term municipal bonds. With these funds, the MSU can provide 20-year loans to solar system purchasers. The MSU can also handle maintenance and repairs for several years after system installation, thus providing added assurance to the purchaser.

Safe Energy Fund. The Safe Energy

Table 8-1
Financing of Energy Conservation and Renewables for Individuals
(government and private)

Financing or Assistance Programs	Populations for Which Program/ Financing Is Intended	Conservation	Renewables	Owners	Tenants	Amounts and Terms of Assistance or Financing	Availability	Specific Legal Require- ments
Grants and Rebates								
Appropriate Technology Small Grants Program (DOE)	Individuals, nonprofit organi- zations, small businesses	X	X	X	X	Grants up to $50,000 for innovative applications of renewable energy systems and energy con- servation technology	Program run through regional DOE offices	None
Housing improvement Community Development (Block Grant Program)	Usually low- and moderate- income property owners of 1- to 6-unit structures	X	X	X	...	Varies but usually a grant, grant-loan or rebate; typi- cally less than $1,200, but could be up to $7,500	In desig- nated neigh- borhoods through municipality or housing improve- ment office	None

Fund is a program developed by a solar organization and a savings and loan association in San Francisco. Depositors at this bank can designate their savings to be used for financing of renewable energy systems. Deposits are invested in money market funds. Loans from this fund are made for 20 years, at interest rates 1.5 percentage points above those being paid to the depositors.

Sunny Mac. The Sunny Mac, or Solar Secondary Loan Corporation, is modeled after existing secondary home mortgage corporations and is still in the proposal stage. It buys packages of solar loans made by banks, savings and loan associations and credit unions. By doing so, the corporation returns to these financial institutions funds that can be used for more loans. Capital for the corporation can come from state pension funds, private investors, lending institutions and federal, state and local governments.

Other Processing Requirements	Typical Time to Process	Waiting after Approval	Other Work Requirements	Contractor Selection	Do-It-Yourself	Other Characteristics	Additional Information
Funds for 3 categories of project: (1) concept development; (2) project development; (3) project demonstration	3½ months	5-6 months	...	Own selection	OK	Only about 10-15% of applicants funded; 1 cycle per year	Regional DOE offices (program may have its funding cancelled for 1982)
...	1-2 weeks	Usually none	Perhaps some programs incorporate code enforcement	Usually own selection	Encouraged by most organizations	Good financing terms and quick; often not enough for all repairs needed; sometimes direct rebates to homeowners	Local housing office or community development office

[*Continued on next page*]

Table 8-1— *Continued*

Financing or Assistance Programs	Populations for Which Program/ Financing Is Intended	Conservation	Renewables	Owners	Tenants	Amounts and Terms of Assistance or Financing	Availability	Specific Legal Require- ments
Grants and Rebates— *Continued*								
Low-income weather- ization (DOE/ CAA weather- ization)	Low-income persons or families	X	...	X	X	Free materials worth up to $1,000 plus CETA (Com- prehensive Employment and Training Act) labor ($240) or contractor labor ($800); no direct cash grants to individuals	Available through CAAs; funds admin- istered through qualifying neighbor- hood group	Verifi- cation of income and residency
Government Loans, Mortgages and Insurance								
Government- insured home improvement loans (HUD Title I program)	Property owners with adequate credit ratings	X	X	X	X	HUD insures loan from designated commercial bank to indi- vidual for home improve- ments, loans are for 15 years maxi- mum; limits are: $5,000 per unit, $37,500 per building; $15,000 per single family	Nationwide	Proof of owner- ship or long- term lease
Government- insured home im- provement loans (Veterans Administra- tion)	Veteran property owner with adequate credit rating	X	X	X	...	Energy- related loans up to maxi- mum of $33,000, at 10-12% interest plus one point for 30 years	Few banks	Proof of owner- ship

Other Processing Requirements	Typical Time to Process	Waiting after Approval	Other Work Requirements	Contractor Selection	Do-It-Yourself	Other Characteristics	Additional Information
If tenant is applicant, owner to agree not to increase rent for at least 1 year	Some CAAs 1-2 weeks; others 4-5 weeks	Depends on workload but typically 1-3 months	Not usually	. . .	Encouraged by some CAAs; discouraged by others	Free materials and labor; risk of poor workmanship	Local CAA or DOE (program may be changed for 1982)
Estimates or bids	1 week or less	None	Not usually	Own selection	OK	List of eligible improvements available at HUD office; simultaneous payments of a mortgage and loan may be high	Local HUD office
Estimates or bids	1 week	Usually none	. . .	Own selection	OK	Processing speed depends on commercial lender; good terms; few banks will honor	Local Veterans Administration office

[Continued on next page]

Table 8-1—*Continued*

Financing or Assistance Programs	Populations for Which Program/ Financing Is Intended	Conservation	Renewables	Owners	Tenants	Amounts and Terms of Assistance or Financing	Availability	Specific Legal Require-ments
Government Loans, Mortgages and Insurance—*Continued*								
Government-insured home mortgages (FHA/HUD-203)	All	...	X	X	...	Home mortgage with increased limits for solar heating w/o Solar 1 fam. $ 67,500 2 fam. 76,000 3 fam. 92,000-107,000 w/Solar 1 fam. $ 81,000 2 fam. 91,200 3 fam. 110,400-128,400	Only FHA-approved mortgage	...
Government-insured home mort-gages—experimen-tal homes (FHA/HUD-233)	All	...	X	X	...	Government insurance up to 90% of loan on experi-mental homes that include advanced housing technology	Only FHA-approved mortgage	...
Residential Conservation Service (RCS)—new	Customers of utilities and heating supply companies that partici-pate in RCS	X	X	X	X	Provides for energy audit that will prob-ably not exceed $15; loans will be made avail-able by private banks through RCS	Admin-istered by each state and run through gas and electric utility com-panies, heating oil suppliers	None

Other Processing Requirements	Typical Time to Process	Waiting after Approval	Other Work Requirements	Contractor Selection	Do-It-Yourself	Other Characteristics	Additional Information
Good credit rating	Varies	...	...	Own selection	OK	Processing speed depends on commercial lender	Local HUD office
Good credit rating	Can be extensive	...	...	Own selection	OK	Program rarely used, so much paper work; generally applicable to new housing	Local HUD office
None	Few weeks	Few weeks	...	...	...	Inexpensive audit; audit includes assessment of home's potential for solar and wind	Local electric or gas utility or state energy office (program may have its funding cancelled for 1982)

[*Continued on next page*]

Table 8-1—*Continued*

Financing or Assistance Programs	Populations for Which Program/ Financing Is Intended	Conservation	Renewables	Owners	Tenants	Amounts and Terms of Assistance or Financing	Availability	Specific Legal Require- ments
Government Loans, Mortgages and Insurance—*Continued*								
3% direct government loan pro- gram (HUD 312 Program)	Low- and moderate- income property owners	X	X	X	…	Up to $27,000 per apart- ment or house at 3% interest for 20 years	In commu- nities with Community Develop- ment Block Grant Program admin- istered by HUD	Clear title to build- ing required
Tax Relief Incentives								
Federal energy tax credit	Middle- and higher- income families	X	…	X	X	15% of first $2,000 spent for qualified energy con- servation measures; up to $300 deduction	All tax- payers who itemize deductions (long 1040 form)	Work must be done to own home or apart- ment
Federal solar tax credit	All taxpayers	…	X	X	X	40% of expenditure up to $10,000 for qualified renewable energy system; up to maximum of $4,000	All tax- payers who itemize deductions (long 1040 form)	Work must be done on principal residence
Municipal property tax exemption	Generally people owning primary residences	…	…	…	…	Generally percentage exemption of property taxes on solar and wind systems	Varies from state to state	Generally applies only to specific conser- vation improve- ments or renewable systems

Other Processing Requirements	Typical Time to Process	Waiting after Approval	Other Work Requirements	Contractor Selection	Do-It-Yourself	Other Characteristics	Additional Information
Preference given to low- and moderate-income property owners	Varies but typically 7-9 weeks	2 weeks	Often building must meet state or local code standards	Lowest bidder in public bid process	Discouraged	Excellent financing terms; may have to make unexpected code repairs; program also provides direct rebates for improvements	Local HUD office (program may be consolidated into a block-grants program for 1982)
...	...	...	Energy conservation measures; must have minimum 3-year useful life	Own selection	OK	OK for the well-to-do; requires very large investment to be meaningful for low- to moderate-income taxpayer	Local IRS office or state energy office (provisions may be amended for 1982); expires in 1985
...	...	...	Renewable energy system must have 5-year useful life	Own selection	OK	Preferential to upper-income groups	Local IRS office or state energy office (provisions may be amended for 1982); expires in 1985
...	...	...	...	Varies	Varies	State energy office or municipal property tax office	State energy office or municipal property tax office

[Continued on next page]

Table 8-1—*Continued*

Financing or Assistance Programs	Populations for Which Program/ Financing Is Intended	Conservation	Renewables	Owners	Tenants	Amounts and Terms of Assistance or Financing	Availability	Specific Legal Require- ments
Tax Relief Incentives—*Continued*								
State energy and/or solar tax credits or deduc- tions	Generally state taxpayers	...	...	...	...	Generally per- centage of expenditure after deduc- tion of federal tax credit	Varies from state to state	Generally work must be done on primary residence
State sales tax exemp- tion	Generally all	...	...	...	...	Generally percentage exemption of sales tax on renewable energy systems expenditures	Varies from state to state	May apply only to specific materials and systems
Private Financing								
Conventional home improve- ment loan	Property owners with good credit rating	...	...	X	...	Up to $30,000 but more likely $5,000; from 12% to 18% interest, 3-15 years	Many banks offer this kind of loan	Proof of owner- ship
Passbook loan	Savings depositor	...	...	X	X	Up to balance of savings account; interest rates and terms vary	Most banks	...
Personal loan	Individual with good credit	...	...	X	X	Depends on personal credit; amount, interest rates and terms vary	Most banks	...

Other Processing Requirements	Typical Time to Process	Waiting after Approval	Other Work Requirements	Contractor Selection	Do-It-Yourself	Other Characteristics	Additional Information
...	...	...	Varies from state to state	Varies	Varies	Eligible systems are usually same as those to which federal tax credits apply	State tax office or energy office
...	...	...	...	Varies	Varies	Exemption taken at time of purchase	State tax office or energy office
Estimates or bids	1 week or less	Usually none	Not usually	Own selection	OK	Don't have to rewrite mortgage; simultaneous payments of mortgage and loan may be too high	...
None	Same day	Usually none	Not usually	Own selection	OK	Almost instant money available; restricts access to savings	...
None	1 week or less	Usually none	Not usually	Own selection	OK	Don't need collateral; very high interest; short term	...

[Continued on next page]

Table 8-1—*Continued*

Financing or Assistance Programs	Populations for Which Program/ Financing Is Intended	Conservation	Renewables	Owners	Tenants	Amounts and Terms of Assistance or Financing	Availability	Specific Legal Require- ments
Private Financing—*Continued*								
Rewrite 1st mortgage	Property owners with good credit rating	...	...	X	...	Up to 80% or 90% of property value; 14-18% interest likely, up to 30 years; equity require- ment from 5-40%	Nearly all savings banks, most commercial banks, credit unions and cooperative banks	Clear title to build- ing required
2d mort- gage	Property owners with good credit rating	...	...	X	...	Where combined 1st and 2d mort- gages don't exceed 90% value of property; amounts, terms and interest vary	Some banks offer this kind of loan	Proof of owner- ship and clear title
Utility financing	Customers of partici- pating utility	X	X	X	X	Varies	Participating utilities	...
Miscellaneous								
Leasing program (Section 8 Moderate Rehabilitation Program)	Property owner with apartments available to low-income tenant(s)	X	...	X	...	Lease with local housing authority	Nationwide and specific neighbor- hoods through local housing authority	Lease between owner and housing authority

Other Processing Require- ments	Typical Time to Process	Waiting after Approval	Other Work Require- ments	Contractor Selection	Do-It- Yourself	Other Character- istics	Additional Information
Estimates or bids	2-3 weeks	Usually none	Not usually	Own selection	OK	Payment of original mortgage and repairs combined in manageable single payment; must give up low interest of original mortgage	
Often- times must hold 1st mortgage	2-3 weeks	Usually none	Not usually	Own selection	OK	Don't have to rewrite 1st mort- gage and lose low interest; simul- taneous payments of 2 mort- gages may be too high	. . .
. . .	. . .	. . .	. . .	. . .	. . .	Fairly new idea; details will vary; presently limited to California and parts of Wash- ington and Oregon	. . .
. . .	Varies	Usually none	Building to meet state or local habita- tion code stan- dard	. . .	OK	Guaran- teed rent; may have to make unex- pected code repairs	Local housing authority or HUD

Table 8-2
Monthly Payments on a $1,000 Loan at Varying Interest Rates and Terms
(in dollars)

Interest Rate (%)	Term of Loan (yrs)*					
	2½	5	7½	10	15	20
10	37.81	21.23	15.82	13.19	10.72	9.62
11	38.28	21.76	16.39	13.80	11.39	10.35
12	38.75	22.24	16.90	14.35	12.00	11.01
13	39.22	22.73	17.43	14.91	12.63	11.69
14	39.70	23.29	18.03	15.55	13.34	12.46
15	40.18	23.79	18.57	16.13	14.00	13.17
16	40.66	24.32	19.15	16.75	14.69	13.91
17	41.15	24.85	19.73	17.38	15.39	14.67
18	41.64	25.39	20.32	18.02	16.10	15.43

NOTE: For monthly payments that are multiples of $1,000, multiply monthly payment by:
$$\frac{\text{loan in dollars}}{\$1,000}$$

*Longer-term loans have lower monthly payments but larger total repayments. For example, a 5-year, 12% loan for $1,000 involves repayment of $1,334. A similar 10-year loan involves repayment of $1,722.

FINANCING FOR GROUPS

Table 8-1 addresses the financing of energy conservation and renewable energy systems for individual homeowners, tenants and landlords. Most of the programs described there are not available to groups such as for-profit developers, nonprofit corporations, local governments, community action agencies, housing authorities and local housing improvement programs. However, these types of groups may qualify for grants if they use such monies to provide aid to individuals.

While there are a number of group assistance programs (mostly federal), the variety of funding sources available to groups is more limited than for individuals. Banks state uniformly that no financing is available for groups interested in energy conservation, but because banks are generally required to invest in the communities with which they do business, special community-oriented energy conservation projects might be funded by some banks. Most energy-related tax credit programs are intended solely for individual use. The one exception is the Business Energy Credit which applies only to renewable energy systems and not to energy conservation property. It is possible that energy conservation tax credits for multi-family housing may be legislated for 1982. The Urban Development Assistance Grants (UDAG) Program and the Section 312 Property Rehabilitation Loans from HUD contain provisions for financing energy conservation improvements by groups. The National Consumer Cooperative Bank may fund neighborhood cooperatives whose business is manufacturing and supplying energy-saving materials to low-income neighborhoods. Some private foundations may be willing to fund small energy-related projects, particularly if they are innovative and encourage community self-reliance.

Tables 8-3 and 8-4 list sources of group financing by government programs and by foundations that may be sympathetic to funding energy-related projects.

Government Programs

In some of the programs financed by the government, a community group or local government acts as a vehicle or sponsor through which money is channeled to the individual homeowner, landlord or tenant. In other programs, money is given directly to neighborhood organizations for use in energy-conserving projects in the form of grants, subsidies, rehabilitation and property-improvement loans, rental and housing assistance, mortgage insurance and awards for project proposals. As with any financing program, you should obtain the regulations and/or guidelines before submitting a request or proposal. Don't be discouraged if a program turns you down. Use the experience you have gained from the effort and try again!

Sources for group financing include the following:

Department of Energy (DOE)
1000 Independence Ave. SW
Washington, DC 20585
(202) 252-5000

Department of Housing and Urban Development (HUD)
451 7th St. SW
Washington, DC 20410
(202) 755-6422 or 755-6270

Institute for Local Self-Reliance
1717 18th St. NW
Washington, DC 20009
(202) 232-4103

National Center for Appropriate Technology (NCAT)
P.O. Box 3838
Butte, MT 59702
(406) 494-4572

National Consumer Cooperative Bank (NCCB)
2001 S St. NW
Rm. 302
Washington, DC 20009
Toll-free: (800) 424-2481 or
(202) 673-4300

Small Business Administration (SBA)
1441 L St. NW
Washington, DC 20016
(202) 653-6887

DOE, HUD and SBA have regional offices. NCCB also has state-chartered branches.

Other sources include the following:

Your state energy office or energy extension service

Your state community development office

Local community development corporations or community action programs

Your city energy office or energy conservation program

Your gas or electric utility or home heating oil supplier

Your state public utilities commission

Your municipal housing office

Your local Community Action Agency (CAA)

Table 8-3
Financing of Energy Conservation and Renewables for Groups

Financing or Assistance Program	Population for Which Program Intended	Amounts and Terms of Assistance	Processing
Business energy investment credit (federal)	Businesses	Credit covers 10% of equipment cost or 5% if financed by tax-exempt industrial development bonds	Business must file forms 3468, 3468-Schedule B or 4255; credit is limited to 100% of liability, except for wind and solar, and is refundable if it exceeds the liability
Community Development Block Grant Program (HUD)	Local governments	Local government receives block grants from HUD for community development activities; cities funded according to entitlement based on need (factors include population, poverty, housing conditions, growth lag); direct grants and loans (or private loan subsidies or guarantees) given to individuals to meet weatherization standards for residential and nonresidential properties; most cities offer rebate program for energy conservation	Local government applies to HUD: disperses funds to eligible groups
Community Energy Block Grants (proposed)	Local and state governments	Several proposals have been submitted to Congress to consolidate a number of formerly separate funding programs into block grant funds to be used specifically for energy conservation projects	Application would be made to state- or locally administered program responsible for disbursing block grant funds
Federal matching grants for energy conservation (HUD)	Public and private nonprofit institutions, including: schools, churches, hospitals, nursing homes, municipal offices	Provides (1) matching grants for energy audits of institutions and (2) matching funds for technical assistance programs and for energy conservation measures (including renewable resources)	Program administered by state energy office
National Center for Appropriate Technology Small Grants Program	Any nonprofit organization that is community-based and endorsed by local CAA (must be in low-income community)	Awards for proposals for energy conservation techniques; designed to encourage local development of solutions to common technical problems	CAAs and development groups or any nonprofit organization that is community based submit proposals to NCAT
National Consumer Cooperative Bank	Consumer cooperatives (mainly); 19% of capital can also be lent to producer cooperatives; cooperatives must be licensed by state; no lending to individuals	(1) Technical assistance department (2) Self-help fund: $11 million; reduced interest loans to low-income cooperatives (people who belong to cooperatives or are served by cooperatives must have low income) (3) Bank-market rate loans to cooperatives	Depends upon the state-chartered branch of the bank

Characteristics	Legal Authority	Additional Information
Credit may be taken on energy equipment used for conservation, recycling, solar, wind and other systems designed to displace oil and natural gas	Energy Tax Act, Code Section 46 (A-2)	Contact the IRS
Local government determines funding allocations; federal regulations authorize local government to make CDBG funds available to nonprofit groups, local development corporations and small business investment companies for neighborhood revitalization projects	Title I Housing & Community Development Act of 1974 (PL 93-383) as amended by Title I Housing & Community Development Act of 1977 (PL 95-128)	Office of Block Grant Assistance HUD 451 7th St. SW Rm. 7182 Washington, DC 20410 or local community development and housing agencies
Local and state governments would use discretion in using funds from program as they see fit	. . .	Local housing or community development office or local or state energy office
Institutions appoint energy conservation manager responsible for monitoring and evaluating energy use; matching funds come from state, local or private sources	National Energy Act	Community Energy Conservation Competition Office of Policy Planning HUD 451 7th St., SW Rm. 7134 Washington, DC 20410 or the state energy office
Grants for demonstrations of energy conservation techniques	Funded through Community Services Administration through the Training and Technical Assistance Program	The National Center for Appropriate Technology P.O. Box 3838 Butte, MT 59702
Consists of two parts: the bank and the Office of Self-Help, Development and Technical Assistance; bank will make loans to eligible co-ops at prevailing interest rates, and Office of S.H.D. & T.A. will provide capital advances and technical assistance to newly forming or low-income co-ops	Public Law 95-351 signed into law 1978	National Consumer Cooperative Bank 2001 S St. NW Rm. 302 Washington, DC 20009 or the state-chartered branch of the bank

[Continued on next page]

Table 8-3—*Continued*

Financing or Assistance Program	Population for Which Program Intended	Amounts and Terms of Assistance	Processing
Section 235 Home-owner Subsidy for Low & Moderate Income Families (HUD)	Builder/developer	HUD subsidy makes up the difference between monthly principal and interest payments at a 6¾% interest rate and the payments at the current interest rates plus the FHA insurance premium; married couples or single people over 65 or handicapped are eligible; applies only to unbuilt homes	Builder applies to HUD office for homes he is rehabilitating; sells homes to purchasers as ones that could be financed under 235
Section 8 Lower-Income Rental Assistance (HUD)	Low-income families via sponsors; profit-motivated or nonprofit or cooperative organizations or public housing agencies	Rent subsidies provided to low-income families who pay 25% of income toward rent; money paid directly to owner of rehabilitated unit; dwelling must meet energy-efficiency standards	Nonprofit and profit-motivated developers either with or without public housing agencies, submit proposals to HUD for rehabilitation projects; individuals wanting assistance contact project sponsors directly
Section 213 Cooperative Housing Mortgage Insurance (HUD)	Nonprofit cooperative ownership housing corporations and sponsors who intend to sell projects to nonprofit corporation or trust	Provides federal mortgage insurance to finance cooperative housing projects; FHA insures mortgages made by private lending institutions on cooperative housing projects of 5 or more dwelling units to be occupied by members of nonprofit cooperative; ownership housing corporations to finance rehabilitation improvement of project already owned; rehabilitation of projects owners intend to sell to nonprofit cooperatives	Apply through a HUD/FHA approved lender
Section 221 (d) (3) and (4) Multi-family Rental Housing for Low and Moderate Income Families	For (3): public agencies; nonprofit, limited-dividend or cooperative organizations; For (4): only profit-motivated sponsors	Mortgage insurance program to finance construction or rehabilitation of multi-family rental or cooperative housing: (1) under 221 (d) (3) HUD insures 100% of project value; (2) under 221 (d) (4) HUD insures 90%—has provided bulk of financial assistance for Section 8 projects	Application made with HUD area office
Section 241 Property improvement loans (HUD)	Owners of projects with FHA-insured mortgages (i.e., apartment houses, nursing homes, etc.)	HUD insures loans by private lenders for improvements to existing FHA-insured properties including multi-family projects and nursing homes; eligible uses for loan proceeds include improvements related to energy conservation	Approved FHA lender can administer program; FHA-insured project contacts local HUD office to discuss eligibility requirements and procedure for applying

Characteristics	Legal Authority	Additional Information
Program designed to encourage builders to rehabilitate housing for low- to moderate-income families	Section 235, National Housing Act (1931) as amended by Section 101 HUD Act of 1968	Subdivision appraiser in local HUD office
Sponsoring organization works through local housing agency; rehabilitation work includes energy-efficiency improvements	Section 8, Housing Act of 1937 (PL 73-479) amended by Housing and Community Development Act of 1974	HUD; housing agencies; community groups can also disseminate information
Eligibility extends to rehabilitation improvements of housing already owned, which must include weatherization	Section 213, National Housing Act	Local office of HUD
HUD Minimum Property Standards apply to housing under this program	National Housing Act of 1934 (PL 73-479) as added by the Housing Act of 1954 (PL 83-560)	Local office of HUD
Maximum loan amounts and interest rates for program are established by HUD and fluctuate over time based on prevailing market conditions	Section 241 (PL 73-479)	Local office of HUD

[Continued on next page]

Table 8-3—*Continued*

Financing or Assistance Program	Population for Which Program Intended	Amounts and Terms of Assistance	Processing
Section 312 Property Rehabilitation Loan (HUD)	Federally aided CDBG, urban homesteading, urban renewable and code enforcement areas; administered through local community development agencies	Long-term (up to 20 years), low-interest (3%) direct federal loans, administered by local government agency, designed to finance rehabilitation for single-family or multi-family housing et al; maximum resident loan is $27,000/unit	312 Loan Office in city hall
Small business energy loan	Small firms engaged in energy conservation activities (renewable energy)	Direct or immediate participation; loans may not exceed $350,000; loans under SBA/Bank guaranty program may not exceed $500,000 or 90% of total loan	Contact participating bank of SBA office
State energy conservation and/or renewable tax credits for businesses and corporations	Businesses and corporations doing business in particular state	Tax credit (or deduction or rebate)	Tax credit taken at time of filing of tax return
Urban Development Action Grant (UDAG) —HUD	Local government (cities must meet minimum criteria indicating physical and economic distress)	Grants assist severely distressed cities and urban counties to revitalize local economies through public and private investments in projects of benefit to low- to moderate-income persons; projects should take maximum of 4 years to complete; energy conservation introduced at every opportunity	Interested communities request determination of eligibility from HUD area offices before applications can be distributed; applications accepted throughout the year during 1st month of each quarter and awards announced during last month of each quarter
Urban Homesteading Program (HUD)	Local governments	Vacant HUD-held properties transferred to local governments which sell homes for token sums to families who pledge to renovate them; weatherization must be part of overall rehabilitation effort	HUD chooses localities after acceptable plans submitted
Weatherization Assistance (DOE)	Local community action agency (CAA), either federally or privately funded	CAA receives funding for insulation and other conservation materials; CAA supervises weatherization for low-income persons' homes (individual eligibility determined by family size and income); grants up to $1,000 available to cover costs of energy audit and weatherization materials; grants up to $800 for labor if not supplied by CETA (Comprehensive Employment and Training Act); no direct cash grants to individuals	Funding goes to state; state energy office subcontracts to groups; individual applies to CAA; applications from CAA due April 30 in DOE

Characteristics	Legal Authority	Additional Information
Weatherization is part of rehabilitation	Section 312 Housing Act of 1964 (PL 88-560)	HUD regulatory offices; local housing and community development agencies (program may be folded into Community Development Block Grants program for 1982)
Businesses must be for profit and engaged in the manufacturing, retailing or wholesaling of energy-conserving materials and devices	. . .	Small Business Administration (Funding for this program may be reduced for 1982)
Qualified measures vary from state to state; credit generally equal to some fraction of purchase and installation cost of measure	. . .	State tax office or energy office
Money provided by HUD is used by city officials to attract private investment in a project in a distressed neighborhood	Section 119, Housing and Community Development Act of 1977	Local office of HUD
Neighborhood organizations can work as partners with local government in planning and implementing programs	Section 810, Housing and Community Development Act of 1974 (PL 93-383) Section 20, Housing Authority Act of 1976 (PL 94-375)	Community Development & Urban Homesteading Division HUD 451 7th St. SW Rm. 7174 Washington, DC 20410
Weatherization work crews paid for by CETA funds from Department of Labor, obtained by CAA through local prime sponsors (i.e., state, city or county); weatherization programs can be established as joint ventures with local community development block grant program (e.g., community development agency payments for activities such as home repair done by CAA weatherization crews)	Published in the *Federal Register*, dated June 1, 1981, Weatherization Assistance for Low-Income Persons Amendment of Regulations 10CFR Part 440	DOE 1000 Independence Ave. SW Washington, DC 20585 or regional office of DOE or state energy office (program may be changed for 1982)

Private Foundations

There are over 20,000 private foundations and charitable trusts in the United States. Many of these foundations are quite small, but a good number have assets in excess of $100,000 and are interested in funding programs related to urban development, energy and the environment. Table 8-4 lists a limited number of national foundations that may be potential funding sources. Before sending proposals to any of the foundations listed, be sure that your project fits the foundation's guidelines and funding interests. Often a phone call to the foundation contact person can clarify the foundation's specific interests. If you are interested in learning more about these foundations, or about others, and how to apply for a grant, there may be a local association of foundations in your area that can provide you with such information. Contact a local foundation to find out whether such an association exists, or consult the *Foundation Directory* for the names of foundations located in your area.

A good source of information about foundation funding opportunities is the Foundation Center, which has offices at the following addresses:

The Foundation Center
Kent H. Smith Library
739 National City Bank Bldg.
629 Euclid Ave.
Cleveland, OH 44114
(216) 861-1933

The Foundation Center
888 Seventh Ave.
New York, NY 10106
(212) 975-1120

The Foundation Center
312 Sutter St.
San Francisco, CA 94108
(415) 397-0902

The Foundation Center
1001 Connecticut Ave. NW
Washington, DC 20036
(202) 331-1400

The Foundation Center also maintains collections of foundation reference materials in libraries in all 50 states, Canada, Mexico, Puerto Rico and the U.S. Virgin Islands. A listing of these collections may be obtained from the Center, which can also provide a fund-raising bibliography.

Table 8-4
National Foundations that May Be Potential Funding Sources for Energy-Related Projects

Name, Address, Phone number, Contact	Funding Interests	1. Assets ($) 2. Expenditures ($) 3. Funding Range ($)	Application Deadline
Ahamson Foundation, The 3731 Wilshire Blvd. Los Angeles, CA 90010 (213) 383-1381 Kathleen Gilcrest	Community funds	1. 107,709,600 2. 4,111,076 3. 50 to 1,800,000	Sept.
Arca Foundation 1425 21st St. NW Washington, DC 20036 (202) 822-9193 Margery Tabankin	Safe and healthy environment	1. 9,000,000 2. 600,000 3. 1,000 to 50,000	Mar. 15, Sept. 15
Atlantic-Richfield Foundation 515 S. Flower St. Los Angeles, CA 90071 (213) 496-3342 Walter D. Eichner	Community development	1. 26,123,451 2. 26,123,451 3. 1,000 to 500,000	None
Brown Foundation Inc., The P.O. Box 13646 2118 Welch Ave. Houston, TX 77219-3646 (713) 523-6867 Merritt A. Warner	Conservation (most grants limited to applicants from Texas)	1. 173,881,552 2. 7,291,685 3. 50 to 1,603,226	Quarterly
Bydale Foundation 60 E. 42nd St., Rm. 5010 New York, NY 10165 (212) 682-4052 Milton D. Solomon	Energy, environment	1. 6,963,382 2. 446,271 3. 150 to 30,000	June, Sept., Nov., Dec.
Geraldine R. Dodge Foundation 95 Madison Ave. P.O. Box 1239R Morristown, NJ 07960 (201) 540-8442 Scott McVay	Energy, environment	1. 29,545,625 2. 1,749,057 3. 5,000 to 60,000	Jan. 15, Apr. 15, July 15, Oct. 15

[Continued on next page]

Table 8-4—*Continued*

Name, Address, Phone number, Contact	Funding Interests	1. Assets ($) 2. Expenditures ($) 3. Funding Range ($)	Application Deadline
EXXON USA Foundation P.O. Box 2180 Houston, TX 77002 (713) 656-3008 Harold A. Reddicliffe	Energy, environment	1. 6,623,011 2. 4,411,139 3. 20 to 345,300	None
Ford Foundation 320 E. 43rd St. New York, NY 10017 (212) 573-5000 Howard R. Dressner	Energy, environment	1. 2,583,000,000 2. 114,000,000 3. 1,500 to 2,963,000	None
Gund Foundation, The One Erieview Plaza Cleveland, OH 44114 (216) 241-3114 James S. Lipscomb	Environment, community development	1. 83,000,720 2. 3,842,447 3. 300 to 300,000	Feb. 1, May 1, Sept. 1, Nov. 1
John A. Hartford Foundation 405 Lexington Ave. New York, NY 10174 (212) 661-2828 Robert F. Higgins	Energy	1. 126,425,765 2. 3,982,647 3. 20,000 to 100,000	Mar., June, Sept., Dec.
Hewlett Foundation 2 Palo Alto Sq. Suite 1010 Palo Alto, CA 94304 (415) 493-3665 Roger W. Heyns	Environment	1. 31,006,724 2. 2,256,273 3. 500 to 300,000	Jan., Apr., July, Oct.
Henry P. Kendall Foundation One Boston Pl. Boston, MA 02108 (617) 723-8728 Robert Allen	Energy, environment	1. 30,000,000 2. 1,500,000 3. 5,000 to 70,000	None
Kresge Foundation Standard Federal Bldg. 2401 West Big Beaver Rd. Troy, MI 48084 (313) 643-9630 William H. Baldwin	Construction and renovation of facilities or purchase of major equipment	1. 655,408,031 2. 29,767,887 3. 10,000 to 1,000,000	Jan. 1 to Feb. 15
John D. & Catherine T. MacArthur Foundation Suite 700 140 S. Dearborn Chicago, IL 60603 (312) 726-8000 Joseph A. Diana	Energy	1. 840,000,000 2. 42,788,000 3. 1,000 to 1,000,000	Monthly

Table 8-4—*Continued*

Name, Address, Phone number, Contact	Funding Interests	1. Assets ($) 2. Expenditures ($) 3. Funding Range ($)	Application Deadline
A. W. Mellon Foundation 140 E. 62nd St. New York, NY 10021 (212) 838-8400 J. Kellum Smith, Jr.	Energy	1. 776,376,000 2. 43,967,262 3. 10,000 to 7,500,000	Quarterly
Charles Stewart Mott Community Mott Foundation Bldg. Flint, MI 48502 (313) 238-5651 William S. White	Community development, energy	1. 428,261,241 2. 28,385,881 3. 107 to 1,559,500	None
Musicians United for Safe Energy, Inc. (MUSE) 72 Fifth Ave. New York, NY 10011 (212) 691-5422 Susan Kallam	Safe energy/Anti-nuclear projects	1. 1,000,000 2. 1,000,000 3. 500 to 25,000	None
Needmor Fund 136 N. Summit St. Toledo, OH 43604 (419) 244-4981 Karl N. Stauber	Environment	1. 3,787,878 2. 890,959 3. 20,000 to 30,000	Varies
Orleton Trust Fund 1777 Borel Place Suite 306 San Mateo, CA 94402 (415) 345-2818 Mrs. Jean Sawyer Weaver	Environment	1. 7,793,849 2. 752,940 3. 10 to 100,000	None
Ottinger Foundation Inc. 370 Lexington Ave. New York, NY 10017 (212) 532-0617 David R. Hunter	Environment	1. 2,695,884 2. 220,850 3. 50 to 15,000	Feb., June, Oct.
Rockefeller Brothers Fund 1290 Ave. of the Americas Room 3450 New York, NY 10104 (212) 397-4818 Russell A. Phillips, Jr.	Environment	1. 177,087,767 2. 22,894,957 3. 5,000 to 3,000,000 (average grant 25,000 to 35,000)	Mar., June, Nov.

[*Continued on next page*]

Table 8-4—*Continued*

Name, Address, Phone number, Contact	Funding Interests	1. Assets ($) 2. Expenditures ($) 3. Funding Range ($)	Application Deadline
Rockefeller Family Fund 1290 Ave. of the Americas New York, NY 10104 (212) 397-4844 Robert W. Scrivner	Energy	1. 11,780,870 2. 1,370,101 3. 2,000 to 50,000	Apr., Dec.
Rockefeller Foundation 1133 Ave. of the Americas New York, NY 10036 (212) 869-8500 Lawrence D. Stifel	Occasionally funds projects in energy and environment	1. 1,000,890,354 2. 47,902,311 3. 5,000 and up	Semi-monthly
Scherman Foundation, Inc. 1740 Broadway New York, NY 10019 (212) 489-7143 David F. Freeman	Energy	1. 26,871,350 2. 1,578,351 3. 1,000 to 100,000	None
Schumann Foundation 33 Park St. Montclair, NJ 07042 (201) 783-6660 Harold S. Merrell	Community development	1. 41,800,000 2. 1,866,600 3. 3,000 to 50,000	Jan. 15, Apr. 15, Aug. 15, Oct. 15
Shalan Foundation 2749 Hyde St. San Francisco, CA 94109 (415) 673-8660 Drummond Pike or Yolanda Adra	Environment	1. 650,876 2. 617,656 3. 3,000 to 15,000	None
Stern Fund, The 370 Lexington Ave. New York, NY 10017 (212) 532-0617 David R. Hunter	Social change	1. 2,660,569 2. 523,000 3. 1,000 to 35,000	Quarterly
Youth Project Foundation 1555 Conn. Ave. NW Washington, DC 20036 (202) 483-1430 William Mitchell	Energy, environment	1. 4,000,000 2. 4,000,000 3. 3,000 to 7,000	None

Sources: *Energy Link,* "Special Private Funding Issue" (Washington, D.C.: Lieutenant Governor of Massachusetts O'Neill's Washington Office, 1980); M. O. Lewis, ed. *The Foundation Directory,* 7th ed. (New York: The Foundation Center, 1979).

Note: This list should not be regarded as a complete one. Foundations change their priorities from time to time, and some of the foundations listed may no longer fund energy-related projects while some foundations not listed may have made energy a new priority.

For Further Reference

Barret, D.; Epstein, P.; and Haar, C. M. *Home Mortgage Lending and Solar Energy.* Washington, D.C.: Department of Housing and Urban Development, 1977.

Energy Link. "Special Government Grants Issue." Boston: Commonwealth of Massachusetts Office of State-Federal Relations, 1980.

_____. "Special Private Funding Issue." Washington, D.C.: Lieutenant Governor of Massachusetts O'Neill's Washington Office, 1980.

Federation of American Scientists. *Energy Conservation and Renewable Energies: A Summary of Legislative Progress.* Washington, D.C., 1980.

Lewis, M. O., ed. *The Foundation Directory.* 8th ed. New York: The Foundation Center, 1980.

Glossary

absorptivity—a number that indicates what fraction of sunlight falling on a material is absorbed by the material

active solar—a system that uses pumps or fans to move solar energy from the point of capture (in the collectors) to the living space or to a storage component

air conditioner—an appliance used during hot weather to cool a building; air conditioner is a cooling-only heat pump

airstat—a regulator that controls the operating temperature (normally 90 to 150°F) of the warm air distribution system in furnaces

alternating current (AC)—ordinary household current, 120 volts, 60 cycles per second

anemometer—a device that measures wind velocity

anticipator—a control subsystem on a thermostat which operates pumps, fans and fuel burner independently of one another

aquastat—a regulator that controls the operating temperature of the hot water distribution system in boilers and hot water heaters

aquifer—an underground body of water or river

ASHRAE—American Society of Heating, Refrigerating and Air-Conditioning Engineers; a source of data and handbooks on energy that are available in many libraries

automatic damper—a motorized door inside an air duct that controls airflow to different parts of the building; is controlled by a zone thermostat

back-draft damper—a cover over a vent that allows air to pass in only one direction

barometric damper—a delicately balanced air inlet shutoff that is found on flues or furnaces and boilers; purpose is to control the air supply for combustion and keep it constant in all wind and temperature conditions

306

baseboard convector or heater—a heat delivery unit made of pipes with finlike extensions (finned tubes) located in slotted sheet metal covers; pipes carry hot water or steam and run along the building perimeter at the height of the baseboards

base load demand—the amount of utility-supplied electricity that is always being consumed, 24 hours a day

biomass—renewable fuels derived from plants; for example, wood, grain alcohol and methane from sewage

boiler—a heater that either heats water or converts water to steam for use in space-heating systems; is usually made up of a burner (through which gas or oil enters), a combustion chamber and a heat exchanger (through which heat is transferred to the water from the flame)

Btu (British thermal unit)—the amount of heat needed to raise 1 pound of water 1 degree Fahrenheit

building envelope or shell—the parts of a building which separate the heated or cooled space from outdoors

burner—the part of a heating system that provides the flame for combustion; typically includes a pump, blower, air inlets, nozzle and either a spark ignition or a pilot (for gas)

calorie—a measure of heat energy; the amount of heat required to raise the temperature of 1 gram of water by 1 degree centigrade; 1 calorie = .004 Btu; 1 Btu = 252 calories

caulking—flexible material that seals cracks or seams in the building envelope, such as where window frames meet walls; typically made of gumlike petroleum-based substances

clerestory—a vertical window located on a flat or pitched roof

coefficient of performance (COP)—a number applicable to air conditioners and heat pumps, which measures the number of Btu's of cooling or heating provided for every Btu of energy used by the appliance; electric air conditioners typically have COPs of 2.5, while heat pumps have COPs ranging from 2.0 to 2.4

cogeneration—the simultaneous production of heat and electricity by one system; typically system is a gas or steam turbine or a diesel engine, but solar cogenerators are also possible

collector—a glazed device, wall or window that captures sunlight for the purpose of providing space heat and domestic hot water

combustion chamber—the part of a boiler, furnace or woodstove where the burn occurs; normally lined with firebrick or molded or sprayed insulation; heat exchanger, which transfers heat to the air, water or steam distribution system, forms part of its walls

combustion efficiency—a measurable number that indicates the percentage of energy content in a fuel that is converted to heat; number is measured when heater is running in a stable, steady state (*see also* seasonal efficiency)

compressor—a mechanical device that pressurizes a gas in order to turn it into a liquid, thereby allowing heat to be removed or added; compressor is main component of conventional heat pumps and air conditioners

conduction—the direct transfer of heat energy through a material

conductivity—the rate at which heat is transmitted through a material

convection—the transfer of heat by movement of a fluid, usually air or water

cooling load—the amount of cooling required to keep a building at a specified temperature during the summer, usually 78°F, regardless of outside temperature

cooling season—that time period each year during which a building needs to be cooled

cut-in velocity—the wind speed at which a wind machine (usually a horizontal axis machine) begins to turn and generate electricity

cut-out velocity—the wind speed at which a wind machine stops turning in order to protect against blade damage and generator burn-out

dead band control—a control subsystem on a thermostat that allows the heating or cooling system to remain on until room temperatures exceed the set temperature by several degrees

degree-day (DD)—a measure of climatic severity used to estimate heating or cooling energy consumption; for heating, if average outdoor temperature for a day is 10° below 65°F (or 55°F), the day has 10 heating degree-days

delta T (Δ T)—the difference between two temperatures

design-day heat load—the total heat load of a structure under the most severe conditions (temperature and wind) likely to occur; estimates of these conditions generally based on 30 years of weather records and quote a figure which will be exceeded only 1 percent of the time

design temperature—the most severe temperature likely in a given location (*see also* design-day heat load)

direct current (DC)—electricity generated by solar cells and many wind machines; does not cycle like alternating current

direct-gain system—a passive solar heating system in which the collector is a window opening into the living space

distribution efficiency—the efficiency with which a heating system provides heat to a building

district heating—heating of buildings by hot water or steam produced at a central boiler and distributed through a network of pipes

dormer—a small pitched structure projecting from a roof, usually with a vertical window

double-hung—a window with two parts that move in a vertical direction, counter-balanced by weights hung inside the window frame

downsizing—measures taken to reduce a heating system's capacity to make it more compatible with a building's heating requirements; often done following major weatherization

drain-back—an active liquid solar system that empties the collectors and pipes, storing the liquid in a reservoir; system avoids freeze-up problems

drain-down—similar to drain-back systems, except the liquid is thrown away (to the house drains) each time the collectors are emptied

duct—a tunnel, made of galvanized metal or rigid fiberglass, which carries air from the heater or ventilation opening to the rooms in a building

eave—the projecting overhang at the lower edge of a roof

electric resistance coils—metal wires that heat up when electric current passes through them and are used in baseboard heaters and electric water heaters

emissivity—the efficiency with which a body or material warmer than its surroundings emits radiation

energy efficiency ratio (EER)—a measure of energy efficiency of an air conditioner, namely the Btu's of cooling provided for every watt of electricity used by the appliance (*see also* seasonal energy efficiency ratio)

eutectic salts—a phase-change material (*see also* phase-change material)

flame retention burner—an oil burner, designed to hold the flame near the nozzle surface; generally the most efficient type for residential use

flue—a metal or ceramic pipe, running from the basement or living space to the roof, that vents combustion gases from a stove, furnace, boiler, fireplace or water heater

flue damper—an automatic door located in the flue that closes it off when the burner turns off; purpose is to reduce heat loss up the flue from the still-warm furnace or boiler

furnace—a heater that warms air for use in the distribution system; is usually made up of a burner (through which gas or oil enters), a combustion chamber and a heat exchanger (through which heat is transferred to the air) packaged in a cabinet

gable—that part of a vertical wall that projects above the lowest point of the roof

generating capacity—the total amount of electrical power that a utility can produce at any one time

geothermal energy—energy from hot water or steam warmed deep inside the earth's crust

Glauber's salts—a phase-change material (*see also* phase-change material)

groundwater—water from an aquifer or subsurface water source

heat capacity—the quantity of heat that a given volume of a material can hold for each unit increase in temperature, usually given in terms of Btu's per degree Fahrenheit per cubic foot (*see also* specific heat)

heat exchanger—a device, usually made of coils of pipe, that transfers heat from one medium to another; for example, from water to air or water to water

heating load—the amount of heating required to keep a building at a specified temperature during the winter, usually 65°F, regardless of outside temperature

heating season—the time period during which a building needs to be heated

heat loss—the rate at which a building loses heat, usually expressed in Btu's per hour, Btu's per degree-day or Btu's per square foot per degree-day

heat of fusion—the quantity of heat released when a material freezes or absorbed when it melts (in Btu's per pound)

heat pump—a device that mechanically removes low temperature heat from a source, concentrates the heat and delivers high temperature heat to a sink; source can be indoors and the sink outdoors, or vice versa

humidity—the quantity of water vapor contained in air (in pounds per pound of air)

hybrid system—a solar system that combines both active and passive elements, for example, a passive system that contains fans or blowers to aid heat circulation

hydropower—energy produced by water, for example, at a hydroelectric dam

indirect-gain system—a passive solar system in which the glazing is separated from the living space by a heat storage wall that may be masonry or containers of water

infiltration—the passage of air from indoors to outdoors and vice versa; term is usually associated with drafts from cracks, seams or holes in buildings (*see also* convection)

insolation—the amount of sunlight falling upon a surface, usually measured in Btu's per square foot per hour or Btu's per square foot per day

insulation—material used in a building's walls, ceiling, floor or roof to hinder the flow of heat

internal gains—sources of heat within a building that are not part of the heating system, for example, people, animals, lights and appliances

inverter—a device that converts direct current to alternating current, either mechanically or with solid state circuitry

isolated-gain system—a passive solar system in which the collector, storage and living space are all physically separated from one another

kilowatt (kw)—a unit of power (energy consumed per unit time) equivalent to 1,000 watts or 1.3 horsepower or 3,413 Btu's; this unit is usually applied to electrical power

kilowatt-hour (KWH)—a unit of energy equal to the energy consumed by 1 kilowatt of power acting for 1 hour

lath—any material fastened to wood framing that is designed to hold plaster

life-cycle cost—the total cost of purchasing, owning and maintaining a device, including the cost of energy used to operate the device, over the lifetime of the device

liquefaction—the process of converting a gas to a liquid, either by removal of heat or an increase in pressure

load management—control of energy consumption at any instant through the use of mechanical or electronic devices or conscious consumer limitations on the use of energy-consuming devices

luminaire—a glass or metal fixture that reflects or diffuses light from a bulb

marginal cost—the cost of one additional unit of something, for example, the cost to find, pump, ship and deliver the next new gallon of oil

MBh—thousands of Btu's per hour; a measure of the heating capacity of a heating system

nozzle—the part of a heating system that sprays the fuel or fuel-air mixture into the combustion chamber

off-peak—periods other than peak electrical demand periods

on-peak—periods when energy consumption is highest and the most expensive energy-supply systems are operated; generally peak period refers to period of greatest total demand on an electric utility

passive solar—a system that uses natural heat transfer processes to move captured solar energy (in the collector) to the storage or living space

payback period—the time it takes for the fuel savings realized by some action to equal the cost of that action

peak demand—the greatest amount of electricity or natural gas used by utility customers during the day or year

permeability—a measure of the ease with which water vapor penetrates a material

phase change—the change of a material from liquid to solid, or liquid to gas, or vice versa

phase-change material (PCM)—a material that melts at or near room temperature and that can therefore be used for heat storage purposes

plenum—that part of an air heating duct system that collects together all the branch ducts; the main duct from the furnace

process heat—heat, usually steam at high temperature and pressure, used in industrial processes

radiation—the transfer of energy or heat by electromagnetic waves, such as light

rated output—the maximum amount of power produced by a wind machine (or other energy system); a compromise between maximum output and safety of operation

rated velocity—the wind speed at which a wind machine reaches its rated output

refrigerant—a substance that remains a gas at low temperatures and pressure and can be used to transfer heat

roof ridge—the line that runs along the peak of a roof, where sloping parts of the roof meet

R-value (thermal resistance)—a measure of the ability of a material to resist the flow of heat through it

sash—the part of a window that holds the glass

seasonal efficiency—the average efficiency of a heater or cooler over the heating or cooling season (*see also* combustion efficiency)

seasonal energy efficiency ratio (SEER)—the average energy efficiency ratio (EER) achieved by an air conditioner over the cooling season; this number tends to be lower than rated EER (*see also* energy efficiency ratio)

seasonal storage—systems that capture and store energy when it is not required (for example, heat in the summer) for use when it is needed (for example, space heating in winter)

sheathing—the layer or structure, under the exterior finish of a roof or wall, upon which the finish is mounted; sheathing found in wood frame buildings

sill—in a window, the bottommost part of the frame; in a house, the lowermost beams above the foundation

sink—the point to which a heat pump transfers heat, often a living space or thermal storage

skylight—a more or less horizontal window located on the roof of a building

soffit—the underside part of an eave

solar cells—*see* solar photovoltaic cells

solar gain—the heat gained in a building due to sunlight, principally that entering through windows

solarium—a glazed room or structure (also called a sunspace) whose purpose is to capture solar energy to heat the building; atrium is solarium in middle of a building

solar photovoltaic cells—semiconductor devices able to directly convert sunlight to direct-current electricity

source—the point from which a heat pump removes low temperature heat, for example, air or groundwater

space heat—heat supplied to the living space, for example, to a room or the living area of a building

specific heat—the heat capacity of a unit amount of a material, usually given in units of Btu's per pound per degree Fahrenheit

standby loss—the heat lost in a boiler or furnace when it is not in operation; greatest standby loss is due to warm air flowing through combustion chamber and up flue

storage mass—the component in a solar system that stores heat energy for heating when the sun is not shining

subfloor—the layer of flooring immediately over the floor joists or beams

sunspace—*see* solarium

synchronous inverter—a device that converts direct current from a wind machine or solar cells into alternating current matched to the utility electricity and draws electricity from the utility grid in order to make up any household or building power deficits

temperature regulator—*see* airstat, aquastat and thermostat

therm—a unit of heat equal to 100,000 Btu's; frequently used on gas bills

thermal lag—the delay between the absorption of heat by a storage wall in an indirect-gain system and the radiation of heat into the living space

thermal mass—those portions of a building that store significant quantities of heat; these components may be wood or masonry

thermal storage—the place in which energy can be stored for use at a later time, for example, a tank of water (*see also* storage mass)

thermocirculation vent—vents in a Trombe wall that allow air to circulate by natural convection from airspace to living space and back to airspace

thermosiphon—a system in which heat is captured by a collector and moved to the living space or storage by natural convection

thermostat—a device which regulates the temperature of a room or building by switching heating or cooling equipment on or off

TOURS (time-of-use rates)—special electric rates that are high during on-peak periods and low during off-peak periods

Trombe wall—a passive indirect-gain solar system in which the space is heated by natural convection during the day and radiation from the wall at night

turbine—a machine in which the energy contained in a high-pressure gas or liquid is converted to rotational energy, often to turn an electric generator

utility grid—the combination of electric power plants and transmission lines belonging to an electric utility

U-value (coefficient of heat transmission)—the rate of heat transmission through 1 square foot of building envelope for 1 degree Fahrenheit difference in temperature between indoors and outdoors

vapor barrier—a material which impedes the passage of water vapor through building surfaces to insulation; located on the warm side of insulation, and is usually a polyethylene sheet, aluminum foil or coated paper

vaporization—the process of converting a liquid to a gas, either by addition of heat or reduction of pressure

weatherization—modifying a building envelope to reduce energy consumption for heating or cooling; involves adding insulation, installing storm windows and doors, caulking cracks and putting on weather stripping

weather stripping—a narrow strip of material, such as plastic, felt or metal, which is installed around windows and doors to reduce the infiltration of air between them and their frames

weep holes—small holes in storm window frames that allow moisture to escape

window frame—the stationary part of a window unit; window sash fits into the window frame

window sash—the operating or movable part of a window; the sash is made of windowpanes and their rim

wind shear—a change in wind direction or velocity over a very small standard distance; the standard distance is usually equal to the rotor diameter of a wind machine

work—the transfer of energy from one system to another; for example, heat transferred by a heat pump from source to sink

zone—the section of a building that is served by one heating or cooling loop because it has noticeably distinct heating or cooling needs

zone valve—a device, usually placed near the heater or cooler, which controls the flow of water or steam to parts of the building; it is controlled by a zone thermostat

Index

Boldface numbers indicate entries in glossary.